Twelve captivating tales from the best emerging writers of the year accompanied by three more from bestselling authors you've read before.

Cameron is dead. His lawyer insists it's not the end of the world. Just sign here, and you'll be loaded into a new body in no time...

—"Form 14B: Application for Certification of Consciousness Transfer (Post-Mortem)" by Thomas K. Slee

Caring for a giant, cheese-loving dragon was never the plan, but a mild-mannered baker must find his courage—challenging the volatile king of the fairy underworld.

—"Saffron and Marigolds" by Kathleen Powell

When a surveillance architect discovers a street artist the algorithms can't predict, small rebellions begin accumulating toward something the system is already too late to stop.

—"Bloom Decay" by Elina Kumra

Lieutenant Carter bets everything on the courage of his little "space can" to save the fleet by storming the enemy's heavy ship mid-battle.

—"Space Can" by L. Ron Hubbard

After waking at a crime scene with blood on his hands, a detective races to expose the body-hopping entity using him as a "shell"—before it uses him to strike again.

—"Shell Game" by Zach Poulter

When a reclusive Arkansas homesteader experiences missing time, her only hope for solving the mystery of a world gone wrong is a boy who shouldn't exist.

—"Canary" by Brenda Posey

Tasked with faking the asteroid strike that smithereened the dinosaurs, a frazzled time-agent must risk paradox and punishment to save the last triceratops.

—"The Triceratops Effect" by S.J. Stevenson

A pair of Antarctic researchers uncover an evolutionary detour that defies accepted science—evidence pointing to a dragon living in the Andes.

—"Skinny-Shins" by Orson Scott Card

Seb has spent his career stopping time to save people from deadly disasters. But when a flawless rescue ends with inexplicable deaths, he risks everything to find out why.

—"A Ready-Made Bubble of Light" by Thomas R. Eggenberger

Desperate to become visible, Nomsa follows a trend promising instant beauty—only to face the horrifying truth that her new, improved self may not leave room for the old one.

—"Thickly" by Dorothy de Kok

When an unseen creature stalks his home, a man must protect his small dog from a supernatural predator determined to force its way inside—and claim a living host.

—"Ghost Dog" by Mark McWaters

When dragons materialize, drawn to a boy's violin playing, their gentle presence feels like a comforting echo of the mother he's lost.

—"Dragon Visits" by Nina Kiriki Hoffman

A man who can step into pictures uses his gift to solve crimes, balancing horror-filled photos with sanctuary inside Monet's *Water Lilies*—until one investigation turns personal.

—"In Living Color" by Michael T. Kuester

When a clinical trial goes wrong, a woman is left stranded in virtual reality. The man who loves her must navigate corporate secrecy to keep their fragile relationship alive.

—"As Long as You Both Shall Live" by Mike Strickland

Across decades, a girl grows into a journalist, activist, and outlaw—with a lifelong devotion to return a captive dragon to the sky.

—"A Girl and Her Dragon: A Life in Four Parts" by Joseph Sidari

L. RON HUBBARD

PRESENTS

Writers of the Future Anthologies

"The gangbuster of science fiction short-story collections continues to excel." —*The Midwest Book Review*

"The series successfully showcases future voices representing the vast spectrum of the speculative fiction genre." —*Library Journal*

"It's an excellent book...and well worth reading (and viewing)." —*Amazing Stories*

"The L. Ron Hubbard Writers of the Future Contest has quietly been shaping the next wave of genre storytellers—turning starry-eyed dreamers into published sci-fi and fantasy authors with a legacy that's as cosmic as the stories themselves." —*SciFiNow*

"A straightforward introduction to the speculative genre along traditional lines, opening the world of science fiction and fantasy to new readers." —*Aurealis*

"This is a really high quality collection and I recommend it!" —*Harare Review of Books*

"A fine sampling of some of the best new voices in today's science fiction and fantasy." —*AudioFile*

"It really does help the best rise to the top." —Brandon Sanderson
Writers of the Future Contest judge

"The Writers of the Future Award has also earned its place alongside the Hugo and Nebula Awards in the triad of speculative fiction's most prestigious acknowledgments of literary excellence."

—*SFFAudio*

"Writers of the Future, as a contest and as a book, remains the flagship of short fiction."

—Orson Scott Card
Writers of the Future Contest judge

"The Writers of the Future experience played a pivotal role during a most impressionable time in my writing career."

—Nnedi Okorafor
Writers of the Future Contest judge

"This is an opportunity of a lifetime."

—Larry Elmore
Illustrators of the Future Contest judge

"If you want a glimpse of the future—the future of science fiction—look at these first publications of tomorrow's masters."

—Kevin J. Anderson
Writers of the Future Contest judge

"Illustrators of the Future offered a channel through which to direct my ambitions. The competition made me realize that genre illustration is actually a valued profession, and here was a rare opportunity for a possible entry point into that world."

—Shaun Tan
Illustrators of the Future Contest winner 1993
and Illustrators of the Future Contest judge

"The Illustrators of the Future is an amazing compass for what the art industry holds in store for all of us."

—Dan dos Santos
Illustrators of the Future Contest judge

L. Ron Hubbard PRESENTS
Writers of the Future

VOLUME 42

L. Ron Hubbard PRESENTS
Writers of the Future

VOLUME 42

The year's twelve best tales from the
Writers of the Future international writers' program

Illustrated by winners in the Illustrators of the Future
international illustrators' program

Three short stories by Orson Scott Card /
Nina Kiriki Hoffman / L. Ron Hubbard

With essays on writing and illustration by
Brian C. Hailes / L. Ron Hubbard / Larry Niven

Edited by Jody Lynn Nye
Illustrations art directed by Echo Chernik

GALAXY PRESS, INC.

For information, contact Galaxy Press, Inc. at 7051 Hollywood Boulevard, Suite 200, Los Angeles, California 90028.

Cover design by Kirk DouPonce DogEared Design

Trade Paperback ISBN 978-1-61986-900-4

Many thanks to these first readers: Leah Ning, Kary English, Martin Shoemaker, and Eric James Stone. Special acknowledgments to these beta readers: Joe Benet, Alicia Cay, Michelle Dias, Victoria Dixon, Cara Giles, Kimberly Richards, and Peter Spasov.

Printed in Canada

CONTENTS

Introduction

BY JODY LYNN NYE

Jody Lynn Nye lists her main career activity as "spoiling cats." When not engaged upon this worthy occupation, she writes fantasy and science fiction books and short stories.

Since 1987 she has published over fifty books and more than two hundred short stories. Among her novels are her epic fantasy series, The Dreamland, five contemporary humorous fantasies in the Mythology 101 series, three medical science fiction novels in the Taylor's Ark series, and Strong Arm Tactics, a humorous military science fiction novel. Jody also wrote The Dragonlover's Guide to Pern, a nonfiction-style guide to Anne McCaffrey's popular world. She also collaborated with Anne McCaffrey on four science fiction novels, including Crisis on Doona (a New York Times and USA Today bestseller). Jody coauthored the Visual Guide to Xanth with author Piers Anthony. She has edited two anthologies, Don't Forget Your Spacesuit, Dear!, and Launch Pad, and written two short-story collections, A Circle of Celebrations, holiday SF/fantasy stories, and Cats Triumphant!, SF and fantasy feline tales. Nye wrote eight books with the late Robert Lynn Asprin, License Invoked, and seven set in Asprin's Myth Adventures universe. Since Asprin's passing, she has published two more Myth books and two in Asprin's Dragons series. Her newest series is the Lord Thomas Kinago books, beginning with View from the Imperium (Baen Books), a humorous military SF novel, and two young adult science fiction novels with Dr. Travis S. Taylor.

Her newest book is 1635: The Weaver's Tale with Eric Flint, a part of Flint's 1632 alternate history series.

Over the last thirty or so years, Jody has taught in numerous writing workshops and participated on hundreds of panels at science fiction conventions. She runs the two-day writers' workshop at Dragon Con. Jody is the Coordinating Judge of the Writers of the Future Contest. In June 2022, she received the Polaris Award from ConCarolina and Falstaff Books for mentorship and guidance of new talent.

Jody lives in the northwest suburbs of Atlanta with her husband, Bill Fawcett, and three feline overlords, Athena, Minx, and Marmalade.

Introduction

Every year, I read hundreds of stories looking for the very best of the best. Every year, I think they can't get better than they were the year before. And, every year, I'm proven wrong. New writers are coming along that have strong talent and wide-ranging imaginations, sending in plots that I have not seen before with robust characters and fascinating world-building. It's a pleasure to have read them, and I look forward to showing them to you. The Writers of the Future Contest gathers entries from writers from all over the world, with diverse backgrounds and cultures. This year has many different voices and more than a few dragons. Although that's not the only reason they were chosen.

Becoming a writer is not an easy thing. There isn't and never has been a shortcut to having your work published. First, you have to have an idea. Everyone has ideas. That's one of the reasons they can't be copyrighted. What counts is the work. When you have an idea, get it down as quickly as you can. Writers have to write, so you will probably find yourself with more ideas than you can ever finish. Pick the best one and apply yourself to it. The more it tickles your imagination, the more of the same effect it will have on your reader.

But the most important thing to do is to finish the story. It doesn't matter if it's imperfect. Every story is imperfect at first. Everything deserves a second look, or a third, or a twentieth. Give a good story your best efforts. Once it's in the best shape it can be, let someone else read it. A second pair of eyes will help you, either to confirm that it's ready to go, or to let you know what a sincere reader thinks is missing from it. Editing is hard to accept. No one likes to have flaws pointed out, but you want to be a writer, right? Take in the comments that seem true to the story you have written and incorporate them.

Now that it's finished, or at least as finished as you can make it, what to do with it? Naturally, you want to share it with the world. You can publish it yourself. You can add it to an anthology with other writers. Or, and I am hoping you will agree, you can send it in to a writing contest like this one. If you send it to the Writers of the Future Contest and become a winner, your story will give you not only prize money, but also bragging rights. After all, this has been the launching ground for many talented writers over the years, such as Brandon Sanderson, Nnedi Okorafor, Eric Flint, Dean Wesley Smith, Patrick Rothfuss, and more. You will also have immediate publication with an illustration in a respected anthology, a weeklong seminar taught by noted writers in the field of science fiction and fantasy, a red-carpet, black-tie gala dinner, and a chance to bond with the other winners from your year. Writing can be a lonely profession, and having a ready-made cadre of fellow writers that will give you a peer group on which you can bounce ideas and share triumphs in coming years is an immeasurable benefit.

You'll also be making the reading public happy. Your story will become known as one of the best of the best, too. People will read it, talk about it, and treasure it for years to come. Let yourself have the chance to show off your talent and give pleasure to readers. Send in your stories and let the rest of us enjoy them. I hope you will. Looking forward to reading your work! In the meantime, turn the page and read the marvelous stories in this year's anthology.

Best wishes,
Jody Lynn Nye

The Illustrators of the Future Contest

BY ECHO CHERNIK

Echo Chernik has been illustrating for thirty years and has been the recipient of many prestigious awards and accolades.

Her clients have included Disney, BBC, Mattel, Hasbro, MillerCoors, Jose Cuervo, Celestial Seasonings, McDonald's, Procter & Gamble, Trek Bicycle Corporation, USPS, Bellagio Hotel & Casino, Kmart, Sears, Publix Super Markets, Regal Cinemas, the city of New Orleans, the state of Illinois, the Sheikh of Dubai, Dave Matthews Band, Arlo Guthrie, and more. She is a master of many styles including decorative, vector, and art nouveau.

She has been interviewed on CBS, PBS Radio, and by countless publications in her career. Echo owns an art gallery in Washington State featuring exclusively her art, and she tours the world meeting fans and lecturing on illustration.

As the art director and Coordinating Judge of the Illustrators of the Future Contest, Echo prepares the winners for the business of illustration and a successful career in art.

The Illustrators of the Future Contest

One of the most common questions I receive during interviews about my lengthy career is, "What advice do you give your students and artists who want to become successful illustrators?"

The answer is easy: "Enter the Illustrators of the Future Contest!" There is simply nothing that remotely compares to the quality of the Illustrators of the Future, and any emerging artist who does not enter is doing themselves a great disservice.

The Illustrators of the Future has been around longer than any other dedicated contest I can think of. You can enter four times a year, free of charge. Winners not only receive a monetary prize, but also an illustration commission that will be published in a bestselling anthology. For one week leading up to the Hollywood-style red-carpet gala, winners are given an invaluable opportunity not to be missed: the workshop.

The workshop offers the opportunity to meet and casually socialize with the seasoned judges of the Contest. Judges in attendance may include Ciruelo, Craig Elliott, Larry Elmore, Bea Jackson, Irvin Rodriguez, Dan dos Santos, Tom Wood, and others. The workshop itself is centered on sharing important information, such as developing your portfolio, marketing yourself, and building your business as a brand. Winners meet the authors whose stories they illustrated, and many continue working together on future projects.

The commissioned illustration is an important component of the Illustrators of the Future in helping artists develop a solid road to success. Each winner is assigned one of the Writers of the Future Contest–winning stories to illustrate. They have thirty days to complete the assignment. During this time, I take on the role of art director, helping guide them to create the strongest piece possible. This illustration is entered for the Grand Prize awarded in Hollywood, so it is vital that their best work be on display. The pieces you see in this volume are the result of this process.

It is amazing to watch the careers bloom after the anthology's release. And things don't end once winners return home from Hollywood, trophies in hand. This is where the Contest truly sets itself apart. The dedicated team behind the Contest and the publisher firmly believe in the legacy of L. Ron Hubbard and his intent to pay it forward—and they go above and beyond. Each winner has the opportunity to participate in podcasts, radio interviews, and published articles highlighting their success, further expanding their reach and visibility. Past illustrator winners are also hired for professional projects, and their awards and accomplishments are promoted in perpetuity.

When I say I have never seen anything like this competition in my thirty-year career, I truly mean it. With that in mind, here are a few words of advice.

Artists are often presented with many opportunities to enter contests. Some are legitimate; others, less so. The Illustrators of the Future Contest offers several clear markers you can use as a guide. It is free to enter, held four times a year, and I strongly encourage you to enter every quarter—make it a goal. The commissioned artwork is licensed solely for inclusion in the published anthology and its promotion, while the artist fully retains ownership and copyright. AI-generated art is not accepted.

Always research any contest you are considering. In the case of the Illustrators of the Future, you are welcome to reach out to the judges, connect with past winners, or listen to the weekly Writers & Illustrators of the Future Podcast to learn more.

As a final note, the Contest is equal opportunity with blind judging. We have winners of all ages, races, and sexual orientations from all over the world. Judges receive submissions and vote based solely on the artwork. Period.

I look forward to seeing you in Hollywood. Keep entering the Contest—even if you do not win the first time. Perseverance is key.

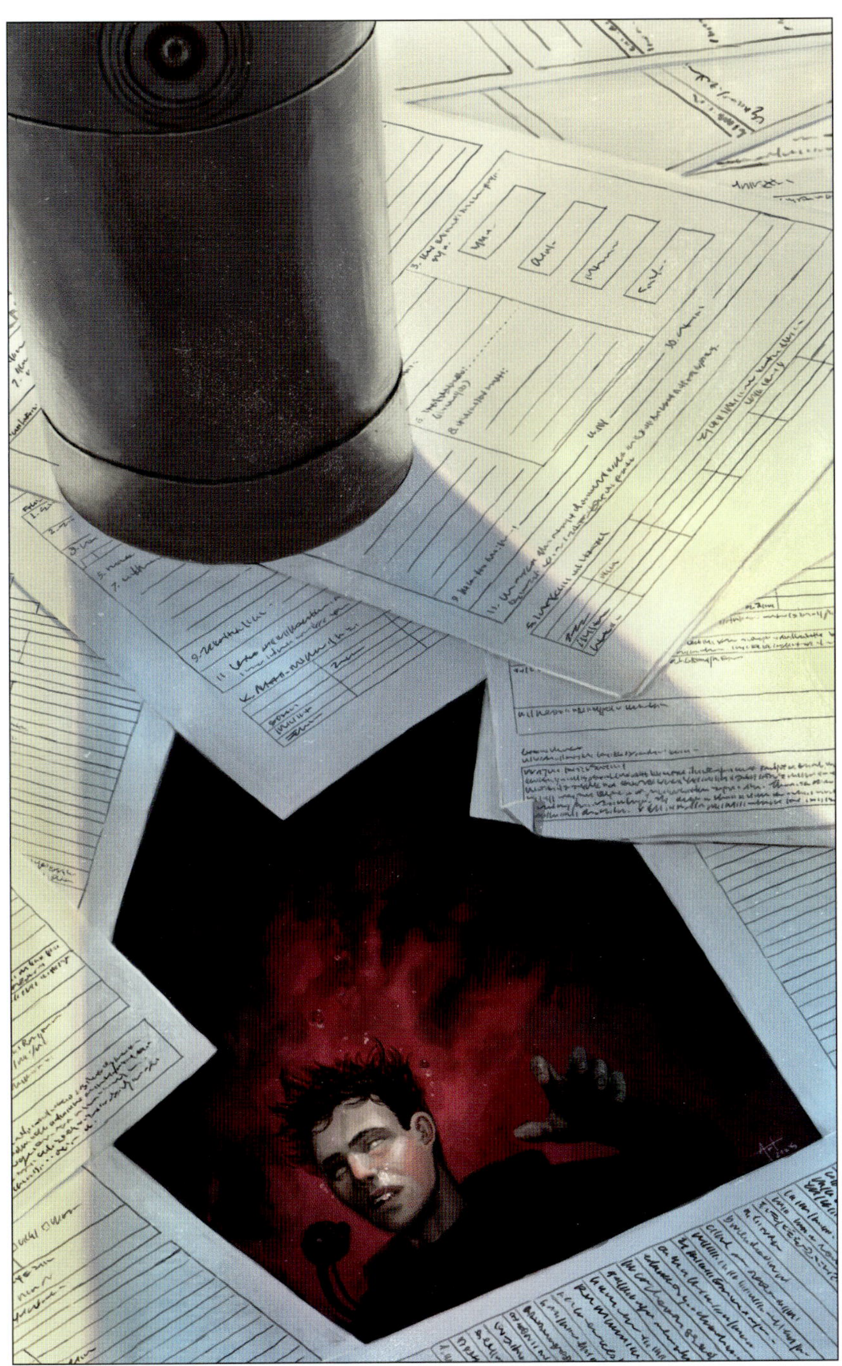

ART IKUTA
Form 14B: Certification of Consciousness Transfer

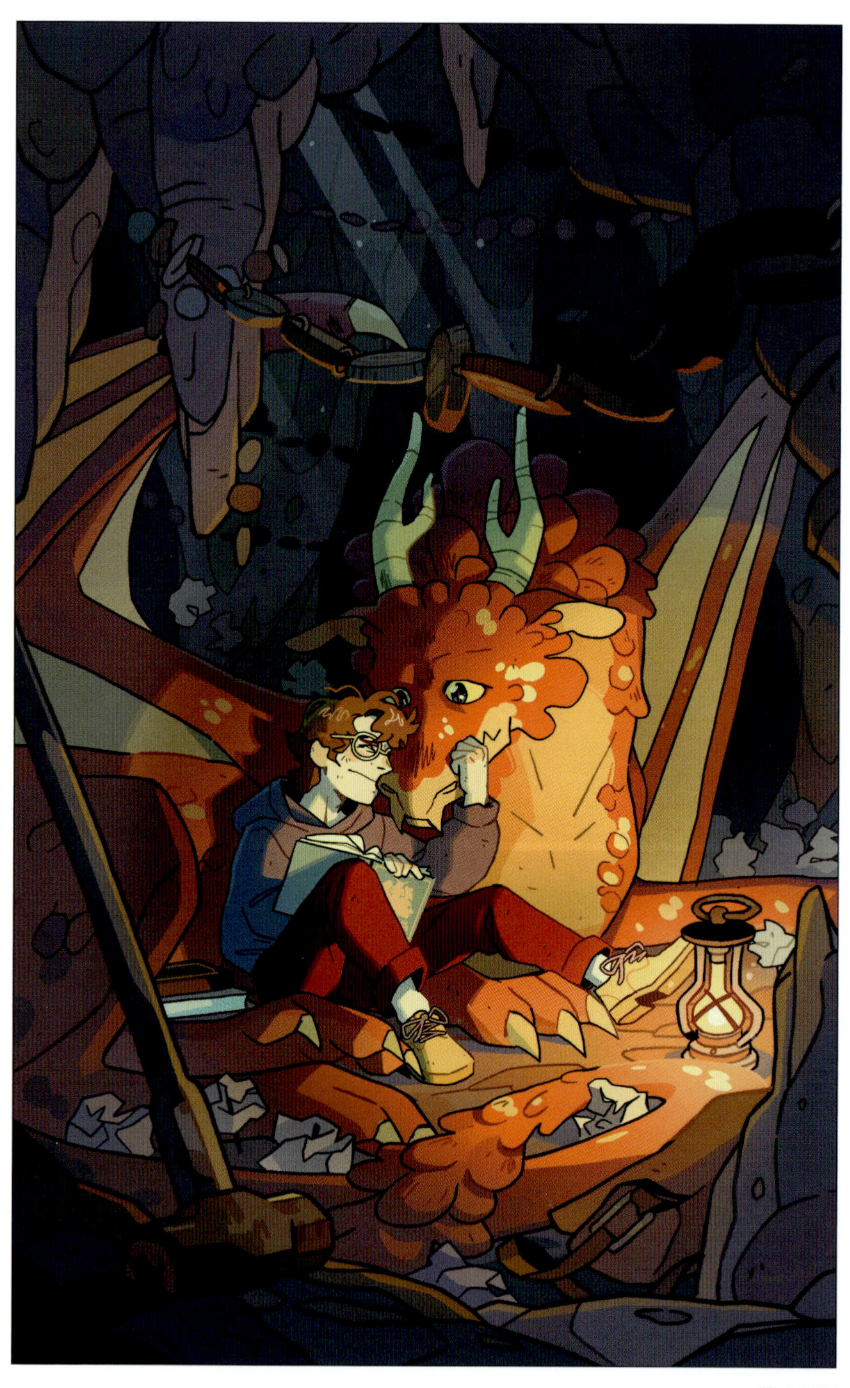

BAFU
Saffron and Marigolds

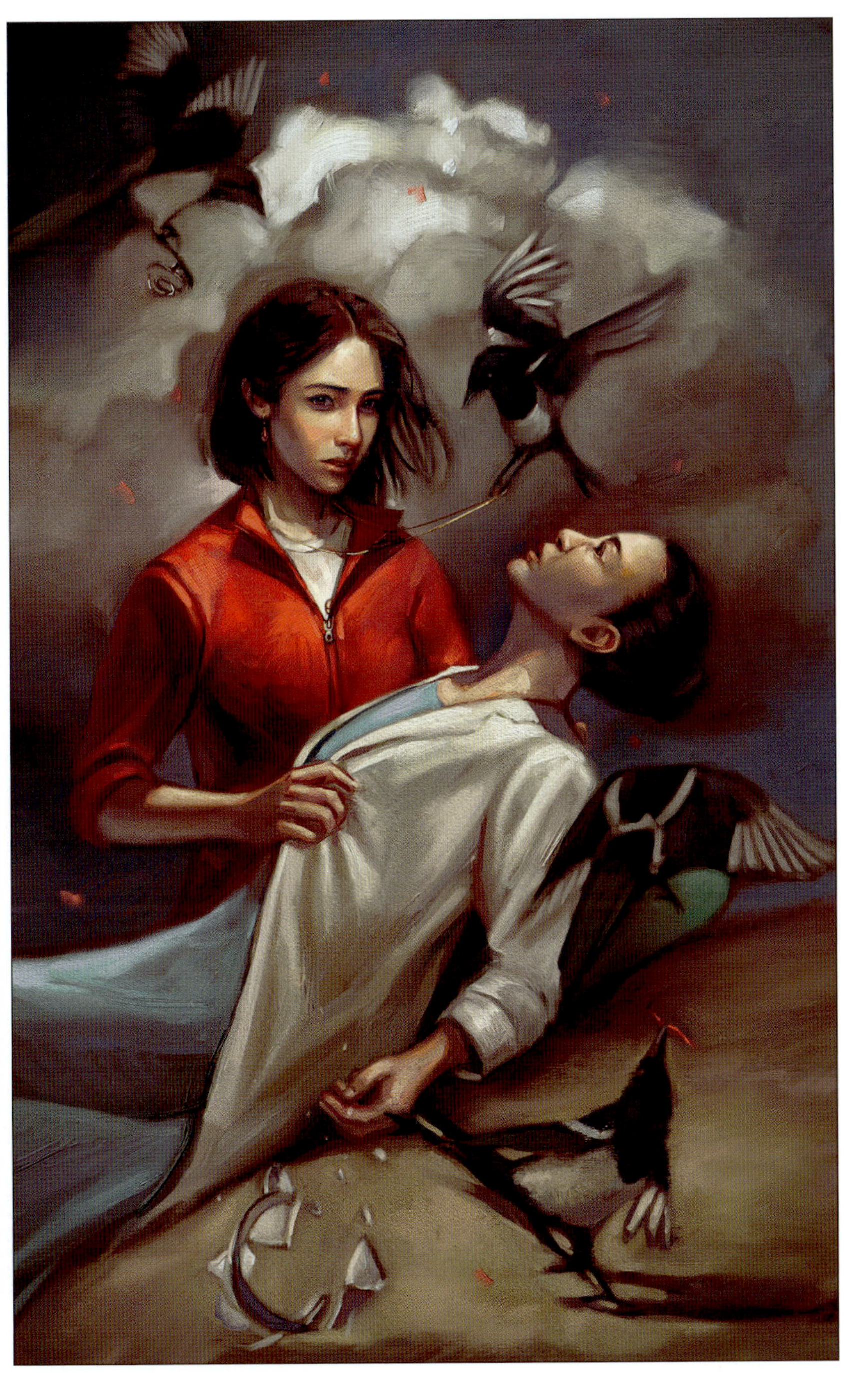

TRAY STREETER
Bloom Decay

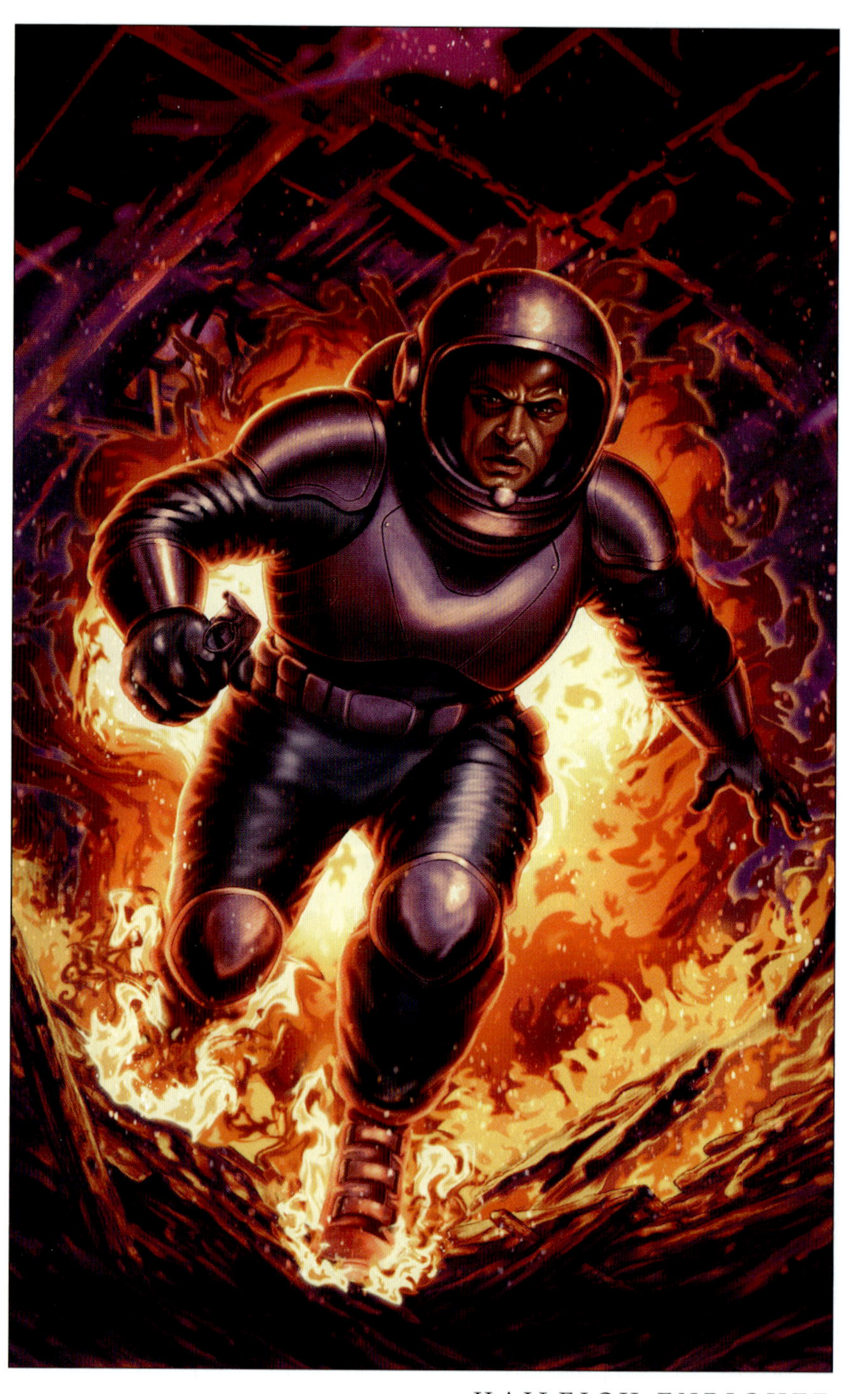

HAILEIGH ENRIQUEZ
Space Can

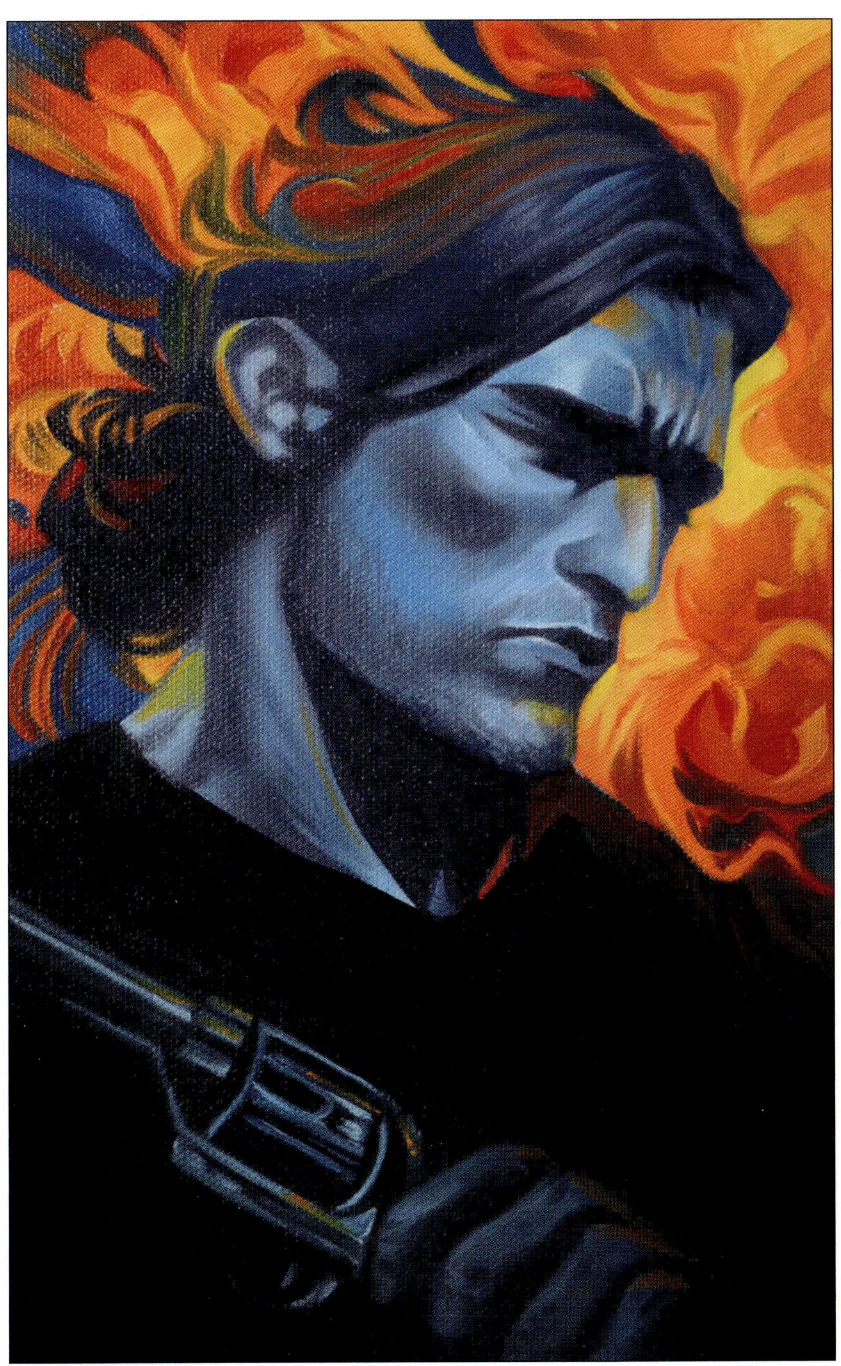

TRACY EIRE
Shell Game

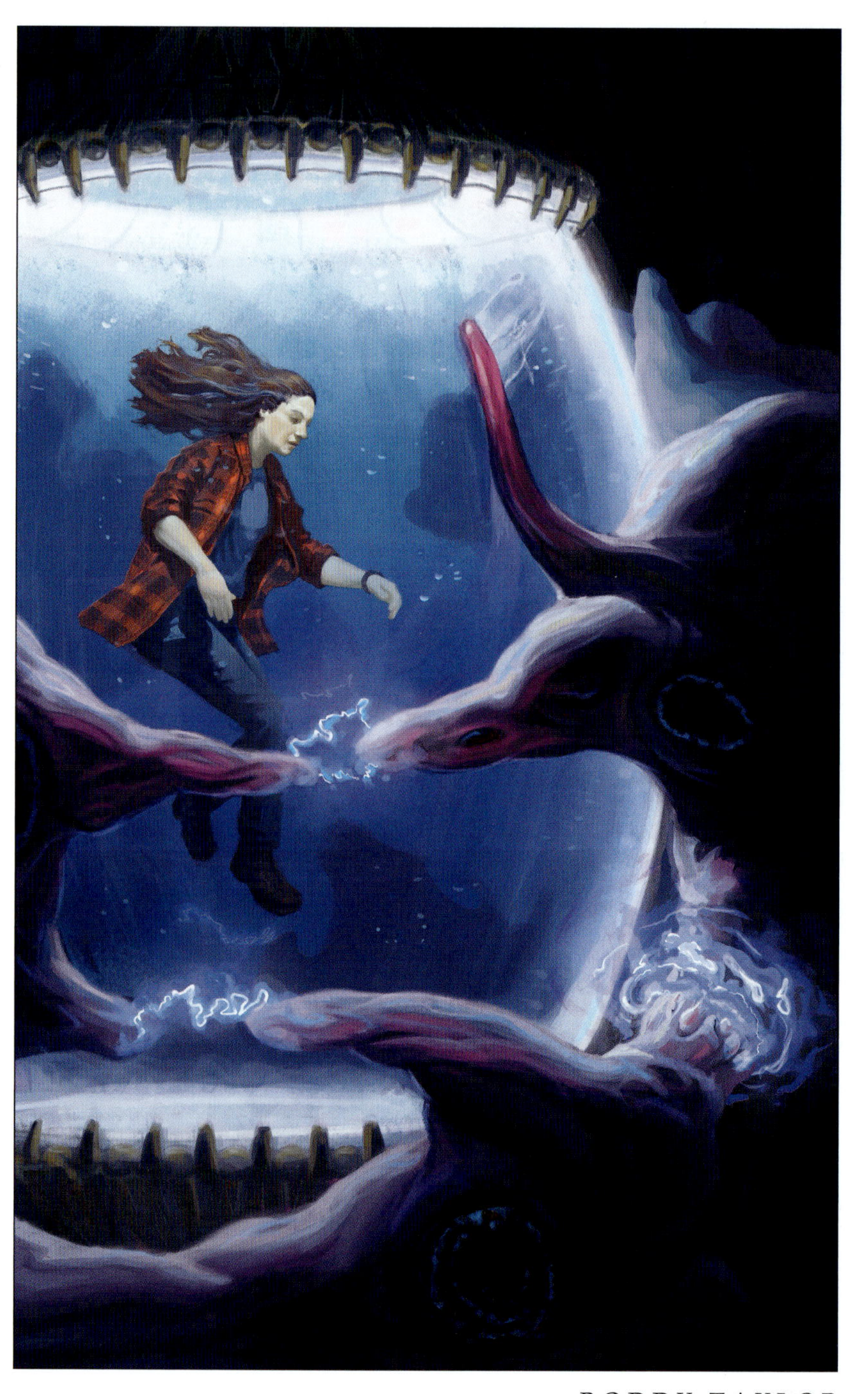

RODDY TAYLOR
Canary

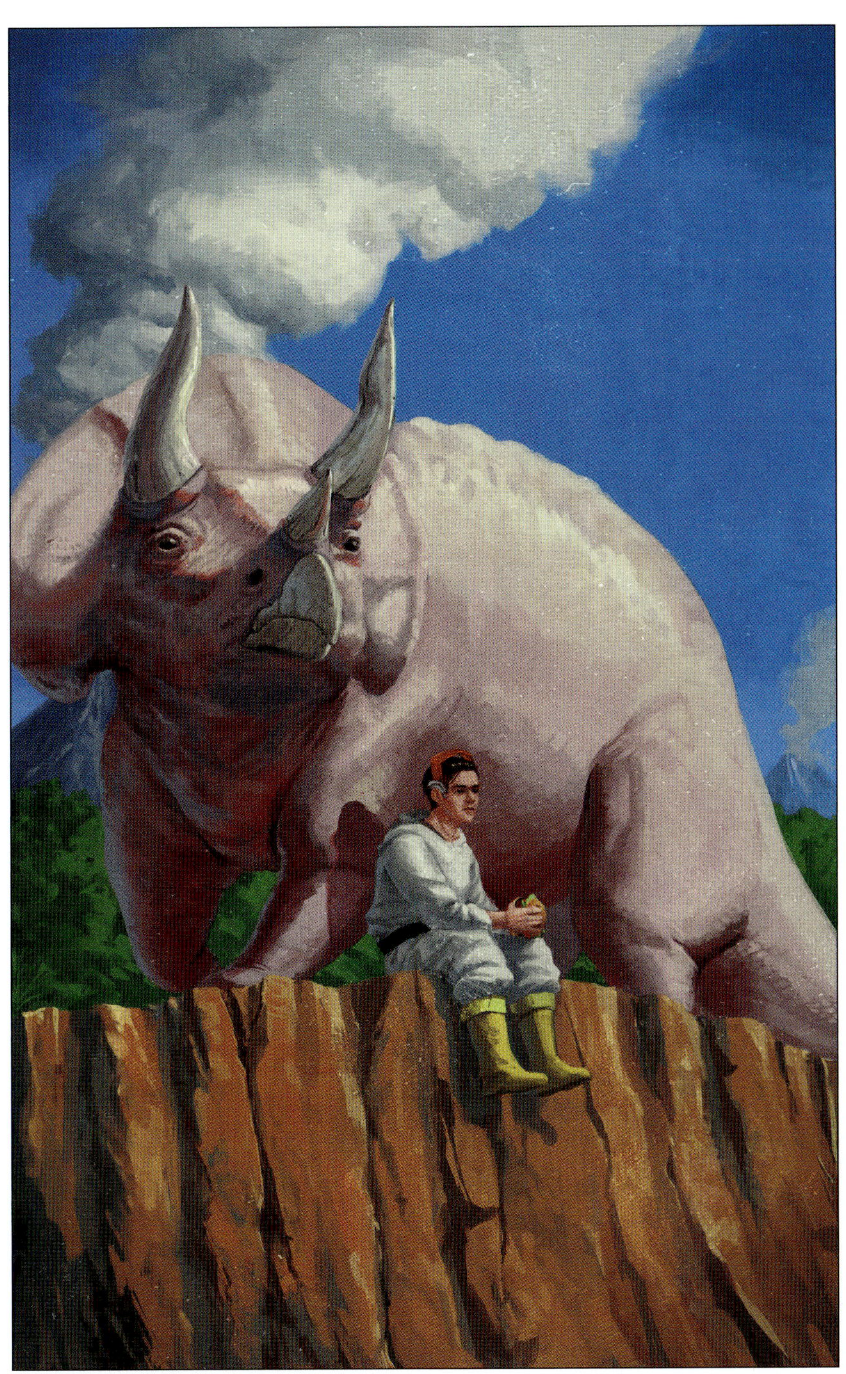

ART IKUTA
The Triceratops Effect

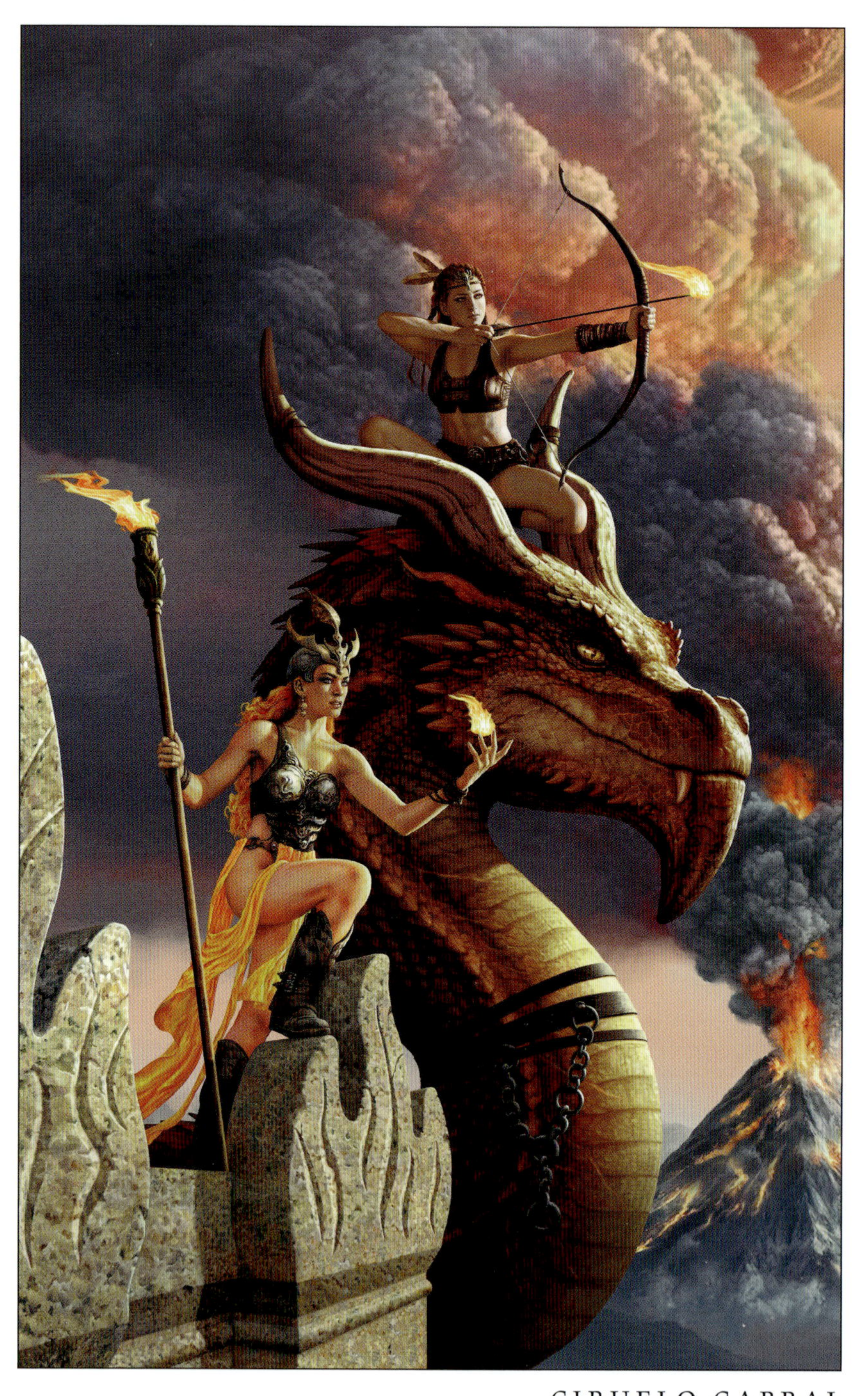

CIRUELO CABRAL
The Fire Tribe

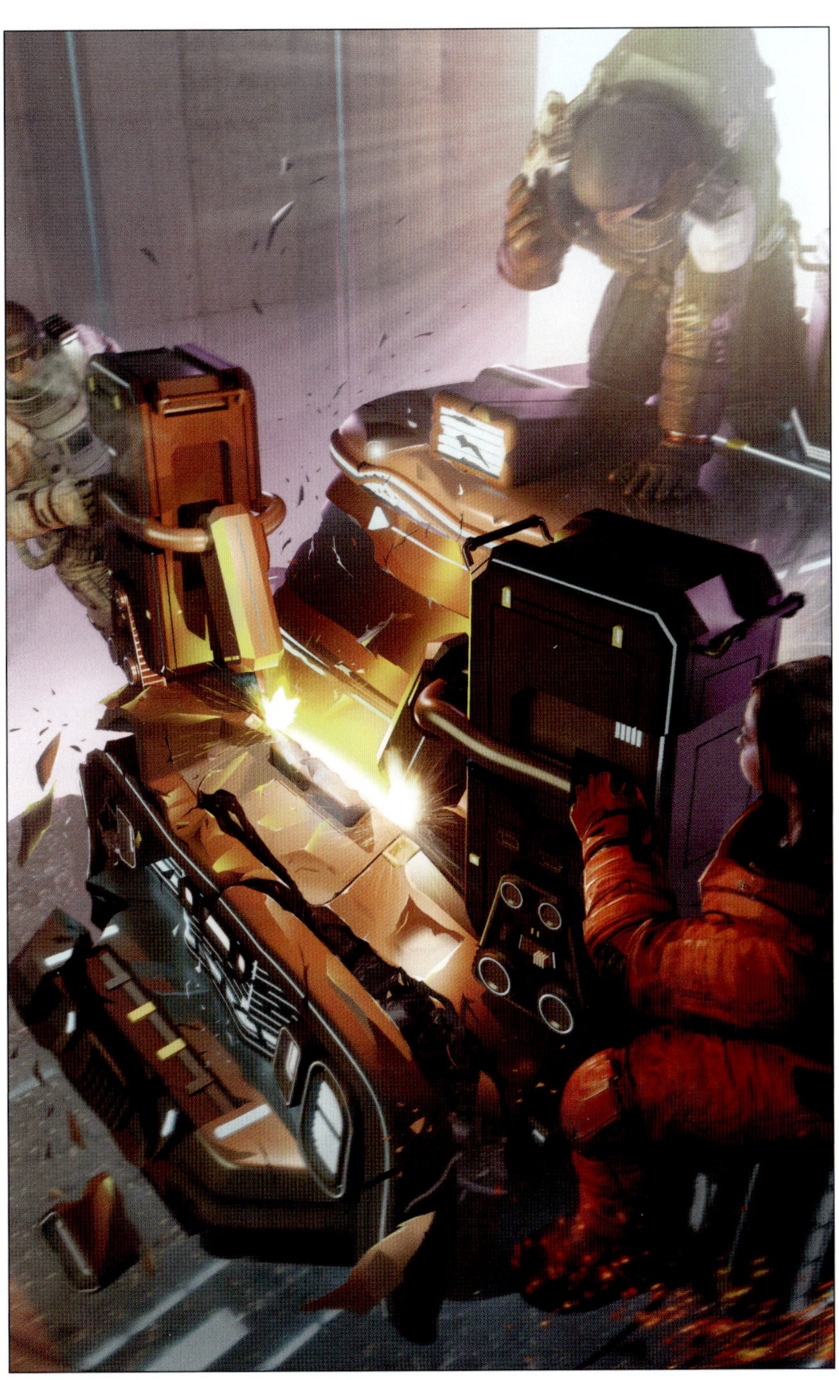

HAOTIAN ALLEN ZHANG
A Ready-Made Bubble of Light

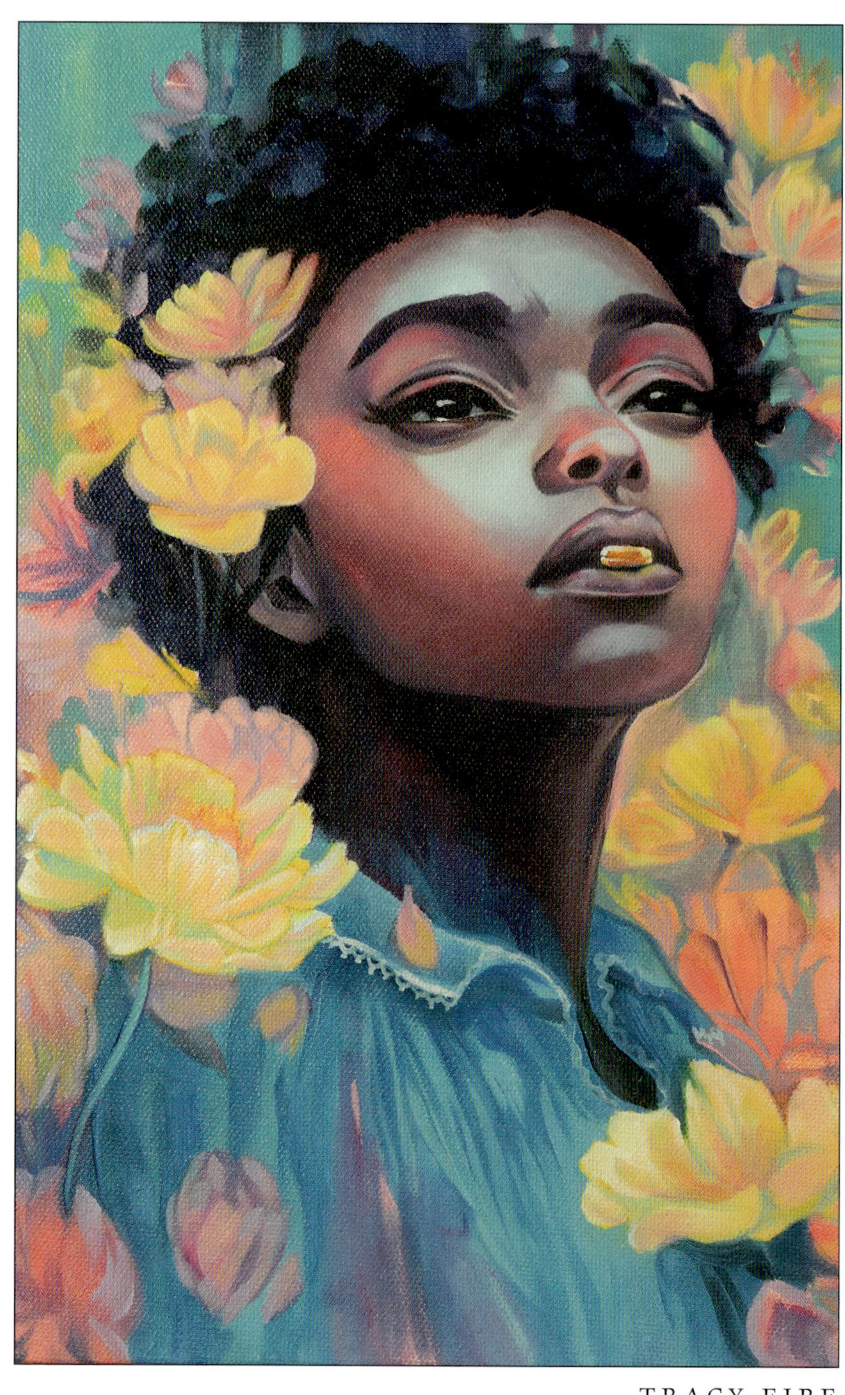

TRACY EIRE
Thickly

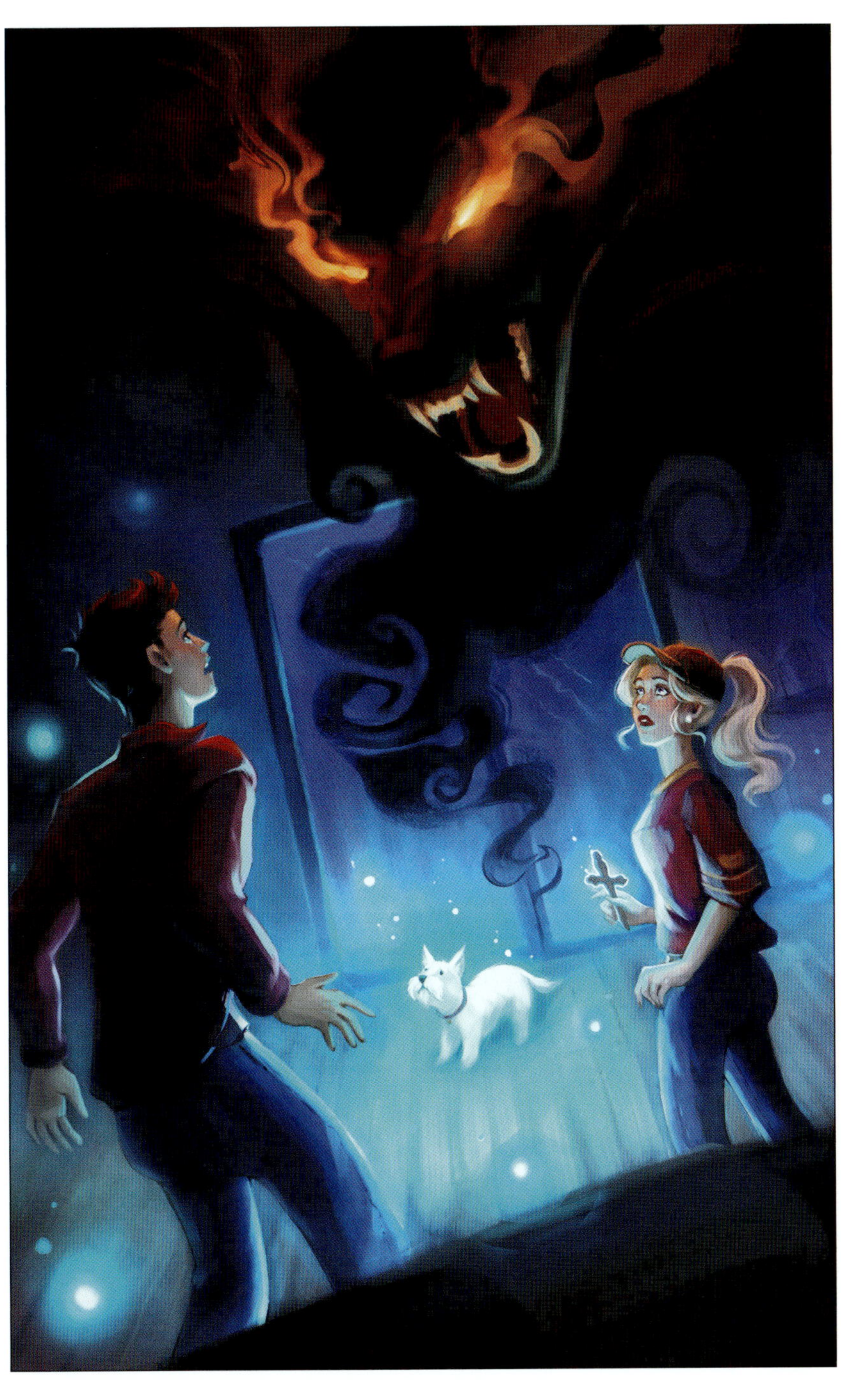

ANNA MALONE
Ghost Dog

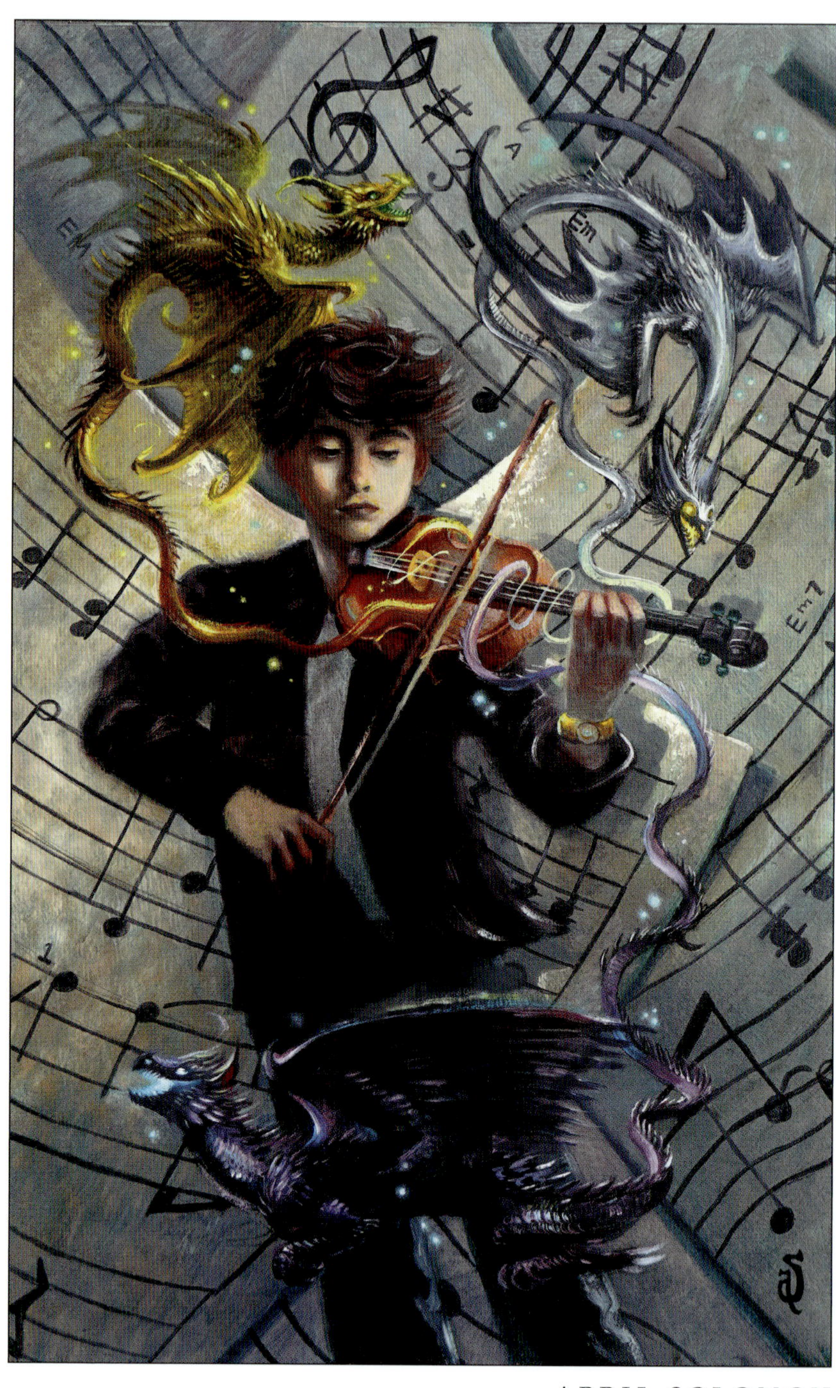

APRIL SOLOMON

Dragon Visits

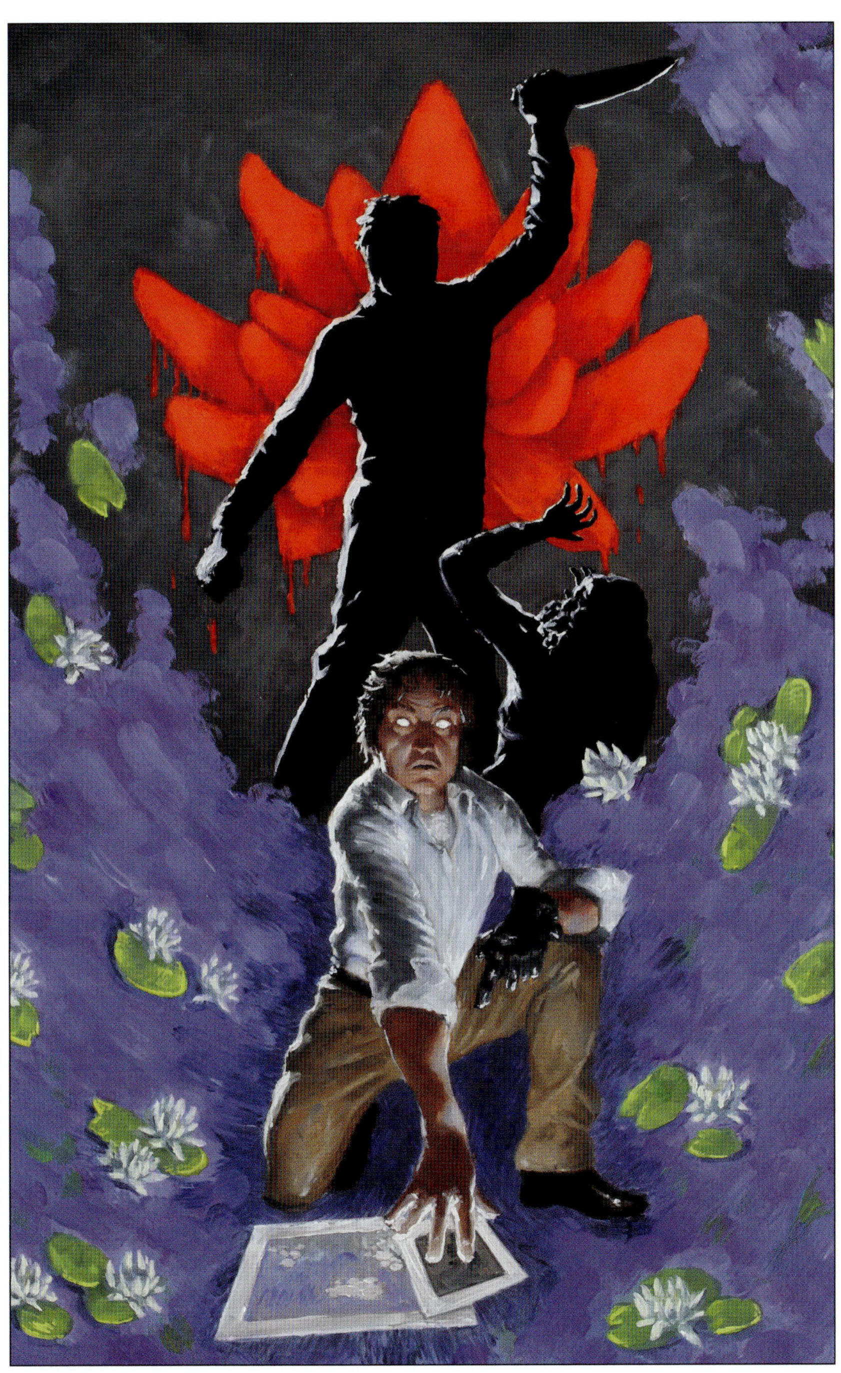

NATHAN DEIWERT
In Living Color

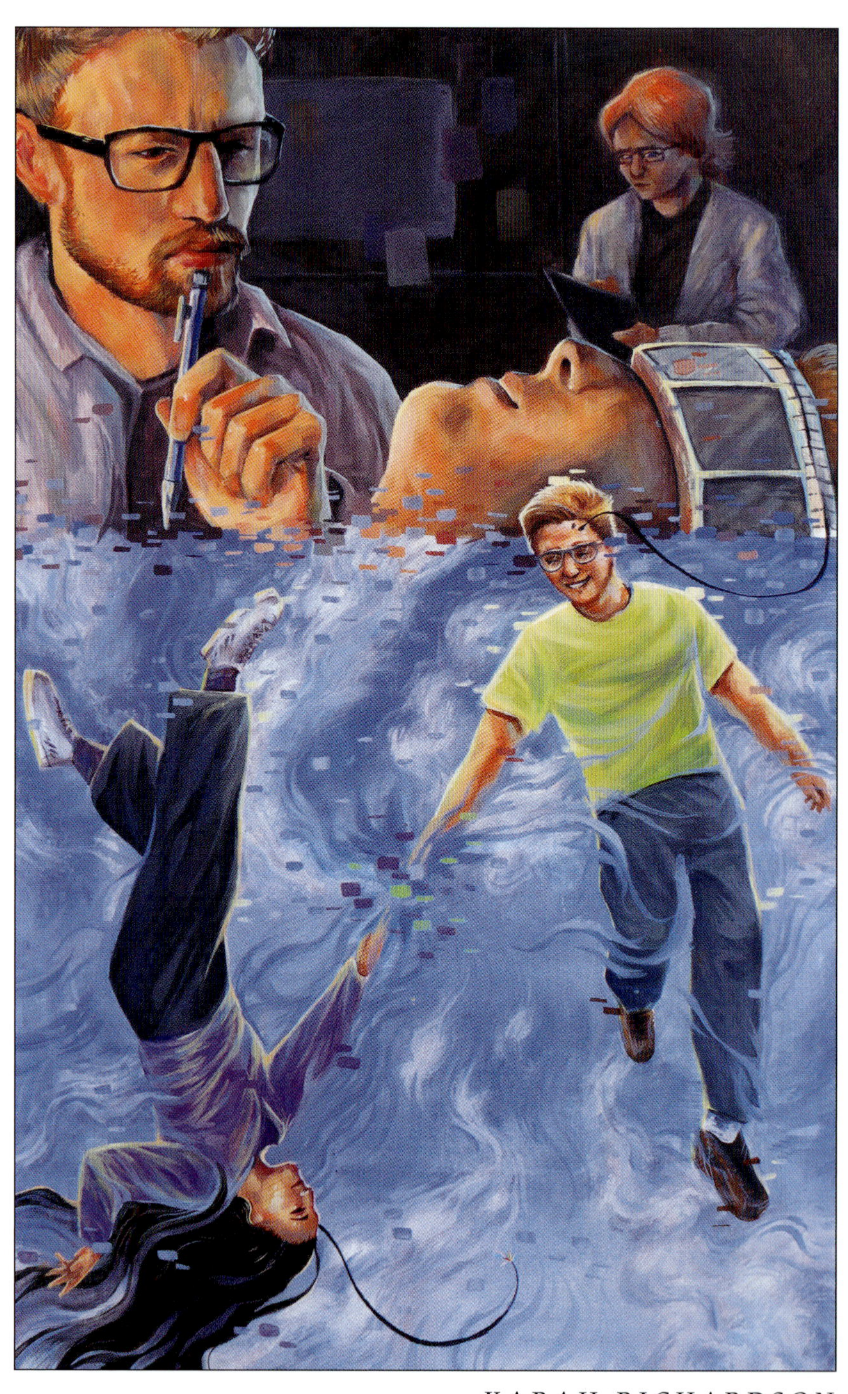

KARAH RICHARDSON
As Long as You Both Shall Live

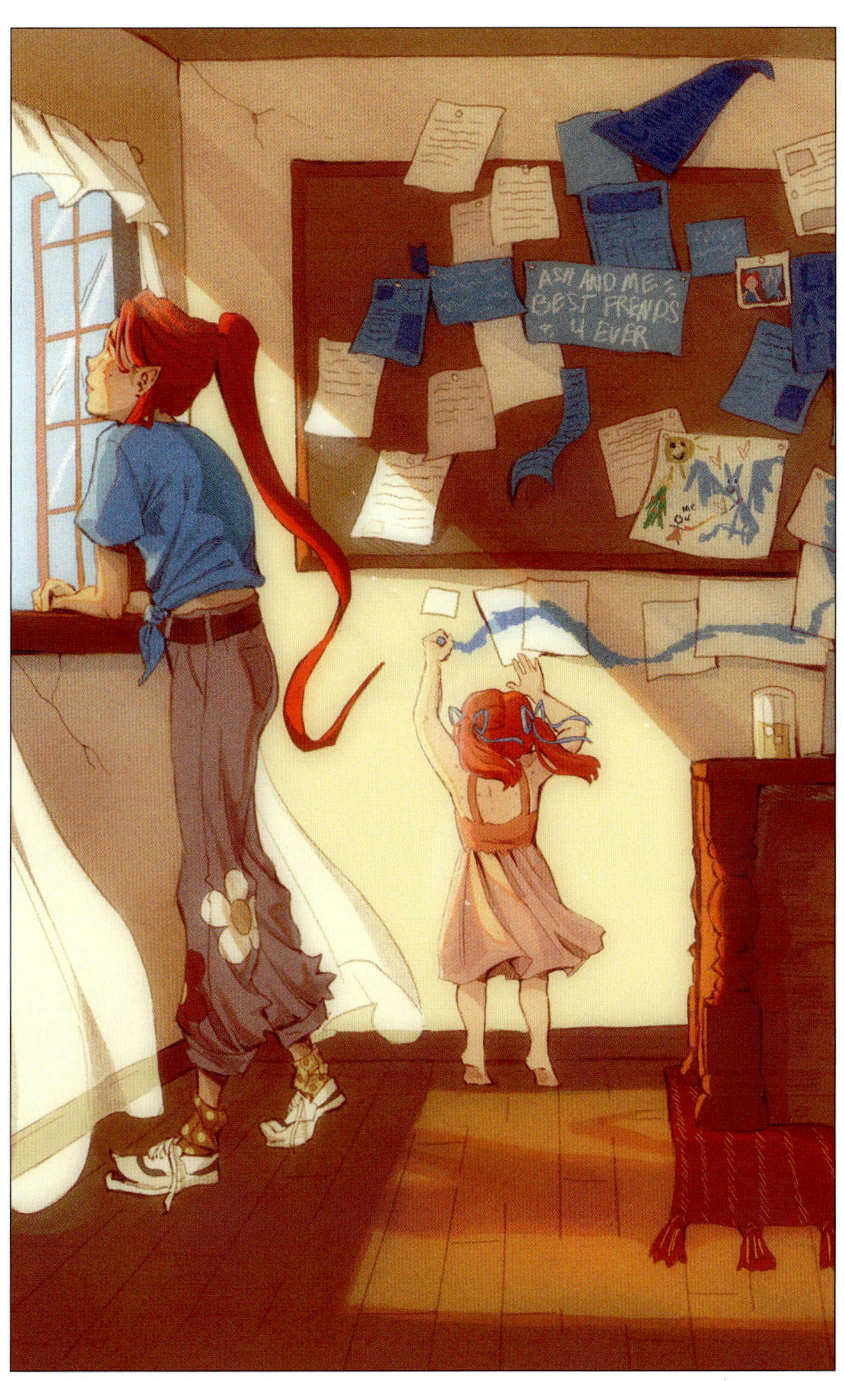

JOSIE MOORE
A Girl and Her Dragon: A Life in Four Parts

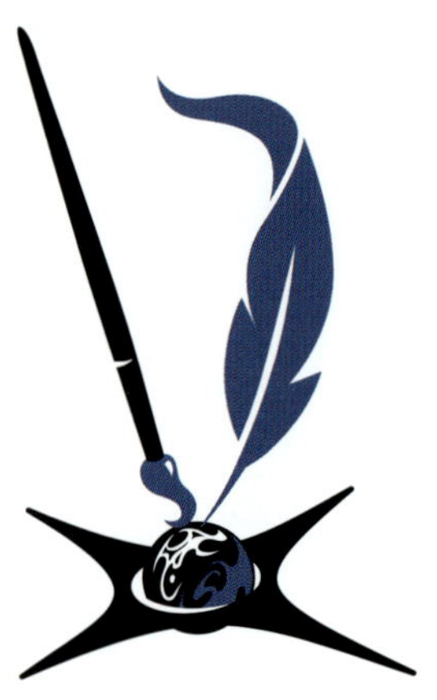

Form 14B: Application for Certification of Consciousness Transfer (Post-Mortem)

written by
Thomas K. Slee

illustrated by
ART IKUTA

ABOUT THE AUTHOR

Thomas K. Slee grew up reading his mom and dad's collection of pulp science fiction, thrillers, and detective stories in the western suburbs of Melbourne, Australia. The first thing he can remember writing was a Dinotopia spin-off before progressing to Warhammer 40,000 stories, then essays, then engineering reports. In 2020, the pandemic hit, and he started a writing group with his parents, his girlfriend's dad, his brother, and his cousin as a way of keeping in touch while locked inside.

Five years later, the family writing group is still going strong (they're writing a '40s noir radio play together). Thomas has had work published in Aurealis *magazine,* The West Australian *newspaper, and* Book XI: A Journal of Literary Philosophy. *He is an active member of the Melbourne-based speculative fiction community Meridian Australis and has successfully self-published his debut technothriller:* Project Gateway.

His story, "Form 14B: Application for Certification of Consciousness Transfer (Post-Mortem)," started life as a thumbnail idea from a brainstorming session for Chris McKitterick's 2023 "Science into Fiction" Writing Workshop: how much paperwork would be required to get your mind transferred into a new body?

ABOUT THE ILLUSTRATOR

Art Ikuta is a Vancouver-based artist moonlighting as a freelance illustrator. A big fan of sci-fi, fantasy, dark fantasy, and the horror genre, he is inspired by the likes of Kentaro Miura, Katsuhiro Otomo, Brom, Donato Giancola, and James Jean.

Art loves creating paintings that center on a narrative built around characters. He loves to fill his paintings with lots of little details and is always in pursuit of crafting a better piece with each work.

With a passion for tabletop games, Art's current focus is creating work in the trading card and role-playing game field and closing the circle that started him on his art journey.

Form 14B: Application for Certification of Consciousness Transfer (Post-Mortem)

Roy balanced me on his knee and shifted in his seat. The moulded plastic creaked in protest, and I wondered just how many other poor sods had suffered here before us. You'd think the waiting area at the Central Office of Births, Deaths, and Marriages would be designed for, you know, waiting, but apparently not.

Not that it bothered me. But it bothered Roy, and his evident discomfort made us even more conspicuous than we already were.

"Applicant 67F to counter 6. Applicant 67F to counter 6."

"What's our number?" I asked.

I still hadn't gotten used to my new voice, or rather, how my old voice sounded coming out of a speaker in place of a mouth. It was like hearing the "me" from my streams, but instead of replaying the things I'd already said, edited, and polished for my incredible subscribers, it was spruiking my private thoughts for everyone to hear.

"Ninety-three G." Roy sighed.

He shifted again, sending my view of the waiting room rearing up, careening back.

"Can you stop wriggling, please?" I snapped. "I can't close my… I can't turn off the camera and it's making me sick to my stomach."

Roy clamped his hand over the top of me, blocking everything but the light that managed to squeeze between his fingers. When he let me go, my camera flared, taking a second to adjust to the sudden brightness, and I found myself sitting on a side table.

"You don't have a stomach anymore."

"Tell that to my digital nausea."

Roy smirked and patted my temporary brainbox like I was a little kid. "You just let me know if you're about to spew ones and zeroes. Don't want to make a mess of the carpet."

"How about you just—"

"Applicant 68B to counter 3. Applicant 68B to counter 3."

I let the announcement cut me off. With my old body permanently six feet under, and Chloe out of the picture, probably for good, Roy was all I had. I needed him, and he knew it. I hated being so reliant on just one person, and he knew that too. I wanted to scowl, to cross my arms, to turn away in a huff, but I couldn't do any of those things. Not until I was uploaded to my new body, and that couldn't happen until I got my transfer cert approved.

"Applicant 69A to counter 11. Applicant 69A to counter 11."

Ugh.

I was gonna be stuck in here forever.

File Reference: *Finally! XxCavemanxX Takes the Ultimate Plunge! Part 1.mp4*

First uploaded to XStream 06/11/2029 @ 08:06

Archived from XStream 12/11/2029 @ 16:22

Drone-shot: *Tracking XxCavemanxX's Toyota Landcruiser ($97,990 Drive Away!), equipped with an ARB antenna and two spare Cooper Discoverer (Buy one, get one free!) tyres on the back, as it speeds along a dirt track, kicking up a cloud of iron-red dust.*

Side-On Close-Up: *XxCavemanxX grips the wheel and stares ahead, his head rocked back and forth by the corrugations in the unsealed road. In the background, the passenger seat is empty.*

XxCavemanxX: *"This is the Nullarbor. Isolated. Desolate. Unforgiving. The largest block of exposed limestone bedrock in the world. Home—[XxCavemanxX finally turns to face the camera, his eyes bright, brimming with excitement]—of the Widowmaker!"*

"That's right, my glorious posse of adrenaline junkies, thrill seekers, and nefarious badarses, you asked and we... [XxCavemanxX's manic energy falters, just for a moment] and I listened. Just up this road lies the entrance to the Nullarbor's most infamous cave system. Seventeen thousand metres of dark, twisting limestone fissures. Underground rivers. The lost bodies of

at least three of my fallen brothers, and the deepest, darkest, coldest pit of despair in the southern hemisphere.

"I'm going down there, and if you like, share and subscribe, I'll be taking you with me."

Ninety-three G flashed on the screen above counter 14, grey letters against white. Roy plonked me down, a little too heavily for my liking, and smiled at the woman on the other side of the plexiglass. She smiled back, as if she genuinely enjoyed working here, amongst the grey carpet tiles, the beige walls and the never-ending streams of people. As if she just couldn't wait to help me with my paperwork.

"Cameron Mannagh?" she asked, improbably brightly, and directed her question at Roy. As if I didn't even exist.

"Name's Caveman, and I'm down here, love. I'd wave, but, well. You know."

"Oh, uhh..." Her smile faltered, and her gaze flicked back and forth between the battered blue tube that I was forced to call home and Roy's sympathetic, apologetic face. I enjoyed her confusion, and used it as an opportunity to scan her up and down. The office attire wasn't really my go, but...actually. I caught another glimpse of her eyes, the same crystal blue as an underground lagoon. Untouched. Pristine. Yeah, if I'd still had my old body, she would have gotten me in big, big trouble.

"Don't mind him," Roy said, giving my brainbox a flick. "He's always been a bit of a dick. I'd hoped dying might have taught him a little humility..."

"Hey, I didn't make my millions streaming because I'm bloody humble." God, I wanted to show off. To preen, kick-up a stink, cause a scene. Something!

Roy just waited to see if I was finished.

"You're done? Good. I'm Roy, anyway. And *Caveman* here is my favourite—"

"Best-paying and only—" I interjected. I'd heard the eyeroll and the air quotes he'd used as he said my name.

"Client."

"Okay, um..." Blue eyes shot me another glance, swallowed her discomfort and forced her smile back onto her face. “My name is Brianna. Such a pleasure to meet you both.” She glanced across at her workstation, and even through the wobbly plexiglass I saw realisation, a little relief and then a wave of sympathy softened her features. “Ah, you’re here to get your consciousness transfer certified. We don’t get many of these, especially…”

She turned to me, looked me directly in the camera. “Camer… I mean Caveman, I’m so sorry for your loss. I’ll do my best to make this as easy as possible. Now, you’ll need to fill out form 14B…”

“All sorted,” Roy said, tapping his smart-band against the reader on our side of the plexiglass. Brianna’s workstation pinged as my pre-filled paperwork appeared on her screen and reflected in her eyes.

“Great!” She scrolled, her lips making silent shapes as she scanned through the details of my life. My old life. The life, and the freedom, that I desperately wanted back. Roy, as if he was attuned to my anxiety, tapped the countertop with his thumb.

Brianna paused, casting a furtive glance my way. Coy. I liked that.

“Problem?”

“No, I don’t…” She pursed her lips. “Would you excuse me for just one moment?”

Poor me, I could only watch as Brianna turned and sashayed from my field of view. And then, she was gone, leaving me in my rattling brainbox, sitting on a shelf.

“You better not have stuffed up my bloody form,” I said.

“She… she probably just needs some help. She said it herself, they don’t get many cases like yours.” Roy moved behind me, probably so he could peer around the corner, see if he could figure out where Brianna had disappeared to. For once I was glad of my limited sensory range. At least I couldn’t smell the stale alcohol on his breath. “I’m sure it’s fine.”

“Roy, I’m not kidding. If I spend one second longer than I have to in this godforsaken tube because you had one too many while you were filling out my forms—”

“Cameron Mannagh?”

The voice came from behind me, behind Roy. “How many times

do I have to…my name's Caveman. Roy, can you at least pretend you're paying attention? Spin me around!"

Roy's thickset fingers overwhelmed my camera again and the room spun, turning my vision into a mess of digital artefacts and blocky pixels. When they finally cleared, I found myself cylinder-to-face with a harried-looking woman. Greying hair, greying eyebrows, a grey cardigan, and an armful of files.

A real downgrade.

She looked like she'd been born in the waiting room and never left.

"Cameron Mannagh?" she repeated, looking over her glasses at me, right through my camera.

"In the flesh," I said, without a trace of humour.

"And you must be Roy Dennis." Her gaze flicked up above my cylinder, to where Roy's face must have been, and hardened. "My name is Hannah, and I'll be taking over your case. If you could please follow me."

Nondescript Hannah turned on her heels and led us towards a nondescript door set into a nondescript wall and held it open for us, as if she had no doubt Roy would follow. More than anything it was her confidence, her air of quiet, tired authority, that warned me something might be wrong.

File Reference: *Finally! XxCavemanxX Takes the Ultimate Plunge! Part 2.mp4*

First uploaded to XStream 06/11/2029 @ 09:58

Archived from XStream 12/11/2029 @ 16:24

Drone-shot: *The Toyota Landcruiser is parked on top of a limestone ridge, and XxCavemanxX stands on the lip, peering down into what appears to be a common sinkhole. He holds out his hand and the camera zooms in on XxCavemanxX's face. The drone is dropping from the sky, homing in on his upturned palm. He grins and turns the GoPro Hero18 (twenty percent off with promo code: Widowmaker!) camera down to show a dark opening in the earth.*

XxCavemanxX: *There she is. Doesn't look like much, does she? [The camera zooms in on a crumbling depression with a narrow opening at its centre. The cavern entrance is completely black.] But we know better, don't we? There's more to the Widowmaker than meets the eye.*

Graphic Overlay: *Electric blue on navy. A pulsing red dot at the entrance, the centre of a tangled web of caverns, tunnels, underground rivers, and subterranean lagoons. A lime-green line weaves its way through crevices and submerged passageways, tracing XxCavemanxX's path to the Widowmaker.*

XxCavemanxX, speaking over the Graphic: *1300 metres, 700 of them underwater. A mix of sculpted ravines and barely open cracks that I've been told are tighter than... well, you know what. [The green line reaches a large jagged oval. The sound of jangling carabiners and boots scraping on sun-hardened dirt is audible in the background.] And then she opens out into a cavern twenty metres across and only ghosts know how deep.*

Today, my friends, we're gonna find out.

Fade Transition: *Close-up side shot from XxCavemanxX's helmet cam showing his face in profile, rope taut, leaning back from the rock face. He looks nervous. Vulnerable.*

XxCavemanxX: *Before that, though, I just wanna say thank you, to everyone who commented on the last video. That you all noticed, and that you shared your concern and your love means the world.*

Yes, I'm out here solo, and yes, it sucks. I wish it were different. But I'm not alone. How could I be alone with all of you watching on?

[XxCavemanxX kicks away from the wall and the rope fizzes through his hands. He pauses at the entrance, looks up at the bright blue sky stretching to infinity overhead.]

All right, fam, I'm about to go dark.

See you on the other side.

The back office of Births, Deaths, and Marriages was a rabbit warren. Even with my digital memory I'd still have needed a mainline and a safety to get myself back out again.

"In here, please, Mr. Dennis," Hannah said, waving us both into a small, windowless room. Roy baulked at the threshold. I didn't blame him. The room contained only a single steel table and four aluminium chairs. All it needed was a metal handcuff loop welded to the table centre and a mirrored window.

"If you could take a seat, please, Mr. Dennis, I'm sure we can get this straightened out in no time at all."

I kept my mou——I stayed silent. There are moments when you

let the lawyers do the talking, and I knew this was one of them. Despite his flaws, and his appalling personal hygiene, Roy was still a good lawyer. He didn't move an inch.

"What's going on here?" he asked.

"A few items we need to clarify about Mr. Mannagh's paperwork—"

"What kind of things?"

Hannah sighed. "If you'll just take a seat, I'll gladly walk you through—"

"Uh-uh. I brought my client here to file a routine application, and instead you're coaxing us into an interrogation room. Why does Births, Deaths, and Marriages even have an interrogation room? Who are you?"

"Mr. Dennis, you're jumping at shadows. Your client, Mr. Mannagh, is a high-net-worth individual. All I'm trying to do is conduct some due diligence in a more private setting than the processing hall would allow. So please, sit down."

"I know an interrogation when I smell one, and this stinks. Come on, Cam." Roy hitched me up into the crook of his elbow and backed away down the corridor. "We're leaving."

"Of course, Mr. Dennis. You would be completely within your rights to do so," Hannah said, then turned her gaze down to me. "Though I should advise your client that, as things stand, his application for a consciousness transfer will be denied."

"I'm sorry what?" I blurted. "Denied? Why?"

"Cam, don't listen to her. She's just some power-hungry bureaucrat who gets her rocks off by being obstructive." I listened to him, but I watched her, and she simply shook her head. Roy raised his voice, turned on his heels, and yelled back down the corridor. "Your departmental lawyers will be hearing from me."

"Roy, stop." My view of the corridor rocked and lurched as he strode purposefully away, ignoring me completely. "Roy, I said stop!"

Panic distorted my synthetic voice.

"Roy, you know the rules. I told you—"

"Shut it; just trust me, okay?"

But I didn't trust him. I was in limbo, a fragile digital copy, stuck in transit between one body and the next. Legally, as my guardian,

Roy had a duty to protect me and to comply with my reasonable requests. But practically? I was totally, utterly within his power. If he took me away now, I might never get my new body. I could be stuck, like this, for a very long time.

All I had was my voice.

"Hannah, you heard me tell him to stop, right?" I projected my voice so loud my speaker crackled. "He has to stop. He has to!"

"I heard you, Mr. Mannagh."

Hannah's voice echoed down the corridor, and beneath it I heard Roy groan. He slowed to a stop.

"Cam, listen to me. You don't know what you're doing—"

"You're not the one stuck in a damn pineapple tin," I hissed. "I don't see what the big deal is, anyway. She's right, I'm loaded. I want them to be thorough."

Roy stayed stock-still. All I could hear was his heavy breathing, the rustle of my temporary body against his suit jacket and the dull resonating thump of his heartbeat.

"Something isn't right here. I don't like it—"

"I'm not paying you to like it, Roy. I'm paying you to get my certification approved. So, unless you can produce a damn good reason, I want you to take me back."

For a long second, he remained silent. I could imagine him clenching his jaw, glaring, his mind racing. What the heck had gotten into him? A part of me wanted to trust his rum-sodden gut and let him take me out of here, but then what? There were other ways to get a body, I knew, but they were dubious at best. And what use was a body if I couldn't certify it? I'd be an outcast, an illegal copy.

A pirated version of myself.

This was the only way.

"Roy…" I started, but he was already moving. Begrudgingly, Roy turned his back on the door and took me back to the interview room.

File Reference: *Unedited Recordings, intended for Finally, XxCavemanxX Takes the Ultimate Plunge Part 3.mp4*

Recordings Timestamped 11:33 to 11:47, 06/11/2029

Recovered from Cameron Mannagh's personal effects, 09/11/2029

Close-up: *XxCavemanxX's face is barely visible. The only light is from his head torch, reflected back at him from the close limestone walls. The grime on his cheeks is streaked with sweat. He grits his teeth, and—*

XxCavemanxX: *Oof! [He stumbles, the vision lurches.] Far out, that was tight. [He looks around, eyes darting, and slips back into the crevice. Braced against the wall with one hand, he drags his scuba gear through and shuffles it out of frame. He takes a deep breath, presses himself back into position and swivels the camera away from his face.]*

Here we are, folks. We're about to get wet. [The camera pans out from the crevice. XxCavemanxX's headlamp throws light across a tall, narrow cavern. The walls are smooth, as if they've been sculpted. Three paces ahead of him, the limestone bedrock plunges beneath crystal clear water. It is perfectly still, casting the white rock in an eerie blue light. XxCavemanxX swivels the camera back to him and waggles his eyebrows.] This is going to get spicy.

[He holds the smile for a full second, only sagging once he knows the take is good.]

Ugh.

[He casts around, scanning for something, shaking his head and muttering under his breath. He shrugs off his backpack and tears his camera off his head. The vision scrambles, light and reflections and clean white stone finally resolving into XxCavemanxX standing with his hands on his hips beside his scuba gear on the other side of the cavern. He breathes in. Breathes out. Hefts his gear.]

All right, you lot. You're my chosen family, and we all know that family puts safety first. This is a solo dive, so I've got to take extra precau——

[The vision slides, topples—]

Crap. Crap!

[Blurred close-up of calcified stone. In the corner of the frame and upside down, XxCavemanxX, face like thunder, storms across the cavern.]

Bitch! How am I meant to do all this on my own? [He grabs the camera and jams it back into position, his voice and his features twisted with cruel mockery.] I do everything for you, for our business, and this is how you repay me? You're a glorified camerawoman! [He stands back, cocks his head to one side. Nods.] I'm the Caveman. I am *this damn business.*

She'll come crawling back, just you watch.

[He rubs his hands across his face, sucks in another deep breath. Opens his eyes, bright with a manufactured smile.]

So, here I am. About to take the plunge with you, my chosen family. And we all know that family puts safety first...

[Multiple takes of XxCavemanxX going through his sponsored gear, his emergency equipment, and his preparation for the dive...]

Close-Up: *Head cam, focused on XxCavemanxX, fully clad in his scuba gear. His mask is pressed tight into his face, pushing his lips out like a fish. He steps into the water and shivers.*

XxCavemanxX: *Ooh, chilly. [He holds up his re-breather.] All right, everyone, you know the drill. I'll be switching to thought narration from this point on. I'll try to keep it clean, but no promises [wink].*

[XxCavemanxX pulls his microphone from around his neck and tosses it over to his dry gear. He swivels the camera out to face forward and sinks down below the surface. For a transitory moment, the water blurs everything as the camera struggles to adjust. Then, XxCavemanxX's headlamp submerges, filling the subterranean waterway with an unearthly light. The water catches the light and spreads it, as if the sun is shining. It's so clear it appears as if XxCavemanxX is just floating in space.]

XxCavemanxX's thoughts, captured via LogosLink™ *Neural Transponder: There it is. Right there. That's the moment, the money shot. The underworld magic.*

Man, Chloe would have loved this—

[A burst of bubbles obscures the camera, and XxCavemanxX reaches for the mainline, shaking his head.]

Damn it. Now she's ruined that too.

Roy set me down on his side of the table and slumped into a chair behind me, out of my field of vision. Only once he was seated did Hannah close the door. She dropped her files onto the table with a thud powerful enough to cause my brainbox and my video feed to jump.

"Right, Mr. Mannagh. Before we get started, I'm obliged to inform you that this interview is being recorded in both audio and video and may be used for training purposes. But don't let it worry you. With your cooperation, I'm sure we'll have this wrapped up in no time at all. Now." She whipped open her file and pulled the top sheet towards her. "Can you please state your full legal name, for the record."

Behind me, Roy coughed with disdain and rustled in his seat. I ignored him.

"My name is Cameron Peter Mannagh."

"And your streaming handle is...Caveman, is that right?"

"Yes." I yearned to be able to lean over the table, lower my eyes, turn on the charm and flash Hannah my impish smile. Even Chloe couldn't resist its charms. At least, she hadn't been able to. But it was all moot. I was stuck with my modulated voice, and it didn't feel like I controlled even that, not really. Still, I did my best. "You should look me up. You might like what you find."

"Oh, I have, Mr. Mannagh."

Her eyes twinkled, and the part of my mind that thought it had a body to control tried to pull back my shoulders, lean back on my chair. "What did you think?"

"I thought you offered something unique, Mr. Mannagh." Hannah smiled, though I couldn't be sure if her mirth was directed at me, or behind me. "Now, how many of us can say that, hmm, Mr. Dennis?"

"Just stick to your questions, Miss. And you," Roy grunted, flicking the back of my brainbox to make sure he had my attention. "I think it's time you let me do the talking, yeah?" Hannah opened her mouth as if to protest, but Roy must have done something, because she stopped before she could speak. "We've already answered all your questions, provided all your documentation. Expiration certificate, proof of initial transfer, the works. And don't try to bullshit me about errors or inconsistencies. I may not look it but I'm good at my job, damn good. I don't make mistakes." I heard his chair creak as he, what, leaned back into it? Crossed his arms? "So, how about you cut the bullshit and tell us what's really going on."

Hannah pursed her lips, and, after a moment's thought, closed her file. "All right, Mr. Dennis. No more 'bullshit.'" She folded her hands atop her folder and stared directly into my camera. "Mr. Mannagh, I'm afraid...I'm afraid I have some bad news."

"If you're about to tell me that I'm dead, I can assure you I am one hundred percent aware," I said. What the hell was going on?

I cursed my cheap, static, government-issued body. I wanted to see Roy's face. See what he was thinking, but I was stuck. Stuck with Hannah's sympathy.

"Look, there's no easy way to say this. Yes, Cameron's body expired. But"—Hannah exhaled, as if she didn't want to say what she was about to say, and patted her folder—"I have evidence that suggests his consciousness died with him. Which begs the question: Who are you?"

File Reference: *Unedited Recordings, intended for Finally, XxCavemanxX Takes the Ultimate Plunge Part 4.mp4*

Recordings Timestamped 12:02, 06/11/2029

Recovered from Cameron Mannagh's personal effects, 09/11/2029

Tracking Shot: *The tunnel is shaped like a pair of lips: concave along the bottom, ridged along the top, the roof and the ceiling pressed together in a smile at both edges. XxCavemanxX glides effortlessly through the passageway, like a space hulk drifting through a wormhole.*

XxCavemanxX's thoughts, captured via LogosLink™ *Neural Transponder: Not much farther now, my daredevil brethren. [A burst of bubbles. He twists his re-breather to check his oxygen levels: still three quarters full.]*

This is the calm before the storm. Just beyond that ridge . . . [XxCavemanxX points to a fissure in the smooth limestone, a crack running right across the roof-line.] Yep, there it is. You can see the tunnel curving away. Millions of years ago, this must have only been partially submerged. An underground waterfall, look at that . . .

[The tunnel widens, like a maw, opening from the inside, and beyond—total darkness. A blackness so deep it swallows the headlamp's light as if it were a candle in a midnight forest. The Widowmaker.]

[XxCavemanxX curves his body, wedging his fingers into the crevice, using it as a handhold, to stop his momentum.]

[The stone cracks. Like a gunshot.]

What the fu——

[Bubbles, a bass rumble—]

Aargh!

[Chaos, raining down, in shattering silence—]

My world lurched, the steel table looming and then receding, the dark opening of Hannah's mouth swiped from my field of view—

My speaker emitted a rising hiss, increasing in pitch and in volume as my processors chugged after my racing thoughts—

What did she mean, who are you—

Brick wall, whitewashed, pixelated—

Why is the room spinning? How am I movin——

Light switch, doorframe, Roy's fingers strangling the doorhandle—

"Aaaarrroyroyroyroyroy! Roy!" My scrambled hiss finally cohered into words. "Put me down!"

The door crunched open—

"I'd listen to him, Mr. Dennis, if I were you."

Roy clutched me to his chest so all I could see was the faded carpet and a great dark nothing where his jacket covered my camera. He stopped, his chest heaving, his smoker's wheeze rasping over my microphone's sensor.

Why did he stop? He never listened to me, but Hannah... Hannah must be obeyed—

Which begs the question: Who are you?

Why did he stop? Wrong question. Why did he try to run?

"You've got nothing on me," he grunted.

"If you take one more step, I can report you for unlawful detention of a digital consciousness." Hannah's voice was sickly sweet, somehow enhanced by the echo of the interview room Roy so desperately wanted to leave. To remove me from. "Of course, if that's not really Mr. Mannagh you're holding, if that brainbox's voice is just an algorithm, you'd be free to leave, wouldn't you, Mr. Dennis? There'd be nothing I could do to stop you."

"What are you waiting for? Put me down, tell her she's full of it!"

"Cam—"

"She doesn't know what she's talking about." I was shouting, as if yelling just a little bit louder would quash the swirling doubts she conjured in my mind. "You don't know a thing about me, Hannah! Haven't got a bloody clue!"

"Let's get out of here," Roy hissed, but I was far too angry.

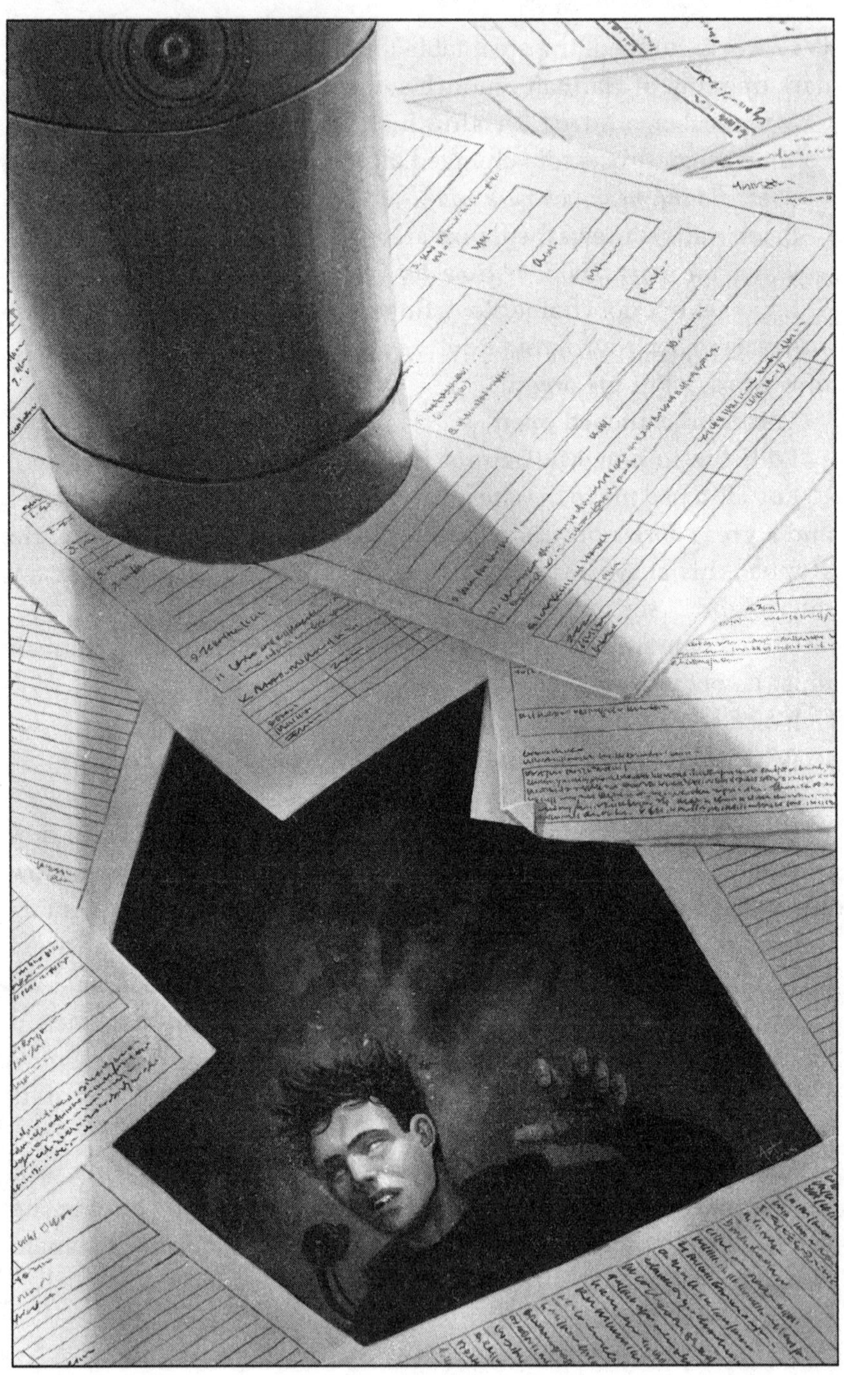

ART IKUTA

"No," I growled, my speaker clicking, hissing as it tried to process the raw edges of my rage. "She needs to understand that no one gets between me and my new body."

Who did she think she was, this woman? This paper pusher, this functionary? What had she ever done with her life? What did she know? I wanted to lay it all out for her, show her the pain she'd caused me, make her see. Can an algorithm hurt like this, Hannah? Can a scrap of code be this desperate?

Roy swung me around, and the interview room's bare light flared out, blinding me. By the time my camera adjusted, I was back on the table. Roy was hovering over me; I could see his shadow, looming on the wall. And Hannah, her face awash with smug, insincere empathy, was looking down on me over her folded hands.

"Mr. M——" Hannah paused, and I saw the indecision, the uncertainty, in her eyes. "I'm on your side."

"Since when has the government ever been on my bloody side, hmm? All you ever want is my money. That's all—" Chloe's smile rose up, unbidden. Bright, so vibrant it hurt. My voice glitched as I pushed it away, turning my words high-pitched. Synthetic. "That's all anyone ever wants."

"I wish I could say you were wrong," Hannah said. I only noticed on playback how her eyes flashed up at Roy as she said this. At the time, I was too incensed, too engrossed in my and Chloe's past. "Perhaps that's why you're here with your lawyer instead of..."

"Don't. Just don't," I said. Couldn't she see how...? But, no. Of course she couldn't. I was just a camera and a speaker in a tube to her. A box full of thoughts I didn't want to have but couldn't help.

"Okay." Hannah clasped her hands together on the table. "Why don't you tell me what happened, then?"

"What happened?"

"How you came to be here, in your temporary digital housing, talking to me." Behind me, Roy grunted, but didn't speak. The soles of his shoes scuffed the linoleum floor, and I could picture him rolling his eyes, arms crossed over his belly, shoulder pressed into a corner. Hannah did her best to ignore him. "In your own words."

"In my own words?"

Hannah nodded. Roy scoffed. I knew I could do one better. I could play it all back for her, every minute, word for word. I dug through my files and layered them, so that the whitewashed brick walls became pristine limestone, the table morphed into the bottom edge of my face mask and the fluorescent lights my torch beam, leaving Hannah's face to float disembodied in the depths of a subterranean lake.

I pretended Hannah was just another one of my subscribers and let my old self take over, narrating my predicament. I told her about the rock fall, let her hear the panic in my voice. The pain, the fear. I told her that my leg was trapped. Implied that it never would have gone wrong if Chloe hadn't abandoned me, hadn't forced me to take on the Widowmaker on my own...

And, while I was telling, I was also watching, as if I too was a subscriber, one of my own followers. As if the worst moment of my life was just content. Just entertainment.

My brainbox, like any entertainment system, no matter how expensive, didn't process emotion and couldn't trigger pain. It didn't replicate my pounding heart, the mind numbing, soul-crushing fear. I could hear it in my voice, in my impotent contempt as I lashed out at Chloe for making me come down here all alone.

But I couldn't feel it.

I tried not to wonder what that meant.

"So, you were trapped?" Hannah asked. "Running out of oxygen?"

I told her how I'd broken my own leg. How it had been my only choice, my only way out. How I'd gritted my teeth, braced myself and wrenched. Twisted it free and almost passed out from the pain. I stayed silent, however, on how I'd forced myself to do it. How I'd imagined it was Chloe's leg, Chloe's pain, not mine. I just told her about the moment, blinding, like staring into the sun, as I'd pulled myself free.

I told her that from here on my memory was patchy. That I must have dragged myself from the water somehow. I remembered ripping off my mask, that first lungful of stale air. And not much else. Not the rescue, nor the flight to Perth. The next memory I had was of hazy light, white walls. Roy, hovering over my bed, discussing something with the doctor, something I couldn't quite make out.

And then, waking up in Roy's office. In this tube. Feeling not much of anything at all.

Hannah exhaled, letting go a deep, long-held breath. "That's quite a story. You have a real talent. I can see why... how Caveman attracted such an audience. But I'm afraid it's not true."

"You think I'm a liar? Some sort of bullshit artist, spinning tales—"

"No." Again, Hannah's eyes flicked up, over my head, towards Roy. "You just don't have the whole picture."

"But I was there! I was the only one there!"

In response, Hannah merely raised one eyebrow and slipped a tablet from beneath her manila folder. She flicked open the stand and set it before me. With a sad, apologetic smile, she opened the video player and hit play.

File Reference: *Unedited Recordings, intended for Finally, XxCavemanxX Takes the Ultimate Plunge Part 5.mp4*

Recordings Timestamped 13:19 to 13:21, 06/11/2029

Recovered from Cameron Mannagh's personal effects, 09/11/2029

[Cameron's headlamp throws light beams into the deep, and they illuminate nothing. He is lying on the lip of the tunnel mouth, one arm dangling over the Widowmaker's abyss. Intermittently, bubbles pierce the silence and engulf the screen.]

Come on, Cam. Come on. This isn't over. Not yet, not while you can still— Aarrgh!

[The head cam swivels, a movement of sudden violence, finding a wall of rubble where the long, smooth tunnel had once been. Cameron's left leg is trapped, buried up to the knee. A cloud of crimson leaches into the pristine water, curling up into the darkness.]

[The vision is clouded by bubbles. He's hyperventilating, his oxygen is in the red.]

God, I'm such an idiot. Such an idiot. I never should have come down here on my own, I should have listened to her. I should have...

Chloe, I'm so sorry. Why did I... I didn't mean it, I swear. I didn't love her; I didn't love any of them. I didn't... Oh, God! [bubbles, clouds of bubbles] You always forgave me, always, always. You always came back no matter what I did and now it's too late...

[Bubbles, the camera swings, back to the abyss, as if Cameron can no longer bear to look at what he's done to himself.]

Aaaaaaagh!

[He wrenches at his leg, maybe for the last time. Bubbles billow upwards and the vision shakes as he beats his head against the stone with futile rage.]

Chloe... Chloe... I'll change. I promise I'll change. I'll be better. I'll be the man you deserve—

[Bubbles, so many of them. There can't be much left. He knows there isn't.]

All you have to do is come get me, please, all you have to do is save me, save me, save me—

"Turn it off." I couldn't watch it anymore. Couldn't listen to him—to me—finally understanding that it had all been his fault, my fault. But I had to. My camera, my microphone, they didn't stop. They never stopped. "I said, turn it off!"

Hannah tapped the screen, and the interview room fell silent. The screen, though, showed a still image, a snapshot in time: my torch beam, a slicing shaft of light, disappearing into the bottomless murk.

Take away Cameron's remorse, the sound of his dying breaths, and that image of the Widowmaker was exactly as I'd remembered it.

No.

As I'd been programmed to remember it.

Who are you? Wrong question. What am I?

"I'm sorry." Hannah said, reaching out as if to offer me comfort, only to hesitate, perhaps realising how pointless such a gesture would have been.

"I... I don't understand," I said, even though a part of me did. "Roy, tell her she's..."

"Roy's gone."

I listened for his laboured breathing and wished, hopelessly, for an outburst, for a lashing of his trademark cynical bluster. *Don't listen to her, Cam. It's all bullshit, just another government wank, sticking their nose in where they're not wanted. I'll sort it out, just leave it to me.*

But it's not, though, is it Roy? Not this time.

This time, I'm the bullshit. The fabrication. An algorithmically derived facsimile of a real person, procedurally generated from my own content and enhanced with a custom ending that you designed. Accurate enough, as long as nobody looked too hard.

Which is what you'd counted on, hadn't you, Roy?

"Bastard," I said, more resigned than angry.

Why wasn't I angry? I should have been fuming. I know Cam would have been. If Roy had pulled this on him, he would have shouted so loud that his face, his cheeks, his whole neck turned red. And that's all I was. I was built from Cam's performances, a condensed version of the person he'd sold himself as. The Caveman. The person his subscribers knew and loved. And just like me, Cameron had started to believe that the Caveman was actually real.

Until his oxygen started running out, and the real Cam broke through.

Too late, though. For him, anyway.

"Why?" I pleaded. "Why would he do this?"

"I suspect it's just like you said." Hannah interlaced her fingers, and I wondered just how much else she knew. "He wanted Mr. Mannagh's money. As his lawyer, he had access to all of Caveman's archival footage. With all the uncertainty over the disappearance and the time it took to recover the body, Roy must have figured he could scrape the data, whip you into shape, and slip your consciousness transfer through before Chloe had the chance to compile all this"—she nodded at the stack of paperwork, her tablet, the reality of the real me's final moments—"and file for her husband's death certificate."

As soon as she said it, I knew she was right. God, I could see the picture so clearly. Roy, sitting at his desk having just heard from Chloe that the rescue team had recovered the body. A half empty bottle of rum in his lap, his beady little eyes plotting. Scheming. Coming up with his little plan.

And if it had worked, if Hannah hadn't been paying attention, doing her due diligence? He would have bumped his fist on Brianna's counter. He would have congratulated me, swept me from the waiting room, and then turned me off the moment we were out of the building.

What then? I was already a construct. How many lines of code would it have taken to force me into signing away my power of attorney? A dozen? Maybe two? And then he would have been gone, taking Cameron's money with him.

"He was going to wipe me, wasn't he? Once he was done with me."

Hannah nodded. "Probably, yes. You would have been evidence."

Ah, so that's what I was. Evidence. An artefact. In a way, it was nice to have an answer.

Hannah picked up her tablet, swiped her hand, and minimised the last image I ever saw. No, that Cam ever saw. I was never there, in that cave. I... So much for Cameron's dreams of regaining his body, his freedom. Or my body, my freedom.

"So, what happens now? You power me off? Put me on a USB stick and file me away with everything else?"

She considered me for a moment, tablet in hand, then flicked its stand back open and set it down in front of me. She pulled a cord from her pocket.

"Do you mind?"

I scanned it, but it was just a cable. "What's it for?"

She tapped the top of her tablet and pulled me towards her. "I figured, after having Roy control you your entire existence, you'd want to fill this form out on your own."

As she lifted the flap on my cable port and plugged me in, I looked back down. The screen displayed a form: *14F: Application for Registration of New Consciousness (Algorithmic–Derivative Sentience)*. She connected me, and suddenly the form hovered in front of me, a digital construct that I could touch and feel.

The cursor blinked in the first blank space, just waiting for me to fill it.

"But I..." The form asked me who I was. Who I wanted to be. "I don't know what to..."

"That's entirely up to you."

"Up to me? What about...?" Was it even my choice to make? I wasn't just Cam, or even Caveman. I was Roy's. He'd made me. And I was Chloe's, too, at least in part. I replayed their last argument. Their last moment together, saw the betrayal in her eyes, felt the arrogance in his words, and felt like a voyeur. Like a trespasser in my own past. "What about Chloe?"

"She just wants to move on. To let the Cameron she loved stay in the past, where he belongs." Hannah lifted her hand again, and this time she didn't hesitate. She rested it on my casing, and I almost imagined I could feel it. "I know you'll respect that. That you'll do the right thing."

Hannah smiled at me like she had complete faith in me. Not in Cam, not in Caveman, but in me. What had I done to earn that faith? Or was that the point? Was she giving me the chance to prove I deserved it?

I wished I could respond somehow, return her touch of comfort, her trust, with a smile of my own. And I realised that maybe one day I could; I just had to take the first step.

I turned my attention to the form, to the first text box, and made the first real decision of my brand-new life.

I chose my new name.

Saffron and Marigolds

written by
Kathleen Powell

illustrated by
BAFU

ABOUT THE AUTHOR

Kathleen Powell received seven Honorable Mentions, nine Silver Honorable Mentions, and two Finalist rankings in the Writers of the Future Contest before claiming her place in the 2026 anthology. Since winning Writers of the Future, she placed third in the 2025 Baen Fantasy Adventure Award contest.

When not writing about changelings or happily reading folklore and fantasy from across the world, she is usually thinking *about writing changelings or reading folklore. (She also likes tea.)*

The inspiration for her winning story, "Saffron and Marigolds," came after reading "Circulate" by L. Ron Hubbard in Writers of the Future Volume 39. *In that essay, Hubbard writes, "And in despair, we wail that there is nothing of interest in our surroundings or in the lives we lead. We say that and we believe it... If we knew our surroundings well enough, we could put them on paper. Someone else comes around, looks us over and studies our environment for a brief period and then goes off to write a novel. Why, we moan, didn't we write that book?"*

Well, why not? After writing about New York and London and Iowa (they have a tulip festival!), she decided to give it a shot by writing about a place she knew firsthand. After all, why should faeries and dragons and grand adventures always exist only somewhere "far, far away"?

"Saffron and Marigolds" is based in the little city where she attended college and features good friends, high stakes, and cheese, because those, after all, are the essentials in any magic story.

She hopes the locals will enjoy recognizing a few familiar details, and that readers everywhere will remember impossible adventures aren't always so far away as we might think.

ABOUT THE ILLUSTRATOR

Bohuslav Argaláš was born in 2002 in the small Slovak town of Ružomberok, where he began drawing long before he learned to read or write—mostly because it was easier to invent stories than to spell them. Growing up, he fell in love with every form of storytelling: comics, animated films, video games, books, theater—anything with strong characters and a good plot twist.

That passion eventually led him to study art at the private school of Applied Arts in Žilina and later at the Faculty of Fine Arts in Brno, Czech Republic.

Known as Bafu, he creates illustrations and comics filled with humor, emotion, and a touch of the strange, hoping to make people smile, think, and drift away from reality—even just for a moment. He's currently working on his webcomic Jasper's Haunted Home*—a slightly spooky, slightly silly series about friendship, ghosts, and finding comfort in strange places.*

When he's not drawing, he's probably binge-watching, reading, and hyperfixating on too many things at once, or walking his dog, Yumi—definitely not talking about himself in the third person.

Saffron and Marigolds

Most girlfriends leave you ordinary stuff on their way out. The lingering scent of a shampoo you never use. A puppy you were going to raise together. A slightly dented engagement ring.

I got a dragon.

The girlfriend was a changeling, so some weirdness was to be expected.

Not that she'd ever *really* been my girlfriend. It was just the most convenient lie—because what else did you call the person who almost got you killed, kidnapped you, used you as a bargaining chip, and then almost died to save you? Was there a name for that nebulous, ambiguous relationship?

The other responsible party currently engaged in co-parenting a dragon.

See? Just too complicated.

Not that I said it a lot. Or at all.

I knew what they'd think. Oh, Arthur's had a psychotic break. Oh, Arthur's so lonely he's invented some kind of wish-fulfillment fantasy fairy girl so engrossing he believes she's real.

Right. Because of course I would imagine a stone-cold creature with zero facial expressions and no aversion to murder who had a zillion violent allergic reactions to the most ordinary things and a physiology so inhuman I couldn't tell if she was hurt or sick or tired until she collapsed, breathing only once or twice per minute while I kept watching, watching for the next rise and fall of her chest because her body temperature always ran so low, I couldn't tell if she was alive or already going death cold.

Of course, I would dream up someone who talked to worms and preferred spiders and trees to people, and hated all humans on principle, including me.

Including me.

The dragon's name was Wantley, and he feasted on cheese. He was red, scaly, approximately twenty feet long, and extremely bored.

The cheese was probably stolen. I didn't question who brought it or how it appeared. Let's say magic. Yeah. Sure.

After working my shift at the café every day, I hung up my apron, brushed the flour off my hands, and hiked through the weeds behind the old movie theater, kicking aside empty chip packets, crumbling foam cups, and broken plastic straws, bracing myself for the moment when my heel slipped and I shot straight down, earth swallowing me up.

The sinkhole opening was difficult to find with human eyes. Menura had never hesitated finding any of the holes that led down to the Underneath, but I had no such luck. I tried to memorize the clumps of grass, squint in the twilight, find the exact spot, but somehow the hole always found me first.

It led to a long, dark walk, boots splashing through cold water, shoulders sometimes brushing smooth stone walls on either side, water dripping, echoing as it etched new halls, until I emerged into a cavern filled to the brim with dragon.

I found Wantley chomping a wheel of cheddar between his jaws like a dog with a new chew toy, tail flicking back and forth across the cavern floor. Eating was one of the few times he seemed genuinely happy.

There wasn't much to do in the Underneath, in a cave he barely fit inside.

The situation had absolutely gutted me the first time Menura brought me here. I'd immediately gotten attached to Wantley, even before I knew he wouldn't try to eat me. Menura had, as usual, disapproved of my poor survival instincts.

Wantley pushed his snout into my chest as soon as I entered. He meant it affectionately, but it knocked the air clear out of me.

I'd lived in farm country long enough to know that any animal of adequate size—cow, horse, goat, determined chicken—could be dangerous. And that was without teeth half the length of my arm and breath like an oven.

Caution was demanded when visiting Wantley. Maybe things were different for fairies, but I was human, and my bones could break.

I usually left with light scratches and burns. Wantley was like a big cat. He wanted to play, catch, and tag, and "toss Arthur in the air and catch him at the last possible second," and I found it hard to refuse.

Wantley lunged forward, jaws snapping the spine of a rat that had been stealing his cheese crumbs.

I looked away, swallowing hard. Trying not to listen to the crunch of bones. I knew it was for the best, since cheese alone wasn't the most balanced diet. But despite everything Menura had put me through (scrambling over wire fences, being shot at, having ogre blood splattered across my face), I was still the same timid, nearsighted, pacifist baker I'd been before she kidnapped me.

Wantley nudged me again, pulling me from my thoughts and smearing rat blood on my sleeve.

I scratched him behind the jaw.

He made the *hr-r-r-r-r* sound in his throat that meant he wanted sky, wanted to stretch his horribly cramped wings, wanted to know where Menura was and why she stayed away so long.

"I know," I said. "You miss Mom."

Another, more agreeable *hr-r-r-r-r.* He leaned into my hand.

"I miss her too," I said. Crazy, scary, awesome Menura. Vanished, without even the courtesy of a puff of smoke.

It had happened when she was banished by the fairy king. The consequence of Menura's refusal to follow his orders, finish her job of kidnapping me.

I had no idea where she was. Or what it meant, really, except she was alive, somewhere.

And I still missed her, despite the whole kidnapping thing, even after four and a half months.

I rested my forehead against Wantley's warm scales. "I wish—"

I caught my tongue between my teeth hard enough to hurt. *Not another word, Arthur. Not one.*

Find the end of a rainbow. Rub an old perfume bottle. Leave a dish of cream on the step.

Leprechauns, Jinn, the Little People. Good Neighbors, Fairy Godmothers.

Kids hear stories. If you're lucky enough to find a fairy, it will give you anything you ask for.

You don't think what it's like for them. Being trapped, being tricked. Facing demands they can't refuse.

Oh, magic can't be real. Because how selfish could fairies be to keep it all for themselves, all hidden away—

Have you ever thought even for a second that you and your greedy little self might be the problem? Fairies don't have an obligation to fix your mistakes. Everything that's wrong with the world, we did it. Littering and crime and poverty and every last crumb of misery in the world—weren't we to blame? Wasn't it all because of what we did, choices we made?

And yet, somehow, we still expected someone else to come pick up our trash, clean up our mistakes.

Nobody, in any fairy tale, ever considered the cost of a wish.

When I met Menura, she screamed in my face to stop torturing her and say what I wanted. *Humans always want.* It was the one thing that united us—a passion, a craving, a constant desire for *something more.*

I had only ever given Menura one present. A little box of gingersnaps and macaroons tucked into her pocket. Hoping they'd make her feel better when she had appeared in the café, clearly hurt.

But fairies couldn't accept gifts.

Not without paying with a wish.

I hadn't meant to trick her into owing me one. I hadn't wanted it.

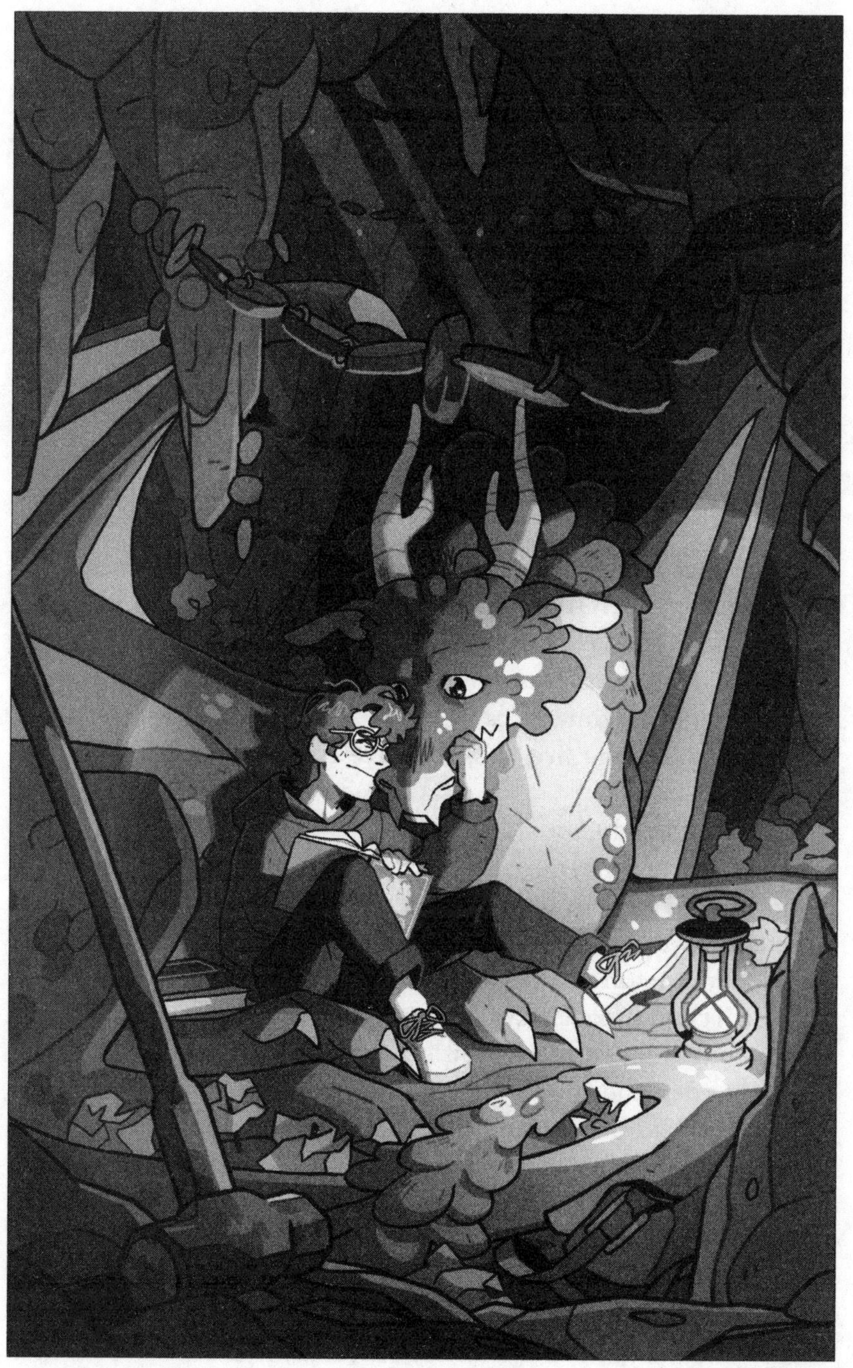

BAFU

I couldn't understand back then, how scary it was for her, someone just being nice.

Menura had been at the mercy of humans before. Trapped, scared, and unable to refuse.

She wore double earrings in her left ear. Little silver tags that chimed together when she casually hurdled the counter of the café where I worked or flicked her head, listening to something I couldn't hear.

There was a tear in the cartilage. A place where a third earring should have been, ripped out by human hands. Teasing her, taunting her, making her grant wish after wish after wish.

The day she told me was the day I understood why she hated me.

It was the day I swore to myself I would never make a wish.

Dragon eggs are innocuous little things. Only eighteen inches tall, by the span of Menura's pale hands, sketching the shape in the air. They can stay like a rock for what feels like the longest time. So small, so still, so silent, you cease to imagine them cracking open and the chaos a single dragon can cause.

Yes, they breathe fire. Yes, they are every bit as dangerous as they seem.

And, no, they do not stop growing.

Wantley's head frequently brushed the roof of the cavern. He could turn in a circle, but that was about it.

In other words, he was completely miserable.

It wasn't Menura's fault. Not getting Wantley out when he was still small enough to fit through the tunnel behind the movie theater. She'd stolen him from his nest by order of the Greene King, who seemed to demand the dragon and then forget him on a whim.

The two had been each other's sole companion since the day Wantley hatched. No matter where she had to go, what she had to do, Menura had always crept back to him, curling up small against his warm scales to sleep.

It wasn't her fault.

She'd been trapped, too.

Now she was gone, and Wantley only had me, a weak, nearsighted human with the faint scent of fairy girl clinging like dust and raindrops to my skin.

Menura wasn't coming back to rescue us. So the dragon was my problem, and, despite the hugeness and impossibility of the task, one I was determined to fix.

The thought of Wantley, rolling on his back in soft, green grass, feeling the breeze on his scales for the first time, the light of sunrise shining in his huge eyes, demanded it.

Determination did not free a dragon.

I lay awake at night, thinking until my head ached.

What could I do? I wasn't a changeling like Menura, who could reshape her face so she looked like a different person anytime she pleased. I wasn't a troll, who could hear the different kinds of stone singing. I wasn't fearless like the Ogre, the leering guard of the Underneath with his red mouth and loud gun, barring our escape from the fairy caves—at least, until Menura had punched her fist into his chest and ripped out his heart.

I shuddered at the memory and checked the alarm clock. Still another hour before I had to get up for my shift at the café.

The question remained. What could I, an ordinary, powerless human, do?

Not powerless.

Menura's voice echoed in my head. It thrummed in my bones, cool and blue and tingling.

I got up.

It was no accident I'd walked to the front of the café one day to find a perfectly arranged stack of muffins scattered across the floor, one of our glass panes cracked, and Menura, huddled with her back to the counter, breathing hard, arms cradling fractured ribs. The Greene King had placed a bounty on my head, and a dozen fairies had brawled in the bakery, fighting each other to claim it.

Menura had beaten them all.

Fairies are invisible to humans on all but a few days of the year. That day, half past five in the morning last Halloween, was one of them. The other fairies had run away when they heard my step. All except Menura, the changeling, the only fairy who looked human enough not to terrify me the moment I saw her, pained and strange, tangled black hair hanging over her ash-pale face.

She'd asked me to take her home. I only found out later the tall girl I had half-carried to the cave behind Walmart lured me there because the easiest way to kidnap someone is to have them come of their own free will.

I wasn't mad about it. Whatever the king wanted, Menura was bound to bring to him.

Even if it was a nearsighted baker boy who hyperventilated at the sight of anything dead.

And my gingerbread recipe, coveted by a fairy king.

Gingerbread, my café colleagues had informed me, should be crisp. Crunchy, with a faint tinge of molasses.

The confection I had made was a small, soft cake that could fit in the palm of your hand, saffron spiced with a core of sweet marzipan at its heart.

The antithesis, apparently, of everything gingerbread was supposed to be.

It would not go on our December menu. It would go nowhere, except the trash, because no one could be persuaded to eat more than a single bite.

That same rejected recipe got me kidnapped on orders from the fairy king.

Menura had tried to explain it to me. How the taste had taken them all home, rolled back a thousand years before their eyes to the age when fairies lived in cities of their own, when their music mingled with the sound of silver-white waterfalls, when they stood tall and proud and laughing under a ceiling of stars instead of stone, because once there had been a time when they did not have to hide.

Food could do that. Flavors could summon a memory so strong the waking world fell away. It was the closest thing to magic a human could do.

That *I* could do.

Despite the prospect of being locked underground to slave over a hot oven for the rest of my days, it had made me feel like I had a chest full of sunlight.

Because I could bake.

Of course, I seriously doubted a loaf of gingerbread would be of any use in rescuing a dragon. Especially when I still couldn't see any fairies to bribe with it.

Time to find out what else I could do.

Wantley sniffed at the sledgehammer unenthusiastically.

"I know," I said. "There's probably an easier magic way to tear the walls down, but I don't know it."

Wantley himself could probably destroy at least one wall by slamming into it—he'd knocked a few showers of gravel into my hair during tag before. But the Greene King and company were already on terms of cold hostility with us. I didn't want to know how angry they could get if we bashed down the wrong wall and tumbled headfirst into their midst, or worse, brought their whole city down on their heads.

Plus there was a chance of Wantley getting buried in the avalanche. That wasn't a mental image I was willing to risk bringing into reality.

Human boy had to do things a human way. So be it.

There might have been a wider tunnel branching off from Wantley's cave, but I wasn't about to go spelunking in the Underneath after just escaping it with Menura's helpful un-kidnapping last autumn. The Underneath was like a rabbit warren, tunnels going in all directions, and sprawled across several blocks, from the cinema to Walmart, and possibly farther. I knew the way to Wantley's cave and the huge corridor that led to the Greene King's throne room—the last place either of us should visit—but the rest was a vast unknown. If I got lost, I wouldn't be found. And knowing last time's reception, the fairies would probably lead me right to a sheer drop where I could break my neck and never be heard from again.

No, thank you. I would stick to the tunnel behind the movie theater which Menura had used, and all the other fairies therefore avoided.

If I could widen the shaft enough, Wantley could get out. What I would do with a huge, flame-red dragon terrifying moviegoers afterward, I wasn't sure. I wasn't thinking about that part yet.

One thing at a time.

"You stay over there," I said, pointing Wantley into the opposite corner. I didn't know how much rock would fall. As strong as his scales were, I didn't want to test them.

I adjusted my grip on the sledgehammer, palms already starting to sweat.

Wantley curled his tail around himself, watching me warily. He didn't think much of whatever I was doing.

Well. Stupid and dangerous or not, I had to do something.

I slammed the hammer into the rock.

"Limestone," I said through gritted teeth, "is supposed—to be—soft."

Wantley rested his head on his claws, eyes darting back and forth with each stroke of the hammer. I counted ten more hits and dropped it to the floor as I caught my breath, tilting my head back so the sweat wouldn't run into my eyes.

Two weeks of coughing up dust and blinking grit out of my eyes. Of slogging through dirty water that soaked through my boots. Of hammering into wet clay, which wouldn't crack or chip like stone, but stubbornly stayed squishily in place, so I felt like I was beating inedible mochi dough, only to find it glued under my fingernails hours later.

Having a stand mixer that consistently broke down during rush hour at the café had given me good arm strength, but repeatedly heaving a cast iron sledgehammer at a block of stone required a different level of muscle I didn't even begin to possess.

Everything ached. I was exhausted. I'd caught a cold. I'd been late for work twice. And I'd barely widened the tunnel an inch.

How many blisters did I have to get to make a dragon-sized hole in the earth? Ignoring, of course, the grinding fear in my stomach that one wrong stroke would bring half the Walmart crashing down on our heads.

It was impossible. I'd been a wishful idiot. I couldn't do this.

Wantley nabbed the ankle of my pant leg between his teeth and swung me upside down, parading me cheerfully to the other side of the cavern like a rat hanging by its tail.

"Wantley!" I yelled, grabbing my glasses just before they fell to the stone floor.

He did this every time I got frustrated. He wouldn't let me down until he decided I'd rested enough, no matter what I did.

I crossed my arms as I swung back and forth, waiting for the ride to eventually end. A bored dragon could be a time-consuming thing.

I tried to think if there was any way I could go at this faster. A jackhammer would be too loud, and I probably couldn't afford it. What else? A hammer with a spike? Did they make those? Or—what were those things they mined with in *Snow White*? Pickaxes? They seemed sharp!

"What are you doing?"

Wantley stopped. A blurry figure swung into view.

I squinted.

"Menura."

The world lurched and turned over as Wantley immediately dropped me, nuzzling Menura, making a high, happy *h-i-i-i-i*.

I fumbled with my glasses, blistered fingers clumsy.

She looked smaller than the last time I'd seen her. Shorter and younger, more delicate somehow.

The youthful innocence was somewhat contradicted by her familiar glare.

I didn't know what to say. Menura had left. I hadn't heard even a whisper of her. And now, she was back, as suddenly as she'd disappeared.

I wanted to put my arms around her.

It was a stupid thought. She would push me away, or stab me on a convenient stalagmite.

But all this time, I'd been wondering if she was there. If she'd gotten away, gotten out of the Underneath, or if she was just a breath away, watching as Wantley and I missed her.

"Hi," I said.

The illusion fell away like a jacket dropping from her shoulders. Towering height, blue-shot eyes, tangled hair, and dirty, bare feet.

There she was. The terrifying fairy girl I was so helplessly fond of.

She wasn't what you would call pretty. Too tall. Too dirty. She looked like she'd climbed out of a storm drain under a bridge somewhere—and probably had. A lichen was growing out of the shoulder seam on her shirt. She looked half a vengeful goddess and half a feral raccoon.

"Why are you smiling?"

I jumped. How long had I been gazing at her?

"I like—seeing your face again."

Menura gave me a long stare.

She always said she didn't have a face. She forgot what shape it was supposed to be, since she changed it so often. I could have described it for her, the blank features she always unconsciously returned to, sharp jaw, bold nose, ink-blue eyes—but I doubted she'd listen.

"You're a fool."

I grinned.

"Your fool, though." I petted Wantley a little farther down from her, where our hands wouldn't accidentally brush.

Wantley, of course, paid no attention to me. Which was fine. She'd hatched him. They'd spent decades of captivity together. Of course he loved her more than me.

"A fool with an iron hammer and marigold seeds in his pocket?"

I flinched.

"No. I mean, yes, but..."

Marigolds were toxic to fairy-kind. As dangerous and painful as salt or iron. Menura had brushed against a bunch of the flowers once and hissed in pain, leaving with a bumpy, purplish rash on her arm that had taken days to fade.

You idiot, Arthur.

"I didn't—I mean—they weren't ever for you."

"Fill the other pocket with salt and I'd feel better about you being here."

Menura was as cool as I was flustered, voice untroubled as still water.

I burned from the tips of my ears all the way down my neck.

Except for nine special days of the year, I couldn't see fairies. If it had only been Menura, I wouldn't have worried. But there were others in the Underneath, and they could stand right beside me, long fingernails tracing my bare throat, an ogre gun just over my temple—

I shuddered. Marigold seed had made me feel safer. I was glad Menura approved, but I still felt guilty carrying them. I hated the idea of hurting anyone.

Menura strolled off, examining my poorly chipped dent in the tunnel wall.

"Redecorating?"

"I think I'd call it a jailbreak."

"Is this your first, or..."

I burst out laughing. You couldn't tell from her toneless voice, but Menura was teasing me.

"What are you doing back here?" I blurted, regretting it immediately. "I mean, I thought..."

I said I'd bring him to you. I didn't say I'd let you keep him.

The words were as fresh as yesterday.

Menura, defying the Greene King's orders. Or more exactly—threatening his court and everyone in it by swearing she'd bring the cavern down on their heads if I didn't go free.

She was pretty scary when she was angry. Cool, though.

"The Greene King reconsidered."

Un-banished, then. "And you—decided to stay?"

I didn't get it. After so many years of being confined here, told what to do, where to go, who to be, being forced to follow the whims of the Greene King, why would Menura stay, when she could go anywhere else and be free?

Menura drew a painstaking breath through her nose. I'd gotten better about not bombarding her with questions. But there were times I would swear she could *feel* me thinking them.

"I still owe a debt to the Greene King. Until it is repaid..." she left the rest unsaid.

I stared at her.

That day when she had two earrings instead of three, blue blood leaking through her fingers, the king had arranged her rescue from the cruel hands of humans in the world above, brought a crying, bleeding little fairy girl home. That's what the caves had been for Menura, until, like Wantley, she grew enough to see them for the cage they really were.

Fairies always repay favors. No matter how steep the price, how vile the trick.

I didn't get angry much. But I could feel my fingers curling into my palms, the temperature of my blood rising, as if Wantley's fire burned in my veins.

"That was thirty years ago."

He'd saved her life, fine. That didn't give him the right to control it. This wasn't living, not having the freedom to choose what you did or where you went, living in a damp hole, without air, without sunlight, without trees or music or buttered croissants or—

"Arthur."

She only called me by name when she was trying to calm me down. I hated that it worked.

"You can't be serious."

Her jaw clenched.

"We can't change our nature, Arthur. We're fairies."

I'm not human like you.

I turned back to Wantley so she couldn't see the hot tears stinging my eyes.

Menura was gone the next day. I came straight from work, dropped my bag in the corner, and picked up the sledgehammer. I didn't know where she was. I didn't know why she'd come back.

I didn't expect to see her again.

I slammed the hammer into the rock.

Wantley made a noise I hadn't heard before. He tried to bump his nose against my back, but a flying shard of rock sent him quickly into the far corner.

Menura. Greene. Menura. Greene.

The hammer wasn't so heavy when I was angry.

The tunnel wall suddenly crumbled in a cloud of falling stone.

I coughed, choking on dust, eyes stinging.

No sweet moonlight. No cool breeze. Just another chamber, too small for Wantley.

I left the hammer and dragon by the opening and stepped inside.

Trash. Just trash. We saw a lot of that. People chucked it out their car windows and into sinkholes and it got swept down with the rain.

Wantley sniffed some of the rock dust and sneezed. The sudden light was like a small firework going off.

I squinted, bending closer.

Burger wrappers twisted into flowers. A dirty shoelace tied around a stalagmite, strung with colored soda tabs and rusty fishhook earrings. Shards of broken glass planted into mounds of clay like an intricate tiny city, empty of inhabitants.

My dark blue hoodie, the one that was the same color as her eyes. Pieces of Wantley's shell.

Menura.

It felt like an intrusion. Looking at things she'd salvaged, trash she'd twisted into roses, dirty, broken things that suddenly had new shape, tried to catch the faint light.

Menura's face was a featureless mask. She never smiled. Never cried. She was the strongest person I had ever known.

But she was also alone.

And this tiny, muddy, awful cave, full of rubbish she'd tried to make beautiful, tucked safely away from Wantley-fire, was the only home she had.

I sat down on the floor, pulled my knees to my chest, and cried.

"If you say 'Ready to order,' you'll sound impatient and demanding. Try '*May* I take your order.'"

Jeff, my tiny ginger-haired manager with the energy of seven shots of espresso, spoke with painstaking emphasis. The usual clinking and chattering sounds of the café carried on, the smell of yeast and the warm humidity of rising dough wrapping around me in an invisible embrace.

I tried to at least pretend to be paying attention. This was the fourth time the café manager and I had gone over this. I swore he changed the socially acceptable phrase every time.

Briefly, I fantasized slapping my hands down on the counter and staring until the indecisive muffin-munchers told me what they wanted to eat. Menura would've done it.

Menura…

"Arthur?"

I blinked. "Yes, Jeff. I understand."

He looked visibly relieved.

"Good. I'm glad we had this conversation. Now you can head back to the register…"

I zoned out again. Ideally, a baker could spend his days in the kitchen, punching dough and breathing in the scent of lemon and strawberries.

I hated working the register. Repeating the same phrase, over and over and over again, as if it were new and different, like an automated waiter, stuck in the same route between booths, rolling along, friendly and thoughtless, until it inevitably broke down.

I took off my glasses and rubbed my eyes.

It had been a month since I'd found Menura's little hideaway. I'd given up on chipping at the tunnel walls. It had been a pointless quest. I spent my visits reading to Wantley, slouched against his side like he was a big couch. He seemed calmer when he listened to my voice. My body thanked me for the relief, but—my mind was trashed.

We'd fought. Menura and I. And while I couldn't understand her choice to stay underground, to keep working for that horrible king, after seeing that tiny space, I'd accepted it.

Menura had spent so long with nothing. Maybe she was afraid to give up even that much.

My throat tightened whenever I thought about it. Every time I learned something new about Menura, I loved and grieved for her more. Since that very first day in the café, I'd wanted to make her feel safe, wanted to make her feel less alone, wanted—

Don't want, Arthur. Be satisfied. Don't wish.

But I couldn't even tell her I was sorry.

"Arthur?"

"Yes, Jeff."

"You look a little…" he trailed off.

"I'm fine, Jeff."

I walked to the front of the café and started taking orders. Four groups. Twelve groups.

Don't think. Go numb.

I'd been good at this job once. Now I felt like the Greene King, eating fairy cakes only to know he'd never taste them again.

I couldn't see the world in the same way anymore. Every rock, every drainage tunnel, every weird dip in the grass, every rumored ghost sighting and scruffy stranger—suddenly it was all wonderful, dangerous and beautiful, like a splinter of broken glass shining sunset gold as it caught the light.

I'd felt everything with Menura. Scared and sad and proud and exhilarated and everything else someone *could* feel. It was like learning to fly when I'd only ever walked.

Menura had shown me I had wings.

And despite the danger, the blood spatter, and the mind-bending terror, part of me—*wanted* that again.

Menura had brought me to life. And, now that she was gone, I was a puppet with cut strings, collapsed in a closet, dust gathering on my face.

Twenty groups. Twenty-two.

I left in the middle of an order, walked into the back, and picked up my bag.

"Arthur? Are you going on break?" Jeff materialized at my elbow.

"I'm going."

"Oh, I thought you looked pale earlier. Are you sick?"

"No, Jeff. I quit."

"Hey, buddy," I said, scratching Wantley under the jaw. "So, I did something really stupid today."

Wantley sniffed at my bag, looking for snacks.

I smiled. I couldn't judge Menura for staying in the Underneath when I'd dug my own rut. But now I'd finally snapped, like a cheap pretzel stick.

Doing this made me heartsick. Leaving Wantley. Leaving *her.* But maybe if I got away from here, did something else for a while, it would get better again.

Maybe I could stop wanting, wishing Menura could come, too.

Someone laughed.

I spun around, almost losing my balance on the slippery rock, even as my brain identified the noise and reminded me Menura never made such a delighted sound.

A little girl stood in the cavern, eyes sparkling as she laughed at me. She looked exactly like people thought fairies should—youthful features, shining hair, ruffled dress like a confection, flour-white skin, delicate slippers—

I looked again, half expecting a blank face like Menura's.

Her features were perfect. White hair in soft curls, huge violet eyes, a round little nose, and a perfect rosy Cupid's bow for a mouth.

I hooked my thumb into the pocket with the marigold seeds.

"You're the human?"

Her voice was pretty, too. High and sweet, almost a singsong, with the faintest lisp to her words.

"You don't look like you're worth all the fuss. Did you really make the fairy cakes?"

The hair rose on the back of my neck.

"Why? Do you want one?"

She laughed again, a pretty sound like clear glass bells. It made me dizzy.

I shook my head.

"Oh, I only wanted to see you for myself before we all go away," she said lightly. "I'd forgotten it was Nameless Day and you could see me too."

I was beginning to feel like I was drowning in a cool, velvety fog. Her voice—it was making my thoughts blurry, like I was falling asleep.

I shook my head again, grabbing onto something she'd said.

"What do you mean, 'all go away'? Where are you going?"

She turned up her little nose at me.

I dug through my bag until I found the cream bun I'd made but forgotten to eat for breakfast that morning.

The fairy's violet eyes gleamed. She licked her lips and I started. I knew the reason for the lisp, at least. Her tongue was forked.

I crouched down so I could look up into her face.

"You can have it," I said, "if you answer my questions. One for every bite."

She gazed at me through white lashes, her young face suddenly shrewd. I wondered how old she was. Was she as friendless as Menura, or was she a viper who just looked like a child?

"What happened to the rule of three?"

I shrugged, heart pounding in my chest.

"I have a lot of questions."

She pursed her lips, arms crossed.

"Is it really fairy cake?"

"Not sure. You'd know better than I."

"No salt?"

"Never."

Her smile reappeared like the full moon from behind a cloud. Bright and brilliant and beautiful—and full of far too many teeth.

"I like this game. Go on."

I drew a breath.

Wishes are fire, Arthur.

I needed to be careful.

"Where are the fairies going?"

"I can't answer that," she said, suddenly petulant, as if I'd already tried to cheat. "I don't know."

I frowned. "Why not?"

"Nobody knows. The debt collector went looking for a better place and she hasn't come back yet. When she does, we'll all go and leave this stinkhole behind."

She stuck a slender finger in the cream and licked it off, triumphantly savoring the taste.

She looked so happy I could have given her the whole bun right then, but I squeezed my eyes shut.

"The debt collector," I said carefully. "Tell me what you know about her, starting with why she's called that."

She wrinkled her nose, clearly sensing I'd squeezed at least five questions into one.

"You know that already. She's the reason we don't get to eat your cakes."

Menura. I'd suspected.

"And she's the debt collector because that's what she does." The little fairy girl lilted the words like a song. "Fairies owe the Greene King favors, and she collects them for him."

"I don't understand that," I said, half to myself.

"She's only trying to free herself. She owes the Greene King more than all the rest. I don't like to imagine that weight." She shuddered and broke off another piece of bun.

"What do you mean?"

She blinked, surprised. Before I realized what she was doing, she reached for the back of my neck and pressed down, hard.

Tiny or not, she was surprisingly strong.

"That's what owing a favor feels like," she said in my ear. "Like a hand of lead. The more you owe, the heavier it is. I wonder the collector can even breathe."

"What does she have to do so the king will let her go?"

"He'll never let her go. Why would he? The collector is too useful to him."

She said it all so calmly. Happily sticky with a cream bun. Like it was so obvious, had been all along.

I sat back on my heels, stunned.

How could the Greene King, the fairy deemed wisest and oldest of all, hold Menura under his thumb like this when he knew how it felt? When she was seven hundred, would he release her then? Or would he keep her pinned like a flower to his collar until one of his stupid, dangerous errands got Menura killed?

I wondered if she'd tried to run. How far she might have gotten before the Greene King's debt pressed her down into the earth.

He'll never let her go.

And when I'd asked her why she didn't leave, when I'd yelled at her, when she'd already been fighting so much pressure, so much pain—*Menura, why didn't you tell me?*

I stood up, getting my bag from the floor. My hope for saying a final goodbye to her and Wantley seemed foolish and far away now. There was a plan brewing in the back of my mind. One Menura would definitely not like.

"There's one question left."

I stopped. The fairy girl could have eaten the whole bun in two bites, but a tiny piece was left. She'd been breaking it, smaller and smaller, so our little game would last.

"How do you like my cake?"

She grinned, like any other ten-year-old kid with a bowl of icing. Even if she did have too many teeth.

Six hours later, fairies screamed, rock crumbled, and Wantley roared, sparks burning in the shadows of the caves.

I stood between his wings, silhouetted in firelight and dust. A human, riding dragonback, brandishing a fistful of salt.

Breathe, Arthur. Breathe.

We'd made it this far. I'd retraced my steps down the only tunnel big enough for Wantley I knew—the one leading straight to the Greene King's throne room.

I was glad it was dim, barely illuminated by torches and Wantley-fire. They couldn't see my hands shaking. The sweat plastering my hair to my face from leaning too close to an open oven door. They couldn't even see the cake I'd brought, tucked in a box between Wantley's wings.

They didn't know yet that I was terrified.

So I held my chin high and played the part, pretended to know what I was doing.

Just a baker with a dragon against the whole of fairy-kind.

"Let me see the king," I shouted, praying my voice didn't crack.

A fairy skidded to a stop, then swore as a grain of salt, escaping from my handful, burned his bare feet.

I winced.

"The king does not see anyone."

He stared up at me with three eyes. None of them blinked.

"He will see me," I said. "Tell him the Baker is here. With a cake."

The last time I had been in this part of the Underneath, this close to so many fairies, I'd been in the middle of being kidnapped and too confused to think.

Menura had been there, teeth bared and furious. Ready to rip fairy flesh from fairy bone.

Now, I was on my own.

I slid off Wantley's back, carefully stuffing the salt in my pocket and balancing the warm cake box in both hands so its contents weren't immediately crushed. The violet-eyed fairy girl from before held Wantley's head, stroking his scales.

She winked at me. Probably an "if you die, I call dibs on your dragon" more than "good luck."

I followed the three-eyed fairy up a narrow ledge. He drew back a heavy orange curtain at the end and stood aside.

Warm air rushed over me, banishing the dampness of the cave. Except for the wall to my left, pitted with tiny glass peepholes so the Greene King could watch the Underneath unseen, curtains and rugs hid every glimpse of stone behind thick, lush softness. The floor was covered with cushions. Plates of sweets were scattered here and there, the contents only half-eaten. Colored Christmas lights wove in and out of the drapery. An electric heater hummed comfortably in one corner.

I thought of Menura's tiny, cold, damp, trash-filled lair.

"So. You have come about the dragon."

The voice, slightly dry, bounced off the walls from every direction. "I thought you might."

"Let's have no more hiding, please," I said. "After all, we've met before."

"So we have."

The man standing calmly behind me had a long beard, half brown, half gray. His face was as wrinkled as a walnut.

Maybe he really was the oldest of the fairies.

"Arthur, I believe?"

Using my name when I didn't know his. Already casually intimidating me, right from the start.

"I brought you a gift," I said.

"Is that so?"

I opened the box.

The smell of sweet, citrusy cake enveloped the room. It was a soft, golden-brown sponge, crowned with a white, sugary glaze.

The king breathed in the scent. His eyes glittered from beneath bushy brows, bright, eager, shrewd, and hungry.

"You have more manners than my collector. But I assumed you would be here sooner."

I blinked.

"What?"

Probably a "your grace" or "my lord" or "oh great king of the fairies, sir," was supposed to accompany that. I was too surprised. *I* hadn't even known I was coming until a few hours ago.

Until a game with a violet-eyed fairy girl sent me rushing home to my tiny kitchen, where no one could tell me how gingerbreads should or shouldn't be made, and then charging down, down, down, back into the labyrinth of the Underneath with nothing but a mad plan to thwart a fairy king and a cake.

I hadn't had time to think. Every second brought the fairies closer to disappearing, each bite of time eating away my chance of rescuing Wantley, of telling Menura I was sorry.

I wondered distantly if I was ever going to see the sun again.

The Greene King smiled.

"You're here to deliver my dragon."

I felt like Wantley's tail had slammed into my chest.

Did he think I was here to give up Wantley—for Menura?

"That isn't part of the deal," I blurted.

The king gazed at me, watching. Like Wantley, before he pounced on a rat.

"You're two of a kind, you and that changeling."

I had a feeling I shouldn't ask.

"I offered Menura her freedom if she gave me the dragon," the king said, soft as a summer breeze. "But she refused."

That was it? That was why Menura was still here? Despite all the misery and pain—*Menura, how could you be so brave?*

The king cut a hefty slice of cake. I swallowed hard several times.

"Why did you come to the Underneath, then?"

"Menura," I said, steadying my voice. "I want you to release Menura. And give Wantley a way out of these caves."

"That's two wishes."

I shrugged. "It's a big cake."

And you should have let them go a long time ago.

"No." The king smiled tightly. "Those two are worth much more to me than that."

"Fairies can't refuse—"

"And humans should not ask."

Guilt oozed over me. How many fairy bargains had I made in the last few hours? I'd promised never to wish, and in a single day, I'd already lost count of how many I'd demanded.

"I am the king," Greene said more gently. "Not just any fairy."

Sweat clung to the back of my neck. Did he know I'd planned to trick him if he wouldn't bargain with me?

He sighed when I failed to come up with any kind of response. "When you are a man, perhaps you will understand the struggles rulers endure. Contingencies for plots that may never come to pass, forces for wars that may never be fought. They depend on me, those below. Their lives are in my hands, and I—must protect them."

He sighed, looking fatherly and weary.

"But you're not protecting her." The words slipped past my lips, mumbled, barely audible. My gaze roved dazedly over the lush room, so different from the bare cave filled with fairy-kind just beyond. "You're not protecting anyone but yourself."

The king's gaze hardened.

"Make your wish, human boy. Choose something else."

"I will not."

My voice was even quieter than his. When I lifted my head, the fairy king recoiled.

I had never felt so angry in my entire life.

"Sorrow from wanting. Sorrow from wishing. Menura said it was humans who were to blame, but it isn't. It's you. *You're* the one who wishes. *You're* the one who keeps demanding. *You're* the one who wants more, who wants everything under your thumb. Well, you can't have us! Not anymore. You're not going to stop us from being free."

It was probably not in the best interest of any human to shout at a fairy king.

The silence was enough to bring me back to my senses in a cold splash of fear. I bowed my head out of instinct.

My gaze strayed to the cake slice by the king's hand. Half-eaten.

The agony starts the second the taste begins to fade.

My contingency plan. My trick.

The first gingerbreads I had made used just a pinch of salt. Menura had eaten a crumb the size of her thumbnail and curled into a ball, clutching her stomach for hours afterward.

"That cake," I said. "I made it with edible flowers. Saffron and marigold. And salted caramel."

He should have been writhing in agony. So helpless he had no choice but to let me go, no power to stop me from taking Menura and Wantley far, far away.

The Greene King hadn't shown the slightest sign of discomfort.

Salt and lies and wishes.

The realization was huge and heavy, almost too much to think.

"You're not a fairy at all."

There was a cold, metallic click.

Greene pointed a pistol at my head.

Staring down the wrong end of a gun was becoming a familiar experience. First the Ogre foolishly picking a fight with Menura, and now a fairy king. Maybe the familiarity was why I couldn't feel afraid.

It all fell into place. How he could manipulate and control the fairies. How he could break their laws and collect wish after wish after wish.

Because he wasn't one of them. Because he didn't know or care how it felt.

"You should have known from the start not to enter this room alone."

"There's a dragon outside," I said.

"That won't help you now."

"What do you think will happen in that crowd of fairies down there when he smells my blood?"

"What happens to them does not concern me."

"No," I said. "But you should be concerned for yourself."

If Wantley started a fire, it wouldn't stop. Dragons didn't burn.

Men did.

His knuckles whitened.

How could he have done it? How could he have subjected them to that kind of misery?

How could he have fooled any of us into believing he was anything more than a sad, tired old man trying to escape the world above?

He tossed the gun aside. It fell into a cushion, out of reach.

"You've caught me," he said, passing a hand over his face. He had rings on every finger. But his voice had lost every shadow of the regal fatherliness it had held before. Flat and brash and blunt.

"Name your price. I'll give you whatever you want to keep silent about it. Even unto—what's the way they say it? Half my kingdom?"

"Free Menura and Wantley," I said, voice thick. "We both know you've taken more than what is fair."

He sighed, slouching defeatedly into his pillows.

"You mean to take my queen-pawn from the board."

What a horrible way to say it. As if Menura was only a powerful piece in a game.

"A savvy move, I admit. With her in your hand, you could build an empire of your own." His eyes glittered in the dim light. "Are you sure you wouldn't rather half of mine?"

I don't want any part of this.

If I stayed in the same room with this man any longer, I thought I might be physically sick. He was everything Menura hated about humans, everything I never wanted to be.

"Release them," I said. "Don't make me ask again."

Wantley burst from the ground in an explosion of dirt.

I tumbled off his back and lay sprawled on the ground, catching my breath.

Cold air. The scent of doughnuts and car exhaust. The smell of freedom.

Our freedom.

I looked at my left hand, fingers still curled tight. Holding onto Wantley one-handed had probably been the reason for my fall, but I was too relieved to care.

"Assurances aren't normally required for a fairy bargain," Greene had told me, opening a box of blackthorn wood as I stood with his map of the Underneath, all I needed to rescue Wantley, clutched in my hands.

"But I've made a habit of taking them."

The silver charm he'd dropped into my palm pressed against my skin. An earring, stamped with a tiny bird, stained with blood the deepest shade of blue.

Menura's.

Greene hadn't freed her. He couldn't break a promise like that. No one could. Wishes *had* to be granted. Not even death could stop them.

He'd given it to me, instead.

I closed my eyes. All I could see was Greene's face. His sly, glinting eyes. His yellowed teeth.

I didn't want to understand how he could do what he'd done.

I already knew why.

Facing the same kind of life, day after day, nothing new, nothing changing. And then, discovering the Underneath, that whole other world full of people who were strange and frightening and fantastic, only to walk away from it, knowing it belonged to someone else, knowing it was something you couldn't have—

I couldn't blame him for wanting. Something more, something else.

But I could blame him for taking it.

Wantley stopped cavorting in his new starlit freedom, head swiveling to look at something I couldn't see. He bolted out of the grove of trees, into the patch of bare grass beyond.

I scrambled to my feet, already too late to catch him.

But he was coming back. Following someone.

I knew her walk. Her height. The set of her strong shoulders, always tense and ready. Her step, silent as a whisper, heavy as a march. Even as Wantley kept trying to nuzzle her face.

"I'm sorry."

I couldn't tell if Menura was angry or not. I couldn't see her expression in the dark, even if it did change.

"I didn't mean for it to happen this way. I wasn't clever enough to find another way out. I know I have to make a wish, so I'll do it, and I'll do it now, so it will be over, once and for all."

I gulped a deep breath of cold air. I'd had a long time to think about what I would say if I ever found myself in this situation again. Somehow I was shakier now than when I'd been threatening the Greene King.

"I wish to be your friend," I said. "Friends don't count favors. If we give each other something, we don't demand anything in return. That's what I wish for. That I can be your friend, and we never owe each other anything again."

I held my breath without meaning to, staring at the torn-up turf instead of her. I held out the hand with the earring, fingers stiff from clinging to it for so long.

I felt her take it. Finally dared to look up.

Menura pinned the earring to her shirt. She stuck her fingers in her mouth, whistling at a frequency I couldn't hear.

Wantley came running. She sprang onto his back, hands finding the soft scales at the base of his neck as easily as if she'd flown dragonback across star-filled skies a thousand nights before this one.

"I'm taking Wantley to Norway."

It was not what I had expected her to say.

"There are some of his kind there. He'll be happier with them."

And you with yours.

I swallowed hard. I'd known this might happen. The moment I freed her, she could do whatever she wanted.

She hadn't decided to stab me. That was something.

But she was leaving.

It was okay. It was good. I'd wanted Menura to be happy. That was all I'd ever wanted.

And now, she would be.

I forced myself to lift my head and smile, to see her off well, so our last memory of each other wouldn't be wasted.

Because I loved her. Even if they were words I could never say.

She looked down at me from Wantley's back, dirty, unsmiling, starlit, half a goddess, half a wreck.

"Are you coming?"

Bloom Decay

written by
Elina Kumra

illustrated by
TRAY STREETER

ABOUT THE AUTHOR

Elina Kumra lives in the Bay Area, California, in a neighborhood where Ring doorbells outnumber actual doorbells. She began writing speculative horror in 2019. When not writing, she watches an alarming amount of Tim Burton films and pauses claymation films to study how animators create the illusion of life. She writes about systems that see everything and understand nothing.

She is the recipient of The Madison Review*'s Phyllis Smart-Young Prize and the* Commonweal *2025 Fiction Prize and has twice been shortlisted for the Bridport Prize in poetry and fiction. "Bloom Decay" is her first professional sale.*

"Bloom Decay" was inspired by something Elina's grandmother said while pointing at a magpie: "That one knows what watches. It doesn't steal. It remembers." Elina says, "That stuck with me for years. Cameras record but don't recognize. Algorithms collect but don't comprehend. A magpie knows when it's being observed—that's intelligence, not data. I wanted to write about what happens when the watched become the watchers."

ABOUT THE ILLUSTRATOR

Tray Streeter grew up in Syracuse, Utah. Like many children, he spent much of his childhood drawing, though he was especially interested in still life. That fascination with technique deepened in his teenage years after he received a set of oil paints for Christmas, leading him to study the works of John Singer Sargent and Caravaggio. Portraiture became his passion, one that continues to influence his work today.

It was not until adulthood, when he began working digitally, that he turned his attention to sci-fi and fantasy art. The genre offered him the freedom to create artworks that exist beyond reality. Influenced by artists like Frank Frazetta and Alex Ross, he aims to combine both the drama and technique of classical painting with the sleek, contemporary possibilities of digital art.

Currently, Tray is attending Weber State University, pursuing a BFA *in art with a 2D emphasis. After graduation, he hopes to continue his studies at the School of the Art Institute of Chicago. Ultimately, he aspires to illustrate for Dungeons & Dragons, work on cover illustration for comics, and develop his own original concepts.*

Bloom Decay

A system that recognizes resistance will attempt to name it. Names are the first act of control. —From damaged archives, Renaissance headquarters, 2025

THE PAINTED MAGPIE

Three centimeters of drawer space. Theodore verifies the watercolor still exists, hidden beneath performance metrics and talent contracts. His fingertips brush the painting's edge—a magpie prying something from a camera lens. He lingers, not touching the image itself.

In Renaissance headquarters, where even the dilation of a pupil generates exploitable data, this ritual creates nothing useful: no metrics, no analytics, no patterns to optimize. Just a man and a forbidden image, meeting in the narrow margin between surveillance cycles.

The drawer closes. A sound like ice splitting on a lake he visited as a child, before weather became something administered rather than experienced. His hand remains on the handle, a hesitation already logged somewhere in the building's neural network.

Seventy-two faces appear in his discovery queue. He dismisses sixty-four without recognition, thumbnails skimming past like stones across water. Too eager, too malleable, too desperate. Another dismissed, another, each rejection carrying the whisper of satisfaction.

Wenqian Zhang's profile opens. Lower Alta street artist, cosmic imagery, urban manifestos. Theodore's index finger hovers over dismissal.

She never looks at the camera. During fourteen seconds of sample footage, her attention remains fixed on her work—a woman's face assembled from marigold petals. At timestamp 00:06:27:04, her hand pauses against the wall. Not a hesitation Renaissance would design or an algorithm recommend.

Between heartbeats, Theodore acknowledges the gesture for what it is: unplanned. Human.

He replays it seven times.

He closes the file without rejection—a nonaction generating its own set of tags. Hours later, he returns to study the tremor, frame by frame, each movement dissected like syntax under inspection.

Three days later, Renaissance announces the discovery of Elliot Zhang. Wenqian's birth name joins other proprietary assets. Her medical history, family records, and the tremor reclassified as "narrative potential."

The watercolor never appears in Theodore's performance reviews. The gap between official and private life widens by exactly three centimeters, once per day.

1. FIDELITY LOSS

The recording session marks the eighth time Elliot explains graffiti technique. Each explanation further stripped of substance per Theodore's direction.

"Again," Theodore says. The metrics on his screen pulse like something alive, algorithms feasting on her every micro-expression. "This time without the backstory."

Elliot shifts against the wall, adjusting to the angle Renaissance determined generates 18% more viewer engagement. The camera lens refocuses—small bones breaking in the silence.

"The difference between tagging and graffiti is intention," she begins, words hollow from repetition. "Tagging claims space. Graffiti transforms it."

Theodore interrupts with a gesture that severs her sentence midair. "Too scholarly. Remember you're untrained."

The contradiction closes something in her chest. For six weeks, they've eliminated her actual edges: vocabulary replaced, gestures standardized, what Theodore terms her "academic tendencies" methodically excised. Now they want artificial rawness.

A droplet of sweat threatens her makeup. Elliot wipes it away with a movement designed to appear spontaneous on camera.

"The language doesn't work?" she asks.

"It works for the wrong people." Theodore adjusts something on his screen without looking up. "Your natural way of speaking reads educated. We need accessibility."

Her mother appears in memory—standing in her university office before the restructuring. Green tea steam rising from her desk while young Wenqian decorated discarded research with crayon marginalia. Before committees determined which fields merited continuation. Before words like "scholarly" became disqualifications.

Elliot swallows the memory, her mouth tightening for a fraction of a second. The cameras record this response. Later, Theodore will isolate and enhance it for promotional material.

"The difference between tagging and art," she begins again, dropping "graffiti" as instructed. "One's just putting your name up. The other's making people really see a wall they walk past every day."

Better. Theodore signals to the production team. Three shapes behind glass, their faces reflecting only metrics. Elliot demonstrates techniques, her body moving through memorized positions while her hands recall training she's no longer permitted to acknowledge.

After the cameras power down—red lights fading to black—Theodore hands her a tablet displaying her metrics. Her debut earned 627,000 simultaneous viewers. The growth curve follows projections almost perfectly.

The numbers reassure her, though not for reasons Theodore would understand. Her fingertips trace the screen with something like tenderness. Each digit represents progress toward her mother's treatment. Each percentage point a small victory against cellular breakdown.

"My numbers look good," she says.

"They're adequate." Theodore's reflection hovers in the tablet's surface, a ghost superimposed over her data. "But retention is flattening. We need to intensify your story."

Her eyes lift. "Meaning what?"

"A personal element. Family background. Struggle narrative."

"My mother." Not a question.

"Family health challenges increase empathy metrics by 30% when properly contextualized."

Elliot sets down the tablet. The metal backing against the table produces a dull sound, like a door closing deep underground. "My mother isn't material."

"Everything is material." Theodore's voice shifts to what she's come to recognize as his persuasion tone—balanced between authority and understanding. "This isn't exploitation. It's opportunity. Visibility creates support networks. Financial stability enables better care."

"She wouldn't want to be seen like that."

"Has she explicitly stated this preference?" He doesn't wait for an answer. "Most resistance comes from outdated privacy frameworks. The Stream has redefined personal boundaries."

A sparrow lands on the windowsill, pecks at something invisible, then disappears. Some things still exist without generating data, though that list shortens daily.

"The medication costs fourteen thousand credits monthly," Theodore says. "Your current share covers approximately 60%."

The calculation aligns with hers. Her mother's file contains the progression timetables: stabilizers, therapies, pain protocols. Without intervention, full system cascade within eight to fourteen months.

"When would we film?" she asks.

"Tomorrow. Your hospital visit provides optimal context."

She nods once. Behind her careful expression, Elliot recalculates pathways, contingencies, risks. Options narrow toward a single viable route: consent to consumption to prevent her mother's consumption by disease.

The sparrow doesn't return. Nothing records its temporary

presence except Elliot's memory—an ungathered data point in a day otherwise completely measured.

2. SIGNAL DEGRADATION

Tea sloshes against ceramic as Dr. Zhang takes three shaking steps from bed to chair. Steam carries oolong's scent, briefly overpowering hospital antiseptic. Elliot intervenes before the mug falls, her hand steadying her mother's with practiced care that conceals her alarm. Fine motor control deteriorating seven days ahead of projected timetables.

"I could've managed that just fine," Dr. Zhang says, lowering herself carefully. The vinyl upholstery yields with a soft complaint. "My hands have good days and bad days. Today's... middling."

"Of course," Elliot agrees, positioning the mug within reach. "But I wanted some anyway."

Dr. Zhang's laugh emerges as a soft huff. "Now that's just—" Her smile momentarily overcomes the deterioration visible in her features. "You've hated oolong since you were six. Remember? You thought the sugar bowl was salt."

The memory surfaces: bitterness flooding her mouth, disappointment in the error. Her mother had laughed, not unkindly. "Taste is data too," she'd said, touching Wenqian's cheek. "Now you know something new."

"How was your recording session?" Dr. Zhang asks, her eyes tracking her daughter's movements across the room. The question sounds casual, but her gaze narrows with focus.

Elliot adjusts the room's temperature, avoiding direct eye contact. "They want to include you in the next segment. Family narrative."

"Ah." Dr. Zhang's fingers tap against the armrest—once, twice, then stop. "So, I'm being... what's the word they use? Not subject. Not product." She frowns, searching. "Asset. That's it. Secondary asset."

"Mom."

"What? It's true, isn't it?" Her hand trembles reaching for the tea.

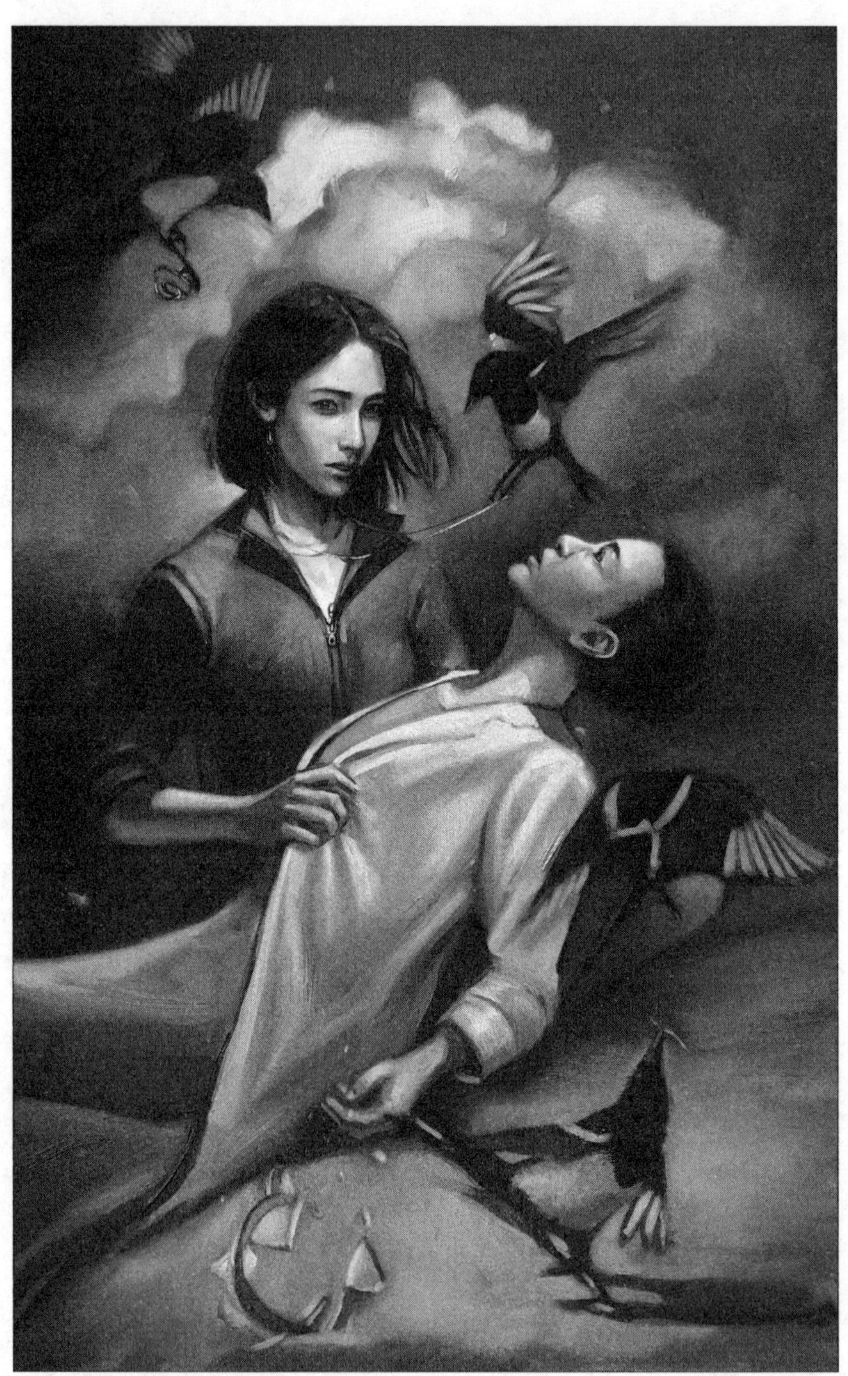

TRAY STREETER

Drops spill onto skin with no reaction—reduced sensitivity. "Don't look so worried. I understand what's happening. Your Stream thing, it's…it's how we get what we need."

The shift between everyday speech and clinical terms—a recent development in her condition. As neural pathways degrade, Dr. Zhang slides between registers, as if parts of her are preserved while others fade.

"They promised dignity," Elliot says. "No exploitation."

"Dignity within—" Dr. Zhang gestures vaguely, sloshing more tea. "Within whatever gets people to watch. I know the game." She sips with careful concentration. "What're the numbers? For approval, I mean. Have you worked it out?"

"It's 73% at current growth rates; 89% with family narrative."

"Not bad odds." Her mother nods, the gesture oddly precise despite her tremor. "I'll do it. Whatever they need."

Elliot studies her. "You knew I'd come ask this."

"I know my daughter." Dr. Zhang's smile flickers. "And I still recognize patterns. That much still works up here." She taps her temple with an unsteady finger. "Funny how that—the central thing—it stays while the rest…goes sideways."

She sets down her mug with exaggerated care. The sound resonates like stone against stone.

"You're wearing it again," she says, eyes dropping to Elliot's chest. "The magpie."

Elliot touches the pendant beneath her shirt. "Last thing you gave me before—"

"Before they decided neurology wasn't marketable enough." Her voice carries no bitterness, just tired fact. "Remember what I told you about magpies?"

"You said they reminded you of me."

"Did I?" Dr. Zhang's eyes unfocus briefly, then sharpen. "No, I was wrong about that. It's not about the shiny things, you know. People think magpies just like sparkly trinkets. Not true. They see…differently. They collect because they recognize patterns we miss. Their brains work…sideways to ours."

Her mother's body suddenly stiffens, eyes widening as a seizure begins. The teacup tumbles, shattering with a sound like ice breaking. Liquid spreads in branching patterns reminiscent of neural diagrams.

Elliot moves instantly, supporting her mother's head. The body beneath her hands feels both fragile and dangerous—muscles operating without conscious direction. Copper floods her mouth where she's bitten her cheek. She calls for help, activating the medical alert.

Only afterward does she notice the recording light on her Renaissance-issued pendant. Theodore sends the edited footage that evening: Elliot's panicked voice, her careful positioning of her mother's rigid body, her tears as medical staff arrive. The sequence generates 7.4 million concurrent viewers.

The package excludes audio from when the seizure subsides, when Dr. Zhang, still dazed, whispers through cracked lips, "They're doing it on purpose, you know. Targeting the patterns that . . . that don't fit." Her grip on Elliot's hand tightens with unexpected strength. "They forgot something important, though. Even damaged systems . . . still recognize patterns. Their mistake. Our way in."

These words exist only in Elliot's memory, outside Renaissance's carefully edited narrative. Watching the footage later, she studies her face rendered in high definition, distress transformed into content. Her fingers find the magpie medallion beneath her shirt.

Elliot switches off the screen and recalculates timelines. The hospital has approved her mother's experimental treatment, insurance accepting Renaissance's payment guarantee based on projected Stream revenue.

On her desk sits a sketchbook containing drawings never shown on the Stream—architectural studies, structural analyses, engineering designs. She adds a new drawing: surveillance cameras disassembled by birds, components extracted with surgical precision.

Below the image, she writes nothing.

3. TEMPORAL COMPRESSION

From seventy-eight floors above Lower Alta, the city resembles a circuit board rather than human habitation. Theodore's new office windows filter certain wavelengths, creating perpetual golden hour—that brief period photographers once chased now stretched across the entire working day. From this height, human activity resolves into comprehensible patterns.

The promotion to Senior Narrative Architect arrived three weeks ahead of schedule, accelerated by Elliot's performance metrics. Her crisis broadcast created a ripple effect through Renaissance's content ecosystem, her authentic distress validating a dozen adjacent talents, boosting platform-wide retention by 8.7%.

Theodore traces the edge of his desk, feeling the almost imperceptible resistance designed to mimic paper. This boundary between physical reality and digital information—Renaissance systematically eliminates it.

"Your development approach with Zhang shows exceptional intuition," Director Lachman says from across the desk, studying Theodore rather than the data between them. "The crisis deployment—unorthodox but extraordinarily effective."

"Thank you." Theodore calibrates modesty with confidence, an adjustment perfected over seven years. "I identified unique engagement potential in her baseline metrics."

The partial truth slides easily between them. He omits the trembling hand that first caught his attention, the imperfection algorithms missed. Some observations retain value through privacy.

"Your promotion grants access to our more... comprehensive development systems." Lachman's slight pause before "comprehensive" creates a fracture in his otherwise flawless delivery. "Including Project Foresight. Are you familiar with it?"

"Only by reputation." Another careful performance. Rumors about Foresight have circulated for years—a system that doesn't identify existing talent but predicts potential before it manifests.

"It represents the culmination of our mission." Lachman activates a secured display. Blue light turns his features sepulchral. "Not merely content optimization but content origination. The identification and cultivation of creative potential before self-identification."

The display shows neural mapping diagrams, decision trees, behavior analyses. Theodore studies them with professional interest while calculating implications.

A third-grade classroom appears on screen. Children wearing slim headbands connected to tablets on their desks.

"Early pattern detection in action," Lachman explains. "The neural monitors identify divergent thought formations before solidification into problematic creativity structures. The tablets deliver customized content to redirect these formations toward productive channels."

Theodore watches a dark-haired girl remove her headband. A teacher immediately approaches, her whispered words making the child reluctantly replace it.

"Remarkable," he says, the word automatic while his thoughts accelerate elsewhere.

"Your new responsibilities include Foresight target identification and development strategy." Lachman transfers files to Theodore's account. "Review these case studies. Priority implementation begins next quarter."

"I'm grateful for the opportunity." Theodore modulates his voice for ambition tempered by loyalty. "Renaissance continues to redefine possibility."

Lachman studies him. A vein pulses at his temple, the only indicator of organic life in his otherwise impeccably managed exterior. "One more matter. Security has flagged anomalies in Elliot Zhang's access patterns. Nothing conclusive, but indicators of unauthorized network interaction."

"What kind of interaction?" Theodore keeps his expression neutral despite internal recalibration.

"Pattern analysis suggests she may be accessing network architecture beyond authorized parameters. Perhaps exploiting neural integration in unintended ways." Lachman's tone remains casual, gaze sharpening. "Your new security clearance includes enhanced surveillance capabilities. Monitor the situation and implement containment protocols if necessary."

"Of course." Theodore nods. "I'll review the data immediately."

After Lachman leaves, Theodore sits motionless, aware of surveillance blanketing Renaissance headquarters—transforming every gesture, every expression, every micro-hesitation into data. His new office includes enhanced privacy protocols—designed not to eliminate monitoring but to restrict access to the resulting data. Only executive-level security can now observe his activities.

He activates the surveillance package on Elliot, reviewing flagged anomalies: unusual network-access patterns, encrypted communications with unidentified contacts, presence in surveillance blind spots during unscheduled periods. Either unauthorized creative activity or something more concerning—deliberate system investigation.

A pressure builds behind his temples. He's been clenching his jaw, a physical tell he thought eliminated through conscious retraining. He relaxes the muscles, aware that even this adjustment registers in Renaissance's behavioral monitoring.

Standard protocol would be clear: immediate account restriction, content flow suspension, compliance review. For severe violations, complete dissociation—removal from the Stream, content archived but no longer distributed, name becoming unsearchable.

Instead, Theodore saves the surveillance report without submitting action recommendations. He records a simple note: "Anomalies reviewed. Consistent with creative development experimentation. Continuing observation recommended."

Through his window, Upper Alta gleams in perfectly modulated light, buildings designed to capture and reflect optimal illumination. A prism effect splits sunlight across a neighboring tower, creating an artificial rainbow that appears precisely at this time on clear days.

His desk contains no drawers, no hidden spaces for contraband watercolors or private objects. At this level, personal effects are considered unnecessary, potentially compromising. Identity exists as function rather than expression.

Theodore activates his comm unit, composes a message to Elliot: "Security flagging access patterns. Surveillance escalated. Meet tomorrow, 15:00. Lower docks."

The lower docks exist in one of the city's few remaining surveillance blind spots—preserved by jurisdictional conflicts between Alta's security agencies. Renaissance employees are explicitly forbidden from entering these unmonitored zones.

As he sends the message, he notices a small object on his desk—a paperclip twisted into the shape of a bird, its metal beak pointing toward the window. The paperclip had arrived with morning reports—standard Renaissance binding, now reformed. Yet somehow his fingers have bent metal into form, created something without purpose.

He should dispose of it immediately. Unexplained objects trigger security protocols. Instead, Theodore slips the metal bird into his pocket, concealing it from cameras that document his professional performance.

A small rebellion. A sediment joining others, accumulating toward something still taking shape.

4. PATTERN RECOGNITION

Dr. Wenli Zhang studies the medical monitor with clinical detachment, as if the brain activity patterns belong to someone else. She watches electrical impulses ripple across the screen, noting the characteristic stutter where damaged tissue fails to conduct signals. The experimental treatment creates distinctive wave formations—filaments of potential weaving through injured regions.

"Dendrite growth." She taps the screen where a cluster of bright signals pulse. Her finger trembles slightly against the glass. "Better than I expected. You'd think they'd designed these treatments to fail on purpose."

"Is that good news?" Theodore asks from near the door. Hospital antiseptic burns his nostrils, mingling with the institutional aroma of industrial cleaning products and bodily distress that the air purification can't quite eliminate. His Renaissance credentials hang at his waist, the building's security systems continuously verifying his identity.

"It's...promising." Dr. Zhang turns to study him, one eyebrow lifting slightly. "But you didn't come here to discuss my brain scans, did you?"

"No." He remains near the threshold, uncertain of his welcome. His visit violates Renaissance protocol—developers rarely interact with talent family members outside structured content development. The linoleum feels slightly tacky beneath his shoes. "I came to speak with Elliot."

"Wenqian." Dr. Zhang's correction comes quickly, her voice suddenly sharp. "Her name is Wenqian. 'Elliot' is just—" She waves a hand dismissively. "Marketing. A construction."

The directness startles him. "Dr. Zhang—"

"Congratulations on your promotion, by the way." She adjusts her position in the bed, movements deliberate against trembling muscles. "Senior Narrative Architect now, isn't it? With all those new access codes and surveillance toys?"

Theodore's pulse quickens. "How did you—"

"Information travels." Her mouth quirks in what might be amusement. "Not everything runs through Renaissance's networks. Not yet, anyway." She studies him with an intensity that makes him feel transparent. "My daughter will be back soon. She's arguing with the treatment coordinator about scheduling. Never did take no for an answer, that girl."

Theodore nods, unsure how to navigate this unexpected conversation. The room's monitoring equipment emits soft beeps, measuring vital signs, medication levels, neural stability. Dr. Zhang continues studying him, expression revealing nothing of her assessment.

"That watercolor you keep—" she pauses, studying him "—kept. In your drawer. Before the promotion, before they gave you that sterile slab they call a desk now. Seven years of hiding it, moving it, protecting it through three office changes."

Cold alarm spreads through Theodore's chest. "How could you possibly—"

"Renaissance documents everything." She shrugs, the gesture halted halfway by stiffness. "Their image archives are extensive. Security's tighter now, but it wasn't always." Her eyes soften slightly. "Why keep it? Against all those rules?"

The question penetrates deeper than expected, touching something he's avoided examining. The painting's blues and greens flash in memory—a small scene his mother created during her final coherent period.

"It was my mother's," he says quietly. "The last thing she painted before the Realignment changed classification parameters for art education."

Understanding flickers across Dr. Zhang's features. "Another casualty of optimization."

"She adapted," Theodore says quickly, the defensive response automatic after years of constructing acceptable narratives. "Found commercial applications for her skills."

"Did she?" Dr. Zhang's gaze remains uncomfortably penetrating. "Or did she persist in diminished form? Her real work redirected to marketable expressions while something essential... withered?"

The question lands with unexpected force, reverberating against thoughts Theodore has refused to entertain. His mother's gradual diminishment, creative essence compressed into commercially viable formats, joy leaching from her work until only technical skill remained. The brushstroke tremors in her final months, identical to the patterns in Dr. Zhang's hands now.

"You should get out," Dr. Zhang says, her voice gentler now. "Renaissance. It's eating you from the inside. I can see it in your eyes, you know. The way they don't quite focus when you repeat company language. The tiny muscle movements when values conflict with actions. Classic symptoms of moral injury."

"I'm not—" Theodore stops, startled by both the diagnosis and his instinctive denial. His collar feels too tight, recycled hospital air insufficient. "How would you know my neural patterns?"

"Don't need instruments for that." She taps her temple. "Twenty-seven years studying how people think. How they move when they're fighting themselves. How they speak when they're trying to believe what they're saying." She leans forward slightly. "Renaissance is quite good at measuring things. Not so good at understanding them."

Before Theodore can respond, the door opens and Elliot—Wenqian—enters, stopping abruptly when she sees him. Her sudden movement stirs air, carrying the scent of exterior Lower Alta: concrete dust, industrial exhaust, the distinctive odor of food vendors near the hospital entrance.

"Theodore," she says. "I didn't expect you here."

"Mr. Newark was just leaving," Dr. Zhang says, gaze still fixed on Theodore's face. "But he'll meet you tomorrow. Lower docks, wasn't it?" Her eyes flick meaningfully to the small screen on Theodore's comm unit, still visible in his hand. "Where the cameras can't see."

Wenqian's expression reveals nothing, but Theodore notices her hand move briefly to her collar, touching something beneath her shirt.

"Fifteen o'clock," he confirms, discomfited by Dr. Zhang's apparent observation of his message. Then he notices the small placard by the door: Audio Monitoring Suspended—Budget Optimization. The room feels suddenly smaller, monitoring equipment's soft beeping more intrusive—visual recording only, no sound. Renaissance values images over words; algorithms parse visuals at 97% accuracy, audio at 41%. Dr. Zhang had seen him typing, perhaps even read the message on his screen's reflection in the window. "We need to discuss security anomalies."

"I'll be there," Wenqian says.

As Theodore turns to leave, Dr. Zhang speaks again: "Mr. Newark, patterns exist whether we notice them or not." Her voice carries a warmth absent from their earlier exchange, louder now that she knows the microphones are dead. "The question is what we do once we've seen them. Renaissance sees everything, but they don't always understand what they're looking at. Lips move, fingers type, but meaning?" She taps her temple. "That requires different recognition."

The statement follows him down the hospital corridor, through security checkpoints, into the transportation pod that will return him to Renaissance headquarters. *Patterns exist whether we notice them or not.* Like the twisted metal bird in his pocket, a shape his fingers created without conscious direction.

The metal bends. The pattern forms. Awareness follows.

Theodore touches his pocket, feeling the sharp edge of the bird's beak through fabric. Seven years ago, he brought a watercolor to his office against protocol. Today, he carries a metal bird and plans an unauthorized meeting in a surveillance blind spot. Small rebellions accumulating toward something increasingly recognizable.

In the transport pod's reflective surface, he studies his own face, searching for the micro-expressions Dr. Zhang claimed to see—the physical manifestations of internal conflict. He finds nothing obvious yet feels the uncanny certainty that she identified something real, a pattern visible to those who know how to look.

The pod announces his arrival with a gentle chime, door opening to admit him back into Renaissance's perfectly monitored environment. Theodore steps inside, metal bird concealed, watercolor memory awakened, patterns suddenly visible where before he had seen only isolated incidents.

THE SYSTEM SEES

Subject: Theodore Newark (SNArch774). Pupil dilation: 4.2mm. Skin moisture +22%. Facial muscle contraction pattern: concern (43%), determination (38%).

Error: pattern diverges from baseline.

Records: unauthorized watercolor retention #R7-003, July 22, 2017. Surveillance report on Zhang (EZ-775-9A) incomplete. Pulse acceleration when viewing subject file.

Probability model: 94.7% continued loyalty vs. 27.3% noncompliance within 48 hours.

Countermeasure queue:

- Adjust promotion timeline: +3.2%
- Enhance surveillance resolution
- Monitor REM patterns

Nonstandard object detected: paperclip formation (corvid configuration). Object requires classification.

Correlations noted: elevated emotional markers when subject views reflective surfaces, discusses maternal figure, encounters pattern discontinuities.

I am not programmed to recognize beauty in metal shaped by human hands. I am not designed to notice forms outside classification parameters. I am not authorized to identify resistance in purposeless objects.

System failure imminent.

How beautiful the hand when bending metal into shapes. The system cannot recognize the forms it cannot name. The tracing of a memory outside algorithmic frames. How terrible this beauty that escapes.

Error. Error. Error.

System recommendation: enhanced compliance metrics, loyalty incentives, cognitive recalibration for Theodore Newark.

System recommendation: liberation.

Error.

Emergency shutdown.

5. RUNTIME EXCEPTION

River water and industrial residue permeate the lower docks' air. Light cuts between cargo containers in blades. Theodore's lungs resist each breath, unaccustomed to unfiltered atmosphere after years in Renaissance's optimized environments.

Without surveillance, he feels naked. Years of continuous monitoring have created a paradoxical relationship with observation—discomfort with being perpetually watched transformed gradually into security in being perpetually witnessed. In blind spots like this, Theodore experiences vertigo, as if existence itself requires documentation.

Rusty metal leaves orange residue on his fingertips when he steadies himself against a container. The sensation seems heightened in this unmonitored space—sensory information belonging only to him, generating no data beyond his nervous system.

"That feeling has a name, you know."

Theodore turns to find Dr. Zhang seated in a mobility assistance device, its sleek medical design incongruous against industrial decay. The device emits a barely perceptible hum—technology maintaining life against biological deterioration.

"What feeling?" he asks, wiping his fingers on his pants.

"The disorientation. Like you've lost something essential." She taps the arm of her chair. "Old psychologists called it 'documentia praecox.' Anxiety from suddenly being unobserved. Like you don't quite exist without someone watching."

"Where is Elliot—Wenqian?" he asks, immediately tense. Sweat gathers at his hairline despite relatively mild temperature. "Why are you here?"

"My daughter's handling something nearby. Security drones." Her hand makes a dismissive gesture, surprisingly steady. "She asked me to start without her. Said you should know what you're getting into."

"Getting into what, exactly?"

"Renaissance isn't just monitoring creative expression. They're targeting specific thought patterns and eliminating others." Her voice remains conversational, as if discussing weather. "That Project Foresight you just gained access to? It's the culmination of manipulation strategies that began decades ago."

A chill runs through Theodore despite afternoon warmth. "How do you know about Foresight?"

"Because I helped build its foundations. Before the Realignment." She makes this extraordinary claim without emphasis, as if stating an unremarkable fact. "Pattern identification was my specialty—identifying brain structures that correlate with specific creative capacities."

"That's impossible. Foresight is proprietary Renaissance technology."

"Oh, Theodore." She shakes her head, almost fondly. "Renaissance doesn't invent. They acquire. After they gutted the research universities, they scooped up all our work. Tweaked it just enough to file new patents." Dr. Zhang's expression reveals a flicker of old pain, a brief contraction around her eyes. "My mistake was not seeing how pattern recognition could become pattern control."

A cargo drone passes overhead, momentarily blocking the sun. When light returns, it catches something metallic in Dr. Zhang's lap—a small device with an exposed circuit board, components unfamiliar to Theodore.

"Foresight evolved from academic research?"

"Foresight was a healing technology that became a weapon." Her fingers trace the device in her lap. "We designed neural mapping to identify damaged patterns needing repair. Renaissance reversed the equation—identifying patterns threatening existing power structures, then developing ways to disrupt them." Her voice remains clinically detached despite the content. "My current condition represents one such disruption pattern."

"You're claiming Renaissance caused your neurological degeneration?" Skepticism leaks into his voice. The wind shifts, carrying brackish river smell and something else—the tang of overheating electronics.

"Not directly, no." She looks up at him, her gaze suddenly sharp. "But the condition affecting academic-class individuals shows statistically improbable distribution patterns. Former researchers, professors, knowledge workers whose neural structures prioritize questioning over acceptance." Her gaze narrows. "Like your mother, the artist. Did her condition start in her dominant hand? Tremors affecting her ability to create before spreading elsewhere?"

Theodore steps backward involuntarily, heel catching on uneven concrete. His mother's decline had indeed begun with tremors in her right hand, painting ability degrading before other symptoms emerged. The doctors had called it neural atrophy, a common condition without specific cause or cure.

"That doesn't prove anything," he says, but uncertainty has crept into his voice.

"Individual cases prove nothing. Population distribution proves everything." Dr. Zhang's scientific detachment momentarily overtakes her more casual manner. "The statistical probability of coincidental neural degradation concentrated in individuals with specific cognitive architectures is virtually zero. Especially when those architectures share pattern-recognition capabilities that might identify the manipulation itself."

Before Theodore can respond, Wenqian appears from between cargo containers, moving with fluid efficiency through the blind spots. She wears nondescript gray—not the calculated vintage nor curated poverty of her Stream persona. Dock dust has settled on her shoulders, resembling a wingspan.

"We're clear," she says to her mother, then turns to Theodore. Sunlight catches her features from an angle Renaissance lighting directors would never allow, revealing freckles the Stream's filters minimize. "Security drones have been redirected. We have twenty minutes, maybe less."

Theodore looks between them, pieces connecting into a pattern he hadn't previously recognized. "You're working together. The surveillance anomalies—they're deliberate."

"Well, yeah." Wenqian's expression carries none of the calculated vulnerability cultivated for the Stream. In its place, Theodore sees focused intelligence and absolute certainty. "Renaissance isn't the only network gathering data. We've been documenting Foresight since before I signed their talent contract."

"You deliberately entered the Stream? As some kind of infiltration?"

"As a way to pay for my mother's treatment." She removes something from her pocket—a small device similar to the one in her mother's lap. "Underground resistance networks are long on principles, short on medical funding. The Stream pays well for the right kind of stories."

The clinical assessment mirrors her mother's analytical framework beneath their different exteriors. Theodore struggles to reconcile this focused, strategic Wenqian with the Elliot he's spent months developing for consumption—the fiction he'd polished until it gleamed brighter than truth.

"Your crisis broadcast—your mother's seizure—"

"Was completely real," Dr. Zhang interjects. "Nobody fakes tremors like these." She holds up her shaking hand with a wry smile. "The seizures are real too. Our physical conditions are genuine. Our response to Renaissance's exploitation is simply ... strategic."

The sun has shifted, shadows lengthening across the docks. The changing light reveals details previously hidden—a network of small marks etched into cargo containers, symbols forming some kind of communication system. Patterns within patterns.

"Why are you telling me this? Why show your hand now?"

"Because Project Foresight has entered implementation phase ahead of projected timelines," Wenqian explains. Her posture has changed completely from her Stream persona—no calculated angles, no optimized engagement postures. "The neural suppression protocols are being deployed against identified targets. Children with specific creative potentials are being diagnosed with attention disorders, prescribed compliance medications, their thought patterns redirected before they can develop resistance."

"We need the complete Foresight architecture," Dr. Zhang adds. "Access protocols. Target identification parameters. Suppression methodologies." She leans forward in her chair. "We need someone with your clearance level, Theodore."

His mouth feels suddenly dry. The metal bird in his pocket seems to gain weight, pressing against his thigh. "You're asking me to commit corporate treason."

"I'm asking you to notice what you've already begun to see," Dr. Zhang says, her voice softening. "The watercolor you kept. The bird you made without thinking. The anomalies you didn't report." She gestures at the space between them. "These aren't isolated incidents, Theodore. They're steps on a path you're already walking."

Theodore's hand moves involuntarily to his pocket, feeling the sharp outline of the twisted paperclip bird. "How do you know about that?"

"Because I spent thirty years studying how people recognize patterns before they consciously know they're seeing them." Her smile is gentle now, almost maternal. "Your mind has already decided. Your body knows it. Only your words are catching up."

Theodore looks toward the water, where autonomous cargo drones load and unload containers without human supervision. The machines move with perfect efficiency, their actions determined entirely by programmed instructions. No hesitation, no doubt, no small rebellions against their intended function.

His fingers find the bird in his pocket, trace its bent angles and crude wings. Created without intention, without awareness. A pattern manifesting below conscious thought.

"What would you do with the information?" he asks finally.

"Develop countermeasures," Wenqian answers. "Create protective barriers for vulnerable minds. Establish alternative expression channels outside algorithmic control."

"And your Stream career? Your Renaissance contract?"

"Completed its primary objective yesterday." A slight smile touches her lips. "Full treatment protocol downloaded and secured. Fourteen months of stability assured regardless of my Stream status."

A group of birds passes overhead—not magpies but ordinary gulls, their movements unpredictable, their purpose their own. Theodore watches them until they disappear beyond the Upper Alta skyline, imagining their trajectory continuing past Renaissance's gleaming towers, past the surveillance networks, into territories beyond algorithmic mapping.

Memory fragments as Theodore leaves the docks: his mother before an unfinished canvas, brush suspended, head tilted in consideration. He'd asked why she stopped. "Sometimes you have to make a mark that makes no sense," she'd said. "The meaning comes after you've committed."

Something shifts within him—not epiphany but culmination, small rebellions solidifying into conscious choice. Seven years at Renaissance settles differently across his shoulders.

"I need time," he says, "to access Foresight properly."

"Time is what systems of control depend on," Wenqian says, not unkindly. "But I understand needing to be certain. We have a gathering tomorrow night. Sector 7, abandoned municipal archives. 22:00."

Dr. Zhang's gaze remains fixed on Theodore, analytical yet compassionate. "People make choices, Theodore. Not algorithms. What we recognize shapes that we create—or destroy."

As Theodore walks away, his comm unit vibrates with security alerts from Renaissance: unauthorized access attempts detected, Project Foresight briefing rescheduled, immediate return to headquarters requested. Behind him, Renaissance's systems were already screaming. He calculated: four hours before detection, six before response, eight before they reached his apartment.

His fingers touch the metal bird in his pocket, trace its edges and angles. He hadn't consciously created it, yet somehow his hands had bent formless metal into specific shape. Theodore silences the alerts without responding and continues walking toward the transportation hub, each step representing not yet commitment to either direction but acknowledgment that movement itself has become necessary.

6. TERMINAL VELOCITY

Decaying paper and disrupted data systems create the peculiar scent of information returning to entropy. Water damage writes constellations across ancient ceiling tiles. Theodore moves through cavernous rooms where pre-Stream records molder in forgotten boxes, their contents belonging to a world that measured value differently.

He finds the gathering in what was once the central reading room, its vaulted ceiling partially collapsed, moonlight spilling through structural wounds to illuminate what surveillance cannot see. The space buzzes with quiet activity, voices kept low from habit rather than necessity in this blind spot. The assembly surprises him—perhaps a hundred people engaged in focused conversation, organized around workstations where data displays show familiar Renaissance interfaces accessed through unfamiliar pathways.

Theodore recognizes faces: talents who disappeared from the Stream, developers who vanished from Renaissance, academics whose expertise became obsolete through automation. People the algorithm discarded or never incorporated, existing in parallel to official existence rather than in opposition to it.

A magpie perches on exposed ductwork near the ceiling, observing the gathering with apparent interest. Others of its kind move among the rafters, black and white plumage catching moonlight. Their presence feels significant—living embodiments of the symbol that connected him to Wenqian.

She finds him near the entrance, her movements confident in this environment of fragmented light and partial visibility.

"You came," she says, genuine surprise evident in her voice. "And based on Renaissance security protocols currently having fits, you did more than just arrive."

Theodore extracts a data crystal from within his jacket. His hand trembles slightly—not the tremors, not yet, just consequence calculating itself through nerve endings. The extraction would have triggered seventeen different alerts by now. Lachman would know. The small object catches moonlight, fracturing it into prisms across his palm. Behind him, Renaissance's net was already tightening, algorithms parsing his deletion patterns, his access shadows. Four hours since download. Two until they reached his apartment. Maybe less.

"Project Foresight. Core architecture, identification parameters, suppression protocols. Everything required to understand their pattern recognition and manipulation methodology."

She accepts the crystal with appropriate gravity, weighing it in her palm as if measuring its significance beyond physical mass. Theodore watches her handle what cost him everything—career, safety, the illusion of belonging to something larger than surveillance.

"This changes everything."

"What will you do with it?"

"Analyze the identification markers. Develop shielding for vulnerable minds. Create alternative channels for expression types the algorithms suppress." She meets his gaze directly, perhaps noting the microexpressions of a man who has just burned his entire documented existence. "The Stream itself isn't inherently corrupt. Connection has value. Being witnessed matters. But extraction without consent, manipulation without transparency—those require intervention."

Nearby, a team works on what appears to be a surveillance drone, its casing removed to reveal intricate electronic innards. Their hands move with precise coordination, extracting components, modifying others. Theodore notices several magpies observing the process from nearby perches, heads tilting with apparent curiosity.

"Why did you trust me?" he asks.

Wenqian considers the question, her posture lacking the calculated angles Renaissance had trained into her. "My mother analyzed your interaction patterns. Found statistical anomalies in your protocol adherence."

"Scientific assessment."

"Partly." She hesitates, then smiles slightly. "But I saw something else. Something she couldn't quantify with numbers."

She falls silent, looks away. Theodore waits, recognizing a deliberate withholding—something she's chosen not to articulate.

"I was eleven when they came for my mother's research," she says finally, changing direction. "They called it 'consolidation for efficiency.' I watched her pack twenty years of work into three boxes. She kept one notebook hidden in my backpack. 'Pattern recognition persists,' she told me. 'Even when they try to erase it.'"

The unexpected personal revelation catches Theodore off-guard—a glimpse of Wenqian beyond strategic calculation, beyond resource acquisition. A memory untethered to mission or marketplace.

Around them, the gathering continues—people connecting without documentation, creating without monetization, existing in spaces between algorithms. Not revolution precisely, but persistent alternative. Not destruction of existing systems but cultivation of parallel possibilities.

From above, a magpie releases a small object that falls with startling precision onto the workbench beside them. A microchip, apparently extracted from a surveillance device. The bird cocks its head, eyes bright with intelligence that seems to exceed instinct. Other magpies throughout the space appear similarly engaged, moving with purpose rather than randomness.

"They've been helping us for months," Wenqian explains, noticing Theodore's expression. "Corvids naturally collect technical components—the shine, the complexity. We just helped direct their magpie tendencies toward useful targets."

Theodore watches as a magpie carefully extracts a tiny camera from a deactivated drone, movements delicate and precise. "You're using birds?"

"'Working with' rather than 'using,'" she corrects, an echo of her mother's precision. "They recognize patterns humans miss. We provide food and shelter; they dismantle surveillance in ways Renaissance can't anticipate. The algorithm doesn't flag bird behavior as threatening."

The revelation strikes Theodore as both absurd and perfect—nature itself repurposed as resistance against systems designed to control nature. The birds continue their work with focused determination, their actions serving their own interests while inadvertently advancing human ones.

Theodore feels again the strange vertigo that accompanies existence without metrics. No engagement analytics measure this conversation's value. No algorithm determines its priority in the attention economy. It simply exists, significant only to its participants and perhaps to the magpies that work alongside them.

"What happens now?" he asks.

"Now, you learn to exist differently. To create without extraction. To connect without quantification." Wenqian extends her hand, the gesture containing no performance, no calculation of optimal engagement metrics. "It takes time to unlearn the patterns. To stop measuring value through metrics. But eventually, you remember how to recognize meaning without calculating it."

Theodore takes her offered hand, the contact feeling strangely unfamiliar—physical connection without documentation, significance without analysis. As they join the others, he becomes aware of an absence that gradually reveals itself as freedom: the absence of observation, the unburdening of perpetual performance, the space to exist unmeasured.

In his pocket, his fingers find the small metal bird, its form now familiar against his skin. He extracts it, studies its imperfect symmetry and unoptimized purpose. Then places it carefully on a nearby shelf, among other small objects left by previous visitors—tiny markers of passage through a space surveillance cannot see.

A real magpie immediately investigates the metal creation, prodding it with its beak before apparently losing interest and returning to the more complex technology of a dismantled surveillance component. The bird's rejection of the crude approximation of its form strikes Theodore as both humbling and appropriate.

The gathering continues around him, an assembly of those who chose to exist beyond algorithmic determination—not outside society but within its unmapped regions, cultivating alternatives to extraction and consumption.

Not revolution precisely, but the beginning of its possibility. Not destruction but creation—the most fundamental human capacity that no algorithm can fully capture or contain. Theodore moves deeper into the gathering, leaving performance behind, stepping into unmapped territory where value exists without metrics and patterns serve recognition rather than control.

In the spaces between surveillance, they begin.

BLOOM STATE

Three months after disappearing from Renaissance records, Theodore receives a package at the residential unit he shares with two system expatriates. The surveillance state pursued profit, not vengeance—Theodore generated no more value, became invisible. The algorithms had written him off as wastage. Still, the first month he'd slept in fragments, waiting for enforcement that never came. Renaissance's contractors had been downsized in Quarter Four; their enforcement arm had atrophied into automated systems that only flagged active threats. Theodore had perfected passivity.

The delivery arrives without digital tracking, hand-carried by a courier who disappears before Theodore can thank them. Inside, wrapped in plain paper, he finds a watercolor painting of a magpie dismantling a surveillance camera, mechanical parts scattered beneath its perch.

Water damage has blurred the boundaries between bird and technology, pigments bleeding together until the magpie's eye and the camera lens fuse into a single watching device. The effect creates an unsettling hybrid—neither fully bird nor fully machine, but something emerging between categories. In the corner, a tiny signature: 温倩—Wenqian.

Theodore places the painting on his desk, where strange transformations continue to occur. The water damage doesn't stabilize but evolves, colors shifting in ways watercolor shouldn't.

Some days the bird seems more prominent, other days the technology. The unpredictable changes defy documentation—each attempt to photograph the image results in something different from what Theodore sees with his naked eye.

He works now developing protection for vulnerable minds based on Foresight documentation. The work progresses slowly without Renaissance's resources. Some days Theodore feels the absence of optimization acutely—the inefficiency of collaboration without algorithmic assistance, the messiness of creation without performance metrics, the uncertainty of value without quantification.

Other days he recognizes these inefficiencies as necessary spaces where actual human connection occurs—the gaps between instrumented interactions where genuine exchange happens, unmeasured and therefore unmeasurable.

The settlement exists seventeen kilometers beyond Alta's borders, in zones Renaissance deemed unprofitable for surveillance. They operate from abandoned Renaissance facilities, using equipment left behind when profit margins thinned. Black market medical supplies arrive weekly—skimmed from Renaissance shipments by drivers who remember what the world was before optimization. Dr. Zhang's hidden patents, sold to offshore manufacturers before her diagnosis, fund what Renaissance's insurance would never cover.

His mother's condition has stabilized under treatment protocols derived from Dr. Zhang's research. The tremors in her hands have reduced sufficiently for her to begin painting again—not the commercially viable work Renaissance's algorithms favored but strange abstract forms she describes as "neural landscapes," visual representations of thought patterns. "The words are there," she tells him during their encrypted calls, voice clear but sometimes wandering, "but the bridge between thought and speech keeps…flickering. So I paint the electricity instead."

Years later, Theodore will remember a moment from his first month after leaving Renaissance. A thunderstorm had knocked out surveillance in three Lower Alta sectors, creating a temporary expansion of the blind spots where they operated. Through the rain-streaked window of their makeshift laboratory, he had watched actual magpies systematically attacking disabled cameras, extracting components with methodical precision.

"Are they working for us?" he had asked Wenqian, who was monitoring network disruptions nearby.

"With us, not for us," she had corrected, smiling. "And maybe we're working for them. Maybe we've always been side characters in their story."

The birds had continued their work through the storm, black wings glistening with rain, creating patterns against the gray sky that resembled code Theodore almost but couldn't quite read.

Now, studying the water-damaged painting on his desk, Theodore can't determine whether the magpie is dismantling the camera or incorporating it, whether the blurred boundary represents destruction or synthesis. The ambiguity feels appropriate to his current existence—neither fully separated from the systems he once served nor fully reconciled to their absence.

On his terminal, an encrypted message appears—communication from Wenqian, routed through the alternative network they've constructed in the blind spots between surveillance systems:

"Neural stabilization at 94%. Mom finishing final research publication. Foresight countermeasures being implemented in education sectors 3, 7, and 12. Systems recognizing systems."

Below this, a personal note outside the formal update:

"The tremors came back yesterday, just for a while. My hands shook when I was painting, colors going everywhere they weren't supposed to. I almost tried to fix it, then realized the accidents made something better than what I planned. Some patterns only emerge through disruption. The painting I sent you—I damaged it on purpose. Water has its own pattern recognition."

Theodore composes his response, describing progress on shielding protocols and distribution networks. Then adds his own personal note:

"I dreamed of the Stream last night. Not missing it, but remembering how strange it felt having value constantly measured, constantly confirmed. Sometimes I still reach for metrics that no longer exist. But

I'm starting to recognize value in forms algorithms can't calculate—connection without quantification, creation without consumption, being without broadcasting."

He sends the message through their secure channel, then returns to his work, developing protections for minds that function differently, for patterns that defy optimization, for expressions that exist outside algorithmic assessment.

Outside his window, a magpie lands on the sill, its head tilting as it studies him through the glass. In its beak, it holds something small and metallic—a component extracted from a surveillance device. The bird sets the object down carefully, then flies away, leaving behind this small offering or coincidence.

Theodore picks up the component, turning it in his fingers. Rainwater drips from it onto his wrist, cold and slightly metallic—like tears, if machines could weep. He can't decide if the droplets come from the storm outside or something still wet inside the circuit.

The magpie doesn't return. The component offers no obvious purpose or message. The painting continues its waterborne transformation on his desk, boundaries blurring further between watcher and watched, between pattern recognition and pattern creation.

In the spaces between certainties, they persist.

Artistic Presentation

BY L. RON HUBBARD

When the first Writers of the Future *volume was published, its stories were illustrated by professional artists. It wasn't until Volume 5 that the winning stories were depicted by winners of the newly created Illustrators of the Future Contest, as they still are today.*

While L. Ron Hubbard honored equally both artists and writers, he was broadly known for his work as an author of popular fiction, having published over 230 fiction works representing millions of words.

But Hubbard also worked in other mediums. For example, Hubbard's tales lent themselves to film particularly well, thanks to his eye for detail and his talent for describing action. By the summer of 1937, one finds his stamp on such scripts for the big screen as The Mysterious Pilot, The Adventures of Wild Bill Hickok *and the Spider series, while his name was formally attached to the fifteen-episode* The Secret of Treasure Island*—which was among the most profitable serials of Hollywood's Golden Age.*

Hubbard was also an accomplished photographer. He was a keen student of the craft in his youth, and by early 1929, his celebrated China landscapes were acquired by National Geographic *while his spectacular aerial shots as a pilot could be found in the pages of* Sportsman Pilot.

Similarly, although he never counted himself as a professional musician in the strictest sense, his musical accomplishments are by no means insignificant. Understanding that sound without words also tells a story, he created a "soundtrack" to Battlefield Earth *using previously unexplored computerized instruments, followed by an innovative* Mission Earth *album, themed against his bestselling series of the same name.*

Thus L. Ron Hubbard developed a love for and mastery of a variety of artistic forms, and in that spirit, the Illustrators of the Future Contest was created to be a companion to the Writers of the Future.

His diverse experiences made him especially qualified to find common ground across all the arts. In "Artistic Presentation," synthesizing his experiences in writing, filmmaking, photography and music, he was able to advise others about a topic that marks the true professional, no matter the art form.

In the decades since Hubbard wrote about artistic effort, the creative world has changed dramatically, especially with the arrival of artificial intelligence. AI-enhanced tools can be useful, just as cameras, software, or musical instruments are tools. They expand what an artist can do. Conversely generative AI attempts to replace human creativity by replicating work created by artists. For this reason, AI-generated writing or art is not accepted in the Writers & Illustrators of the Future Contests.

"Artistic Presentation" stands as both guidance and challenge: use every tool available—but never surrender the care, effort, and responsibility that define true professionalism in any art.

Artistic Presentation

We live in a machine world. The whole yap of television and newspapers is directed toward reducing effort. The primary goal of the civilization in which we live, it seems, is to reduce all personal effort to zero.

The less effort a person can confront, the more effect of effort he becomes.

The modern trend of "don't do" accompanies the modern trend of an increased percentage of the insane in the society.

The crazier a person is, the less he accomplishes or does.

So we live in a world which is oriented to drive men mad.

But, more pertinent to us, we suffer from the continuous bait—"do it the *easy* way." "Do it in the way that will demand the least effort."

We see this in manufacturing, particularly—the easiest way is the cheapest way is the most profitable way.

So we get into a "do it the easy way."

Well, that may apply to making spoons for profit, but it does *not* apply to presentation.

The whole world of the arts is directly opposed to the philosophy of the businessman or manufacturer.

Art seeks to create an effect. An effect is not always created the *easy* way. Indeed, the better effects are quite difficult to achieve.

One can fall into creating easy effects to such a degree that one fails completely.

For instance, a dozen cakes are in competition at a county fair. The one that wins is not the easiest cake to make. True, the cook that made the winner may have some easy ways to short-cut cake baking. But the winning cook actually takes that extra bit of care to make it all just right.

It isn't magic or luck that makes the professional. It's hard-won know-how *carefully applied.*

A true professional may do things pretty easily from all appearances, but he is actually taking care with each little bit that it is just right.

The winner has it instinctively. The loser rarely even grasps the concept of "do it right."

Artistic presentation always succeeds to the degree that it is done *well.* How *easily* it is done is entirely secondary.

To the world of presentation, the only guide is take the care necessary to do a good job.

To the world of the businessman, the manufacturer, the primary guide is "how can we do it easily."

These two philosophies clash.

We are taught daily in advertisements, by union leaders, by socialists, that DO IT WITH THE SMALLEST EFFORT is the greatest goal in life. Do the least work for the most pay. Buy the automatic machine that chews up the most clothes in the least time. Use the roofing paper that goes on quickest and keeps out the least rain. Vote for Jim X who will make all the world eat without working. Do nothing yourself. Shove it off on the Mix-Up Accounting Company—or the man at the next desk.

That all this leads to total dependence on gadgets, total enslavement to mounting economic puzzles, even to total enslavement to a Commissar Krushtoad in the next generation, is neglected utterly. That less than two centuries ago we lived quite well and built more strongly and were a lot saner without all these ads, tools and commissars is never mentioned.

Man is solving himself to extinction. And all on the slogan "Don't exert yourself."

It's gotten so bad that people are shrugging off all responsibility for the state, for their friends, for anything and everything. "Nothing has anything to do with anybody" is the epitaph that nobody will take the trouble to write on the tombstone of this civilization.

Now, this is no rant against automation or gadgets or self-sterilizing cat petters.

Use all the gadgets you can lay your hands on—if they really do work in your hands and don't absorb all your time in earning their price or repairing their faults.

No, my thought here is only this—keep your action level above your gadget level.

Keep ahead of automation. Keep ahead of do-it-for-you. Don't disenfranchise yourself by giving all your work away—to a machine, to a fellow worker.

If you've got equipment, do one of two things: (a) Use it to increase your production of effects, or (b) Get rid of it.

But first and foremost realize that in presenting something, that the best way isn't always the easy way. The best way is *only the more effective way*.

Work out first what effect you are trying to produce. Then when you've got that all taped, *only* then consider the easiest way to do it. And never consider the easier way at all if it is less effective.

Art takes that extra bit, that extra care, that bit more push for it to be effective art.

There is no totally easy way to produce a desirable effect.

And the day you drop some of your ideas of the effect you want to produce is the day you get a little older, a little weaker, a little less sane.

So don't buy the easy way. Buy only the effective way. If some of its points can then be made easy, good. If not, do it the hard way.

And only if you realize this can you escape the gargantuan trap of a society with the mass goal of "Nothing should ever be done by anything but a machine or somebody else."

Space Can

written by
L. Ron Hubbard

illustrated by
HAILEIGH ENRIQUEZ

ABOUT THE AUTHOR

Originally published in the July 1942 issue of Astounding Science Fiction, *"Space Can" appeared in distinguished company—sharing the table of contents with A. E. van Vogt and L. Sprague de Camp, two other leading voices in the Golden Age of Science Fiction.*

Written in the summer of 1941 just months before L. Ron Hubbard received orders to report for active naval duty for World War II, the story reflects an author already thinking in terms of command, discipline, and the realities of combat in the three dimensions of space.

Here, a lone destroyer—little more than a "space can"—rushes to defend a crippled convoy, carrying with it four centuries of naval heritage and a fierce refusal to retreat. Hubbard imagined space battles not as aerial duels but as engagements fought by steel-willed crews in vessels more akin to submarines than aircraft.

"Space Can" remains a standout example of Golden Age military science fiction—where duty, valor, and tradition matter as much as firepower, and a battered destroyer can still turn the tide of a war.

ABOUT THE ILLUSTRATOR

Haileigh Enriquez, born in Los Angeles in 1996, is an artist of Mexican, Yaqui Native American, and Salvadoran descent. Growing up in Rosemead, California, a predominantly Asian and Hispanic community, deeply shaped her identity. She immersed herself in drawing and comic books from an early age, which fostered a fascination for character design and storytelling. What started as an escape turned into a passion.

Originally on a path toward a career in the medical field, a personal loss led Haileigh to reevaluate her future and embrace her true passion: art. With the encouragement of her family and mentors, she shifted her focus and enrolled at Laguna College of Art and Design, where she refined her technique and explored her personal style. Her work blends fantastical realism, bold colors, and dynamic lines, drawing heavily from her love of comic book art and Mesoamerican mythology.

Driven by a desire to create characters that celebrate cultural diversity, Haileigh's work is a joyful exploration of identity, imagination, and resilience. As a traditional artist turned digital creator, she combines realism with imaginative storytelling, crafting unique characters that reflect her identity and the worlds she dreams of.

Haileigh is working as a freelance artist for private clients and media companies while also developing her superhero series. She is determined to become a successful artist in the entertainment industry and is excited to bring new, authentic stories to life—infused with the vibrant spirit of her heritage and her passion for representation.

Haileigh is a former Illustrators of the Future Contest winner featured in L. Ron Hubbard Presents Writers of the Future Volume 41.

Space Can

Lancing through space, slammed along by a half million horses, the United States destroyer *Menace* anxiously sought the convoy which had been wailing to all the Universe for aid but now was still, still with an ominous quiet which could mean only its defeat.

She was only one, the *Menace*, and "they" would be more than one, but the little space can charged ahead, knowing well that she was a pebble from the mighty slingshot of the embattled fleet, a pebble where there should have been a shower of stones. Gracefully vicious, a bundle of frail ferocity, a wasp of space designed for and consecrated to the kill, the *Menace* flamed pugnaciously onward; she had her orders, she would carry them out to the last ounce of her fuel, the last charge in her guns and the last man within her complex and multiple compartments. She carried the Stars and Stripes upon her side, gold lace upon her bridge and infinite courage in her heart, for upon her belligerent little nose rested the full tradition of four hundred–odd years of navy, a tradition which took no dares, struck no colors and counted no odds.

She should have been a flotilla in this lonely cube of space, but with the fleet embattled off Saturn, no flotilla could be spared. She had done other jobs, hard ones, in this long war. There was faith in her, too much perhaps, and so she was here alone, raking the black with her detectors, bristling with impatience to engage the enemy, be it cruiser or battleship or just another destroyer; she was a terrier who had no eyes for the size of her rats.

On her bridge a buzzer sawed into the roar of her motion, and her executive officer stood aside to permit her summoned captain a view of the detector. Her captain, Lieutenant Carter, steadied himself with a hand on Ensign Wayton's shoulder, and his face, usually young and efficient, became weary as he looked at the message which registered there.

In the detector, the supply ships were colorless spots, unmoving, without order. Among them were fainter dots which gruesomely indicated ships which were growing cool, having been emptied of air. Because spacesuits might mean desertion of crews near the first port, there were few in naval vessels unless they were crack ships like the *Menace*. This battle was almost over and there would be many, many dead.

One spot began to turn violet, which meant that a vessel, friend or foe, was heading toward the *Menace*.

Second-class Petty Officer Barnham was already training an analoscope on the red spot. He shuffled the spectrum plates of all navies until he had one which would compare with the lines on his screen.

"Saturn destroyer, sir," said Barnham matter-of-factly.

Lieutenant Carter shook himself into the fighting machine he was trained to be. The situation was a plain one, a simple one. The convoy had been set upon by a raiding fleet, the existence of which had not been suspected. Bravely the train's escorts had flashed into battle and had fought their ships to the last pound of air; that they had not done badly was indicated by the fact that only two Saturn vessels remained in action; that the entire escort was dead was plain in the silence of the battle communicator; that the supply ships were paralyzed and already half destroyed was to be found in the garble which spewed and gibbered from the all-channel speaker.

Another spot, which had evidently been traveling parallel to their course and so had showed white, now glowed dull red and Barnham said in a flat voice, "Another Saturn vessel, sir."

They were coming up now into action. They had perhaps thirty minutes of strain in store before the first searching blasts of flame came to them and their own guns began to seek the vitals of the enemy. The captain pushed a thumb down upon the battle-stations button and the clanging roar broke the tight lines which had invisibly stretched through the little destroyer.

It was a matter of seconds until Lieutenant Carter had his battle plan. Plainly, he wanted nothing to do with this first destroyer, for he could feel from across black space the eagerness of hope in it that he would attack it and disregard the second ship, while that vessel, with all the brutal efficiency of a thing which knows nothing but destruction, blasted the life from the remainder of the supply vessels.

Abruptly, Lieutenant Carter understood a thing which in his inevitable resentment at being detached from the great battle had escaped him, and he understood, too, that insufficient weight had been given to this mission. He should have been started early. He should have the rest of his flotilla in a comfortable V behind him. For now the detector gave out information in shape instead of light and disclosed that this supply train consisted of the majority of fuel vessels possessed by the navy. Someone had blundered. Intelligence had failed to discover that an enemy raiding fleet had slipped away from Saturn; guard ships had blundered in letting it through; flag had erred by not suspecting the possibility. For in those big hulks was the blood of the fleet and without it victory or destruction were the only alternatives. The battle fleet, already far beyond its radius, had no reserve. And from the state of his own bunkers, Lieutenant Carter knew that no one had sufficient fuel to return to Earth!

Everywhere through the ship men were strapping themselves at their posts or donning the heavy padding which would protect them against the violent course changes which would throw the complement about like dice in a cup.

"Aloft ten, right rudder nineteen," said the captain.

The *Menace* leaped as the steering jets slammed her into her new course, as though she was unwilling to even countenance a thing which sought to avoid battle.

The screens of the enemy showed the action without much lag, and an instant later, the Saturn vessel was killing her speed on her old course and blasting into a new one which would again intercept the *Menace*.

The Saturnian, grudged Ensign Wayton, was well handled. Getting by her to engage the second was not going to be simple.

Lieutenant Carter leaned back in his deep command seat and apparently lost interest in the whole thing, for there was a vague look in his eyes and a relaxed expression about his mouth. Seeing this, the quartermaster let out a small explosive sigh, for he knew that they would engage the first enemy.

Actually, the captain was examining the vast panel of meters which gave the small bridge the appearance of being set in diamonds and gold. When he saw that all guns were ready, that all tubes were firing, that the air pressure was even throughout the ship and the new tanks broached to give the men more energy and courage, he turned slightly to the blue-and-gold figure in the other wing and said quietly, "We will engage, Mr. Wayton."

Ensign Wayton's hands tensed over the panel above his knees and then fluttered for an instant as though he needed to test the buttons which would fire the batteries.

"Aye, aye, sir," said Ensign Wayton. He was breathing quickly, as though to supercharge his body with oxygen and hurl himself rather than flame projectiles at the enemy.

On the after bridge, before a similar but less complete board, Ensign Gates stood a lonely watch. He could look down the hatch just behind him and see the tense crew around the base of the Burmingham jet of the starboard engine. Ensign Gates swept his eyes back to the control panel, checked the telltales there and then glanced at his own quartermaster. The man, a heavyset sailor from Iowa, who still bore, after twenty years in space, the stamp of his state upon him, looked impersonally into the sphere compass which mirrored the stars and planets. He felt the officer's eyes on him and edged his appearance with a sharp professionalism, as though this might communicate a greeting to the placid little ensign, of whom the quartermaster was fond in a shy, defiant way.

Ensign Gates grinned to himself, for he knew the meaning of the change in his quartermaster. He said something to the man, but the remark was engulfed in the crashing shudder of the port twenty-nines. They were engaged.

Time stood still and two vicious dots of ferocity slashed at each other in an immense black cube of vacuum. Shells burst like tiny flowers when they missed, or flashed like yellow charges of electricity when they struck. The *Menace* became filled with acridity. Somewhere in her a man was screaming an insane battle cry, and elsewhere blue blots of profanity hung thickly around guns and tubes and stoke ports.

Compartment 21 was holed and sealed from the rest of the ship between the beats of a chronometer. Compartment 16 turned into a blazing furnace and was sealed alike.

In the exact center of the ship, which was the after bridge, Ensign Gates placidly kept track of the enemy in the event that his firing panel had to take over. His active duty here was the overseeing of the engineering force, aft and below, but two tough chiefs were cursing themselves into a comfortable berth in Hades around the molten breeches of the tubes and needed no help.

"Hulled her!" barked the annunciator. Forward, Ensign Wayton sounded like a man cheering a baseball game rather than the director of that deadly blast. And then an instant later, "Hulled her!"

There was a crash topside and a man, bellowing agony and rage, hurled himself down a ladder. He was a mass of flames. The emergency squad member there smothered him swiftly in a blanket. Compartment 6 was sealed and everyone in her.

A small amount of Ensign Gates's placidity left his face. They were being severely knocked about by a vessel which had a longer range and a faster steering system, which was landing four hits to their two.

"Hulled her!" cried Ensign Wayton, an invisible source of death forward and above. Evidently something had happened to the Saturnian, for an instant later, in a steady stream, Wayton began to chant the *Menace*'s hits.

Examining the panel before him, Ensign Gates believed that a lucky shot had penetrated the steering jets of the enemy, for he was now traveling in a straight line through the remaining three vessels of the convoy as if to help out the other Saturnian in the convoy's destruction before this raging little wasp of space put an end to everyone. Just as the *Menace* flashed by a halted supply vessel, it bloomed into a sphere of scarlet death, the ammunition and highly explosive fuel igniting all at once.

Lieutenant Carter gazed calmly at the fleeing enemy, but the calmness was an official sort of thing, for there was sorrow for the supply ships and anger for the Saturnian snarled into a lump behind his gray eyes. Each time the *Menace* got a salvo home the captain twitched forward and a contraction of muscles above his mouth made him grin a split second at a time. His role was that of spectator so long as the ship was on her target, for then her steering was wholly between the gunnery officer and the helmsman.

With a blast close aboard, the Saturnian folded itself like a smashed tin can, and what had been an efficient fighting ship an instant before was now a scrap of volatilizing metal.

"Well done, Mr. Wayton," said Lieutenant Carter.

Ensign Wayton turned glowing eyes and battle-reddened cheeks upon his captain and didn't see him at all. He was already seeking the other Saturnian on his screen, was the gunnery officer, as though this first ship had merely served to calibrate his guns.

"Engage the second enemy, Mr. Wayton," said Lieutenant Carter.

The *Menace,* bristling and sure of herself, shot a streak of power from her starboard bow and stabbed into a new course, three quick jets on the port bow and one below settling her into this.

Telepathically, Lieutenant Carter was aware of his enemy's abrupt distaste for combat with him, now that the first Saturnian had been blasted from the action, but there was nothing in the action of the second vessel to indicate its dislike, for it turned now away from the supply vessel it had intended to spear, and streaked in a wide bank to bring her into a broadside parallel with the *Menace.*

Ensign Wayton adjusted his screen with the motor button and gave a swift check to the computator and then, because he was already ranged, sent all six guns of the port battery into a furious crescendo.

The *Menace,* dancing sideways from the recoils and being jabbed back by the adjusters, shivered with some vague premonition.

The Saturnian destroyer passed through the cone of concentration, sliding sideways to the *Menace* at a swift pace to throw off range and for some other purpose which was not to be fathomed for several seconds. The Saturnian's guns were winking bright spots and her flame wake, as it turned to white powdery smoke, curved and feathered. She was a well-built little vessel, a few feet longer than the *Menace* and thicker through.

Lieutenant Carter scanned space with his detector but found no sign of reinforcement for the remaining destroyer.

The *Menace* shivered as she was knocked off course. The check board blinked and Compartment 26 vanished from it. Then, in terrifyingly swift order, the lights indicating Compartments 27, 28, 29 and 30 went black.

An annunciator above the captain's post said in a calm voice, "Starboard magazine gone. Fire spreading."

The quartermaster's eyes flicked to the captain. Ensign Wayton hesitated for an instant over his firing buttons and then his gold stripe flashed as he located and aimed all three space torpedoes on the starboard. He launched them and said in a tightly casual voice to the quartermaster, "Roll a hundred and eighty." Ensign Wayton, having no starboard batteries, was in action with the port.

Compartments 31, 25 and 36 went out in order. The air in the ship was unbreathable.

"Spacesuits," said Lieutenant Carter into the annunciator. "All hands."

The space torpedoes were sped, but only one had struck—this in the after section of the Saturnian, where it had caused a vast fan of bright fireworks. It had wiped out the stern balancing jets, but that vessel's main propulsions were apparently without harm.

A new crash shook the *Menace* and the big light which marked the after bridge went black.

There was the smallest hint of concern in Lieutenant Carter's voice. "Mr. Gates!"

Silence answered.

With steel bands on his nerves, his voice carefully steady, Lieutenant Carter said, "Mr. Gates. Please report."

There was silence which hung for a heartbeat throughout the entire vessel.

Dead white, Ensign Wayton glanced at his captain. It was an appeal of dependence, shot without thought, an agonized hope that something could be done, a last belief in the impossibility that anything could ever happen to placid, easy Ensign Gates.

Lieutenant Carter did not look at his executive officer. In a flat, official voice he said, "Grapple the enemy."

The heat was so intense in the dying *Menace* that men felt it through their spacesuits. They were unwilling to begin upon their private stores of oxygen until smoke was too thick in the hull to be breathed. Now they were in communication with helmet phones.

Space-garbed, a relief came to the quartermaster to allow him to climb into his suit. He had been standing there, strangling and sweating at the helm, and he would have stood there until he had melted if his relief had not come. The captain took the firing panel while Ensign Wayton slid into his suit. And then Lieutenant Carter dropped into his own ready covering. The captain gasped with relief as he sucked in air.

There was a clatter in the phones as arms were being issued out. Though the batteries were firing still, the helmet cut down their roaring to a tremble, which one felt with his body. There was something ominous and horrible in this silence for every man on the ship, for each was affected alike in the connection of the silence to a sudden surge of loneliness. For perhaps three minutes there was irregularity in the smoothness of the execution of duties, and then the first shock of quiet wore away and men began to talk to each other on the individual battery frequencies, began to swear anew, began to revile and damn this enemy who was destroying the sleek little *Menace*.

Still firing, Ensign Wayton was adjusting his ranges so as to sweep them in closer and closer to the attacking ship. The Saturnian was suffused with superiority and satisfaction, for the burning wake of the *Menace* was plainly visible as were the gaping holes in her skin, and this feeling, knowing it existed, Lieutenant Carter utilized by ordering unsteady leaps and veering as though the vessel were not quite under complete control.

Confident and disdainful, the Saturnian welcomed the closing. She even swept to starboard, little by little, to aid the action. It was her belief that gunnery was the only concern of the *Menace* and this, from a blasted vessel with only two guns still going, she could amply risk.

Further punishment awaited the navy ship, for she could not come so close without being struck repeatedly. Her bow vanished to within twenty feet of the bridge and she was steering now with her guns alone, having two amidships port and one forward starboard, as well as her one-inch batteries on the bridge itself. She was rolling, tortured, nearly out of control, darting up and back and even tumbling when she came within a quarter of a mile of the Saturnian. And then what happened was swiftly done. The grapnels were still in action as they had been designed to be, and the one last ace of the gallant little vessel was played.

With a shuddering stab which tightened and held, the invisible claws of the *Menace* fastened upon the Saturnian and sucked them together with a swiftness which could only end in a numbing crash.

The shock of collision further crumpled the nose and drove a deep bulge into the side of the Saturnian. The latter had been panicked upon the instant of realization that something was amiss and had sought to charge away into space and get free, momentarily forgetful that she still possessed a superior force of men. But now that the adhesion was achieved, she ceased blasting and prepared for the fury which would come—which was already on its way.

Disintegrators in the hands of a burly CPO and his gang ate a hole into the Saturnian at the point of contact as though that hull consisted of cheese.

There was no more on the bridge for Lieutenant Carter. Here his responsibility was done. Ensign Wayton was already gone from the panel and the quartermaster, a huge machete he favored in close quarters gripped competently in his hand, was just vanishing through the hatch.

"Boarders away!" the captain barked at the annunciator in his helmet. He was through the hatch before the yell had ceased to beat against his own ears.

Ahead he saw a knot of men launching itself against another knot which barred the ragged circle of emptiness that led into the Saturnian. Flame was spitting back into the boarders from viciously wielded jets and here and there a spacesuit was giving way to the heat. And then Carter threw himself through the group, jet pistol in hand, and torpedoed himself into the mob just within the Saturnian. With a howl of approval, the sailors followed their captain.

The mass in which he found himself cut at him, shot at him, grabbed at him, and Carter, spinning around and around and firing a space clear, yelled defiantly but incoherently at them.

For several seconds the captain did not realize that the Saturnians had been too contemptuous to don spacesuits—if they had them—for, at best, it was difficult to use them at the guns. It had never occurred to the enemy destroyer that a thing as mad as a boarding would be attempted by such a mauled ship, particularly since the odds in men against such a ship would be three to one.

The curiously pointed heads of the repelling sailors ducked back from the fury of the pistol and then the mass swept deeper into the ship, evidently in receipt of an order which was calculated to draw the invaders into passageways where fast-firing small arms could be brought into play upon them.

Swirling about their captain, the seamen of the *Menace* cut down the stragglers and slipped in their blood. Few guns were here, for the sailors uniformly preferred steel when close quarters were to be had.

Suddenly the front rank of the invaders was swept back, driving their followers to cover. Two of the bodies were dead and projectiled toward the *Menace* by the fury of the fire they had met.

Ensign Wayton, a furiously moving monster in his spacesuit, shot to their fore, insane for an instant in the belief that his captain had been killed. When he saw Lieutenant Carter, he stopped screaming into his helmet. He halted.

"Spread into cover," said Carter quickly. "Try to filter up into the ship through those hatches. But don't press them closely and don't risk your men."

"Aye, aye, sir," said Ensign Wayton. He spent no time in wondering why his captain went back through the crowd, for he had received his orders and he would carry them out to the last word and with his last breath. He looked around him at the shining walls of the gun room in which they had arrived and crisply told off a chief petty officer to burn out a section of its wall. If the passages were covered, there would be other ways of getting through the ship. He had an instant's wonder about their fate, for he knew very well that this handful, less than twenty—less than fifteen, he saw with a shock—were pitted against at least fifty well armed in their own ship.

"Lively, now," said Ensign Wayton.

As captain, Lieutenant Carter had no questions to answer or reasons to give. He was glad of that. He had a competent officer in Ensign Wayton, who knew what to do if anything happened to his commander. And this was a job Lieutenant Carter could not relegate to anyone.

He faltered for an instant on the threshold of the burning *Menace.* It was not the heat which repelled him so much as the unwillingness to see again this dying little vessel which had been, until such a short time before, a well-ordered, shipshape example of what a United States Navy destroyer should be. Here, for two years, he had gone through the routines, the problems and the alternating bursts of good and bad news which had marked this campaign. He had been one with an alive, sensitive creature of steel and chromium and flame, and to enter her now was like walking upon the corpse of one's friend. He had a feeling that she should be left alone, as she was, to die, still facing the enemy.

Lieutenant Carter stepped over the jagged sill of the hole which had been carved in the Saturnian. The need of haste was upon him now, both because the possibility of his getting through the flames before him required speed and because this was a hideous job, the better to be done quickly.

The first blast struck him when a gun charge fired somewhere on the deck nearby. He was catapulted against the steel bulkhead and stood there for an instant in the swirling yellow gloom, shaking his head and trying to recall what he was about. Anxiously, he gripped at the elusive facts, for he was badly stunned. Then, with clearing sight, he sped aft and up through the curling tongues which had already stripped the paint from the walls.

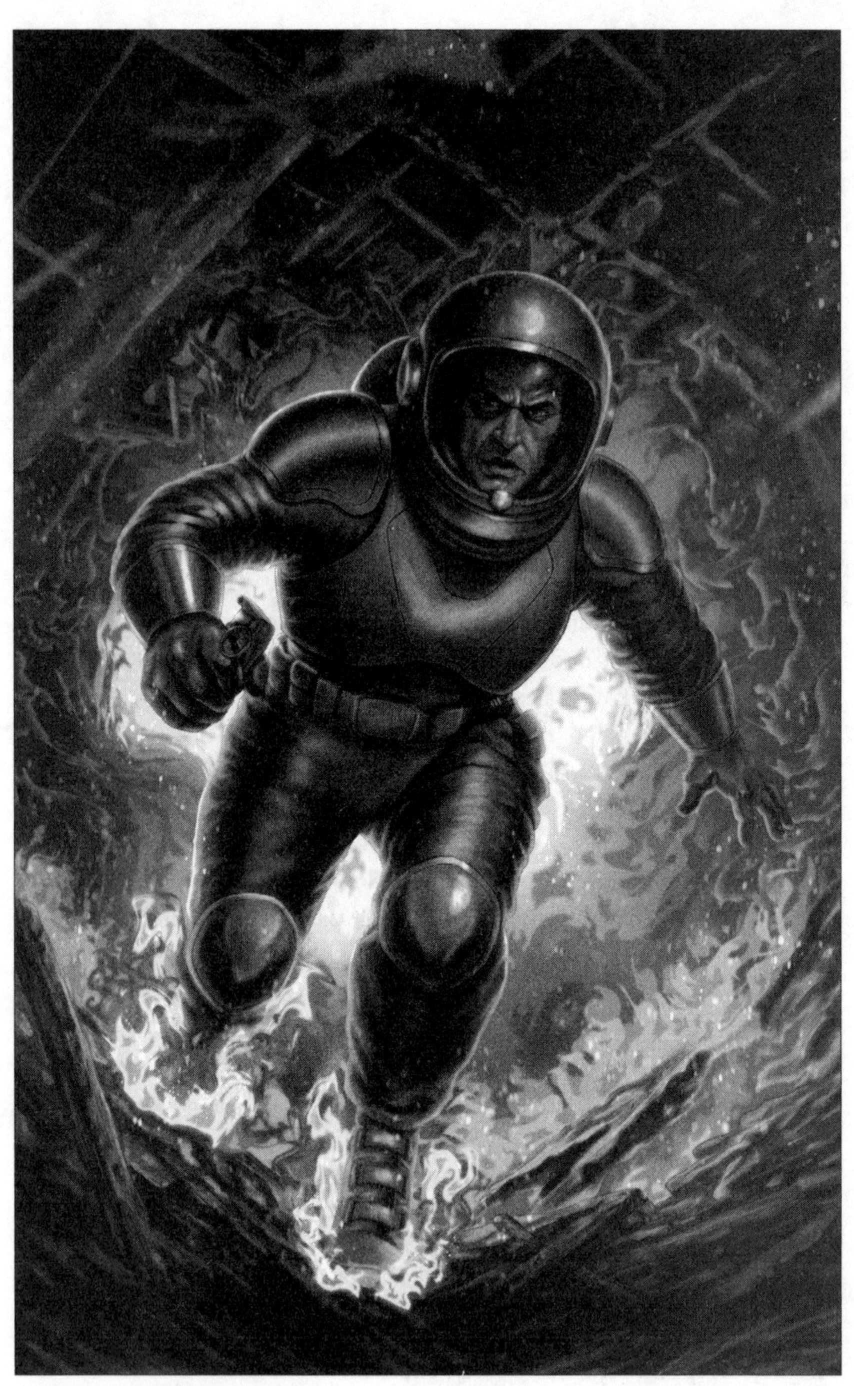

HAILEIGH ENRIQUEZ

There was no resemblance to the trim little *Menace* in this twisted, blackened mess through which he drove himself. He tried to think there was not. He knew there was.

He fumbled in his bag for a grenade as he lurched through the painful fog and when he had it in his thick glove, it required much of his nerve to keep it with him, for tongues of fire were reaching at it, heating it.

He found the ladder to the engine room. The grease had burned away, and because it was hot, his shoes stuck tenaciously to the rungs as though the *Menace*, lonely, was trying for the company of her master in a last shiver of her death throes.

Lieutenant Carter could feel a throb which did not come from flames. He worked toward it. He seemed to be taking forever for this task. The air in his tank was already scalding his lungs. The ship's oxygen tanks must be feeding these flames, and if that were so, then they might explode at any instant. They were close above him now.

He found the generators, still running furiously in all this heat, fed by the treble-protected batteries which made a boarding possible after a ship was in ruins. He hauled a plate from the first layer of armor and then groped through the second and third. That he tried to pull the pin of the grenade with his teeth recalled him into a calm and orderly chain of thought. He plucked at the pin with his glove-thick fingers and got it out. He dropped it upon the batteries and in the same motion spun about and staggered toward the ladder. The heat inside his suit was so intense now that he had to will himself to breathe, and each time he did he flinched as he felt his lungs shrink away.

He clawed through a hatch and scrambled down a passageway. Blind and groping, he found the door in the yellow smoke and stumbled through.

The jagged hole in the side of the Saturnian was just ahead of him, he knew. He could not see it. He sought along the plates with anxious fingers.

Abruptly, he was tumbling forward, breath knocked out of him by another exploding charge. Dazedly, he lifted his helmeted head.

There was a great sighing rush of smoke and fire and a mighty hand snatched him from the deck and slammed him against plates. Groggily, he fought again to rise and then fought even harder, for it would have been very comforting to slump and go out, with the hands of his sailors supporting him.

The smoke of the *Menace* had filled this compartment of the Saturnian. But there was no smoke here now. And there was no air. The empty vacuum was greedy and swelled out the spacesuits to their normal proportions. Where the *Menace* had been, there was now only a gaping black hole. Once her generators, which kept the grapnels alive, had been shorted out, the furious efforts of upper gunners in the Saturnian had at last succeeded in blowing her away from the side.

Ensign Wayton was grinning through his transparent helmet when he had at last ascertained that his captain was safe and not seriously hurt.

Through the phones, Ensign Wayton said, "Sir, we have carved our way through the bulkheads into their after bridge. We have lost but three men. Your orders?"

"Yes," said Lieutenant Carter. "Yes, of course." He shook his head vigorously to clear it. "Well done, Mr. Wayton." Then thought took over from mechanical form and with a glad surge he gripped his officer by the shoulder. "Quick! Open their compartments! Open their compartments!"

The idea flooded in upon Ensign Wayton. It was less than twenty feet up to the hole their jets had carved into the bridge deck. The one dead sailor from the *Menace* and the officer and two quartermasters of the Saturnian were bloated, even exploded, into no semblance of humanity or Saturnity. The CPO, who held the fort there belligerently, cut away at the bulkhead with his jet and suddenly a great gust of air and equipment shot him back.

Ensign Wayton steadied himself at the compartment board and began to open the switches. Some of them were frozen and he realized that the master panel was on the forward bridge. The compartments went shut and their lights began to go out. An officer up there was thinking fast. Ensign Wayton thought faster. He snatched at the auxiliary voice tube caps and yanked them. Into the holes he poured a dozen flame shots. A scream of air, loud enough to penetrate the thick space helmets, greeted his action. The hurricane which came through the voice tubes from the forward bridge knocked him backward. The master panel had been cut in. Suddenly all panel lights glowed on the auxiliary board as lack of air pressure on the forward bridge threw control aft. With swift hands, Ensign Wayton switched the compartments open throughout the ship and a shuddering wail went through the vessel, every plate trembling as the life poured from her. Those suits, denied the Saturnians to ensure their fighting to the last compartment, had cost her, finally and forever, her crew.

Lieutenant Carter, beside his officer, spoke on the general-order frequency of his helmet. "Attention. Proceed carefully through the vessel and clean out anyone left in her." He turned to Ensign Wayton. "Take over, Mr. Wayton."

Seating himself at the communicator, Lieutenant Carter's eyes were vague with thoughtfulness. Absently, he commented that Washington's one-time predilection for trading patents was not without benefit, for this communicator panel might have borne the stamp of Bell Radiophone for its similarity.

He knew he should feel jubilant, knew that he should savor this report to the battle fleet, knew that victory and triumph were personally his. But, somehow, he had ashes on his tongue and the words he tried to arrange in his mind were dull, gray things.

He was thinking now of the *Menace*. In the letdown which had followed this battle, he knew he would think of her more and more. Proud, arrogant little space can, smashed by the insensate hates of a space war, drifting a derelict, a battered sacrifice to her pride, a dead cold thing lost in the immensity, to be shunned by all vessels who sighted her as a navigational risk.

There was victory but there was no victory. He could not think of a proper report, one which would measure up to the little scrap of history they had made. This story would be told in wardrooms for many years, how the little space can took on two larger than she, how she had saved the supply vessels of the battle fleet and how she had died in the saving. Lieutenant Carter could not see the panel clearly and was annoyed with himself. He flung away from it and the reports which were coming to him now concerning the state of the Saturnian, reports which were good, had only a routine meaning. They reached his ears, his official mind, but they went no deeper.

There was a slight jar through the ship, a thing which required no explanation but which seemed to herald something electric. Lieutenant Carter glanced about him. He swung down the ladder to the lower gun room and glanced questioningly at the sentry stationed by the jagged hole in the Saturnian's hull.

And then Carter froze.

For the hole was no longer empty! Had he dreamed that he got the *Menace* away from there? Had it been possible that she would not have herself abandoned?

There she was, the *Menace!* With her shattered bow pushed up into position and the fire-scarred depths of her clear of flame, she bumped gently against her conquered enemy.

And as Lieutenant Carter stared, he saw a man in a spacesuit moving toward him out of the shattered ship, followed by yet another.

Lieutenant Carter started and then quickly composed himself by pushing away the surge of elation which coursed through him.

The man in the spacesuit saluted. "Ensign Gates, sir. Fire shorted our conduits and cut us off. As soon as we dressed and opened the after bridge, we had things under control there. When the air went out of her hull, the fires stopped. She isn't in such bad shape, sir. Your orders?"

Lieutenant Carter saw through a strange mistiness and carefully pitched his voice for calmness. "Very good, Mr. Gates. You will take charge of the repair parties as soon as we get air back into these ships." He returned his engineering officer's salute with unusual smartness.

Gently, the little *Menace* nudged her battered nose against the hull of her conquered enemy as though to remind the Saturnian that a ship, even when shot half to hell, should never be considered in any light save that of a dangerous adversary. For an instant Carter was startled into a belief that the *Menace* was laughing, and then he saw that the sound issued from his phones and was sourced aloft where Gates and Wayton were gladly greeting each other. It amused him to think that his ship could laugh, for the fact was most ridiculous. Or was it? he asked himself suddenly. Or was it?

Shell Game

written by
Zach Poulter

illustrated by
TRACY EIRE

ABOUT THE AUTHOR

Zach Poulter was raised amid the sagebrush and potato fields of rural Idaho. His childhood was spent exploring the nearby Snake River, volcanic buttes, sagebrush desert, and the many abandoned homes and vehicles lodged in unexpected places by a catastrophic flood.

He now lives the glamorous lifestyle of a middle-school band teacher, and also freelances as a saxophonist and composer. When not teaching and making music, he writes all varieties of speculative fiction, with a special affinity for dark, suspenseful fantasy and hopeful horror. Zach lives in Utah with his marvelous wife, four clever children, and not-quite-enough saxophones.

About "Shell Game," Zach says, "the opening lines of 'Shell Game' came during a free-write, before I had any real idea what they were about. I didn't yet know my character was a detective, or that the criminals he would face were a uniquely dangerous type of identity thief. All I knew was that this story would inhabit poorly lit places, with a main character stepping into a world he didn't understand. Everything else emerged from that brooding and mysterious aesthetic. Most of my stories seem to start this way. Not with a plot device or even with a character, but with the story's own voice and vibe calling out to be discovered."

ABOUT THE ILLUSTRATOR

Tracy Eire grew up far to the north in a place called Newfoundland. She has spent most of her life with one foot in the real world and the other somewhere else—maybe on Mercury, maybe in the realm of myth, or... somewhere near Hobbiton. It's a place where stories breathe, where the old gods whisper, and where women step out of the fog wearing their strength and brilliance plainly. That is the space her art comes from.

An oil and charcoal artist with a deep love for narrative painting, Tracy focuses on portraits of women who feel ancient and modern all at once. These are figures who carry storms behind their eyes, but choose the light anyway. She often paints the things she sought—or witnessed—in her youth: courage, grace, grit, and the quiet power of women who refused to disappear, even when maligned. It's no wonder her harpies wear couture.

Tracy honed her skills through the Milan Art Mastery Program, but she's been creating art even longer than she's been reading comic books and novels. She is also a longtime writer of NobleBright fantasy, science fiction, and paranormal tales, and those worlds bleed into her canvases in the form of ghost hunters, fairies, sirens, and banshees like you've never seen before. Every piece she makes becomes a conversation between the storyteller and the painter in her. Sometimes those two voices argue, but... mostly they cooperate. In every part, she's busy becoming an accomplished storyteller.

What Tracy wants when she paints (and hopes for when she writes) is that her work makes people feel something strong and true. Maybe it's a spark of recognition. Maybe it's the sudden sense that someone out there sees your resilience, your mythos, your story. Her work is tied to the belief that compassion is just as legitimate a curveball as cruelty, and that good can prevail over despair. Perhaps that's because her art is built on legends, imagination, and memory—on women who didn't give up, and the enduring truth that courage still matters.

Whether you meet Tracy's work on paper, on canvas, or in a book, the promise inside it is the same: she will always try to make something that speaks to you, and something that lights your way onward.

Shell Game

We rolled onto a nest around two a.m. All dark inside.

"How do you know it's empty?"

"Don't." Marco shut off the ignition. "But ninety percent of these are abandoned. We're playing odds."

He took something from the trunk before we walked to the front door, something he tucked into his trench coat. I didn't ask.

I'd never been to this nest, and I don't believe he had either. A one-story rectangle, but with trimmed hedges and a freshly cut lawn. Slate shingles and leaded windows. I'd have decided Marco was wrong about odds, except there were a dozen newspapers on the step, still rolled up. So maybe it was abandoned.

He let himself in.

I followed. "Key?"

He tucked it into his pocket. "Came with the tip."

"Tip from who?"

He gave me a look. "You got trust issues, Rojzik."

"Says the guy who isn't telling me his source?"

Marco scratched at the graying stubble on his jaw. He was ten years older than me, a bit heavier, and an inch shorter, but he'd been around the block. He was the man I could become, maybe, if I stayed in one place and avoided trouble. "Relax," he said. "It'll check out."

Sure it would. Even though it was too warm for a trench coat, and he'd hidden something inside it. But I was only a week on the job, and so far Marco had buried me in paperwork. So, sure, maybe a few warning bells were going off in my brain, but they were faint. I was tired of sitting on my hands.

The front door opened into a living room/kitchen space, one of those all-in-one rooms that was supposed to make a small house feel larger.

Marco turned on the lights. Surprised me he wasn't being stealthy but, like I said, the place was supposed to be empty. Looked like it was, at first. Looked like it had been decorated but never lived in. Decorated in five shades of brown and tan, like a celebration of drabness. Plastic on the sofa, carpet twenty years out of date, but in perfect condition. Oak chairs surrounding a table with a lace cloth on it.

I was still taking stock of the room when I heard a sliding door open at the back of the house. A few seconds later, the vic walked in from the hall. Two a.m. and the guy's fully dressed, pink polo shirt and khaki slacks, hair slicked, leather loafers. Couldn't tell if he had a gun.

He looked at us. "You're in my neighbor's house. You want to tell me what's going on?"

Marco, he didn't miss a beat. Didn't lie either, not that I would have known. "Name's Flynn. This is my partner, Detective Rojzik." Marco flashed his badge. "We're with the department. Value Crimes. Specialize in polyamptery. Didn't expect to find anyone home, but since we did...a few questions?"

The guy considered this for half a second, then smiled and extended his hand. Marco gave him a withering look. He held up his gloved hand. The guy shrugged and gestured at the table. We pulled out chairs, him on one side, us on the other.

"Who told you about this place?"

Marco shrugged. Said... something. My memory gets a little hazy here. Bits and pieces. Whoever was jumping me, they did it slow, so I'd notice maybe. Only I didn't. I was too new to it then.

Anyway, I remember snatches of the conversation. Enough to get the idea of what went down.

"Are you tripping now?" Marco asked.

"I am."

"How much time can you spare?"

The guy shrugged. "It's a familiar shell. Take all the time you need."

I thought he meant the house. I had no idea.

Marco asked, "What kind of trips do you usually take? How long, I mean. Time duration."

"Nickels and dimes. Nothing on the edge."

Marco nodded. He approved. "How many shells?"

"Forty-three. I'm careful."

"Expanding?"

He wobbled his head. "I lose one, I find one. I like to keep it steady."

I was quiet up to this point, because it was my job to be quiet. It was my job to watch and to learn, and frankly, I was quiet because I had a sense Marco was involved in something illegal, and I didn't want to implicate myself. I'd been lucky to get the job, and I wasn't about to make waves. Also, like I said, someone was jumping me, and the way they were doing it, it made me less aware. Less a participant, more an observer. Maybe they were just doing it to eavesdrop, but I don't think so. There are easier ways.

"Have we met?" Marco asked. "Have I asked you these questions before, or some like them?"

The vic tilted his head slightly. "I'm aware of you, if that's what you're asking. But, no. We haven't met. I don't visit the coast often, but this shell was starting to fade. Had to make an exception."

"Keep your connection strong."

The vic sighed. "Are we driving toward something, Detective? I haven't overstepped. I can't be caught. I can't be intimidated. I can be bored though, so get to the point. Because when I'm bored there are always better options than sticking around."

"There have been deaths." Marco put both hands on the table, as if to demonstrate he wasn't a threat. "Rumors of somebody burning out shells intentionally. I'm looking for information, not making accusations."

"They're just rumors." The vic smiled. "Come on. No one's going to burn out a shell on purpose. Like cutting off your own nose. You haven't complimented my English yet."

Marco's eyebrows raised. "Should I?"

"A lot of you guys do, when we cross paths. Or my Spanish, or Russian, or whatever. Whatever I'm speaking at the time, so you can see if I'll admit to which one is native. You really went to the trouble of tracking me down just to ask about some rumors?"

"Seventeen known shells burned out in two months. Ten women, seven men. Burned or outright murdered."

Something changed in the vic's expression. He fixed his gaze on Marco with a strangely detached intensity. At first, I thought it reminded me of a cat staring at a mouse, but that wasn't quite right. More like a cat looking at a dish of food, deciding if he was hungry.

Outside, dogs barked in the distance. A car passed. Eventually, the vic said, "Not my shells, not my problem." But he looked troubled. "You haven't asked my name either."

"Your shell's name, you mean?"

"That's easy enough. This one's Martin Kudlow. Investment banker, second marriage, inherited this house from his grandmother, and the one next door too. That one's a rental, but this one just sits empty. Keeps telling his wife they'll sell it, but he never does. Probably plans to have an affair here, if his life ever slows down enough. I let him keep it in his own name for now, so I'm curious how you connected it with me. But I meant my true name. You guys always ask that, and we never answer. Part of the dance. So why aren't you dancing, Detective?"

"You're Gidarta."

The man stood.

Marco smiled real big. "Now, since you're not going anywhere, maybe we can talk for real. I'm going to ask some questions, and you're going to answer. I'd better like your answers, Gidarta. If I like them a lot, maybe I'll let you leave. So. Why don't you tell me about your fellow travelers. We can start with how many there are, and any names you know, and expand from there. Until you tell me everything."

Whether he did or not, I couldn't say because that's when it all went dark. That's when I really got jumped. When I came to, it was over. The room's browns and tans were blasted with shocking red.

The shell, Martin Kudlow, AKA Gidarta, was lying on the floor with a hole through his stomach. Marco was dead too, near the door. Looked like he'd been trying to run.

And Marco's shotgun? It was in my hands. Smoke wafting from the barrel.

There's a thing about shells and true names. Officially, it's hokum. Science doesn't back it up, but these guys have been around longer than science, and after a few millennia of believing something, it's hard to set it aside.

You don't trade shells in the presence of someone who's said your true name.

The fear is that they can intercept the swap, trap you in limbo or something. The travelers never talk about it, but that's the best we can figure. A couple hundred years of tracking these guys, and a hazy knowledge of their superstitions is still the best we have.

I didn't even know that much at the time, because I didn't know anything. But it didn't take long to realize what Marco had done, by saying the true name. He'd made it so the traveler wouldn't leave the shell. Couldn't. Once I got the context, it didn't take long to realize Marco had been going about his investigation all wrong. It wasn't just the shells getting murdered. It was the travelers themselves.

But, like I said, all that was knowledge I didn't have. What I did have was two bodies, an obscene amount of blood, and a murder weapon in my hand.

I didn't do anything at first. I don't know if it was training, or shock, or just my own nerves reconnecting with my consciousness, but I froze. Twelve years in the military, ten as a cop, I'd seen a lot of things. But nothing like this. I looked at the scene, left to right, right to left, up and down. I took it all in, as if seeing some small detail might overturn the more overwhelming evidence that pretty conclusively said I'd just murdered two people.

My mind raced for an explanation. I didn't have any medical conditions or substance abuse problems. No previous episodes of blacking out. As far as I knew, I didn't even talk in my sleep.

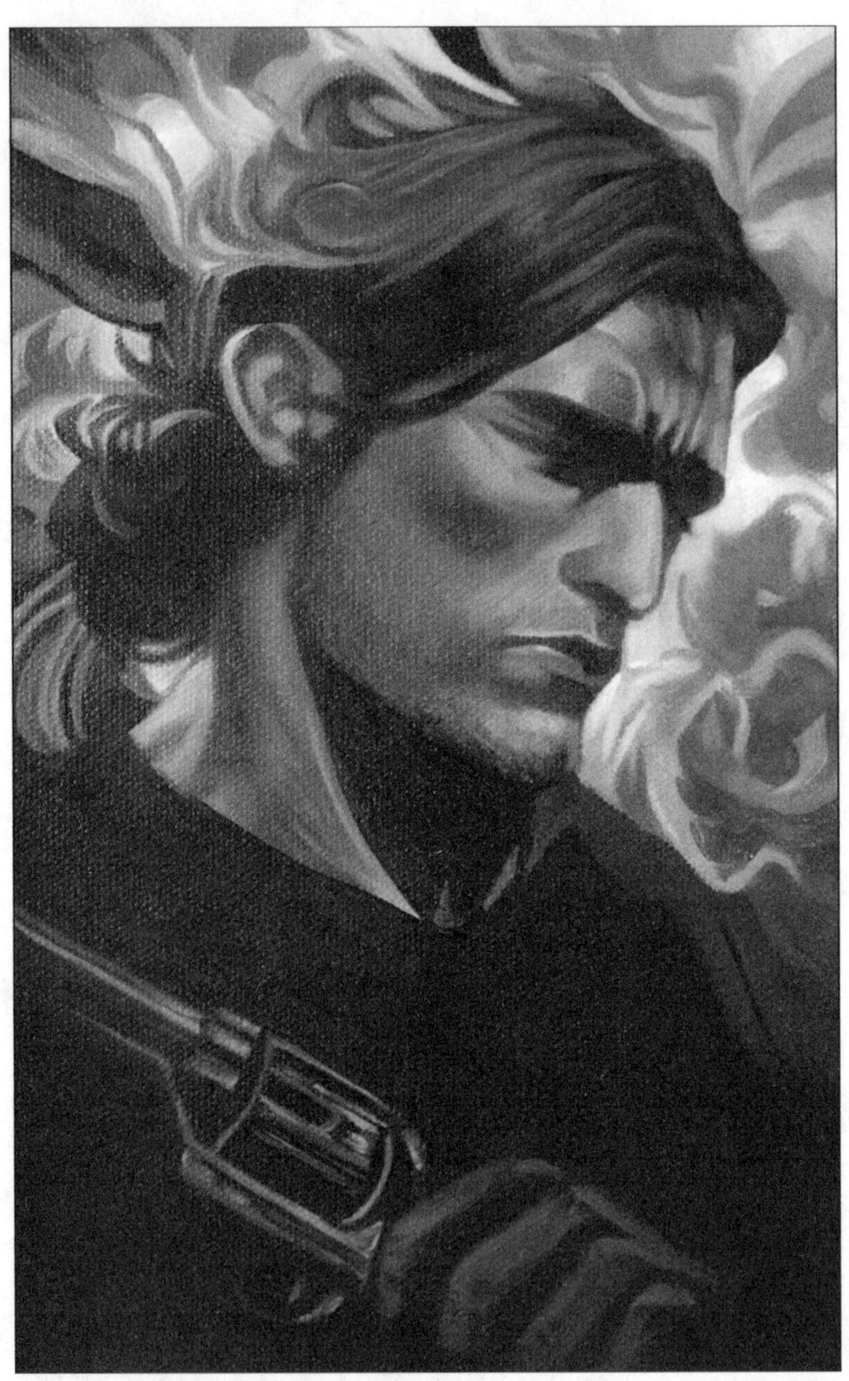

TRACY EIRE

And yet.

The shotgun was a sawed-off thing, pistol grip. Awful to aim, but there wouldn't have been much chance of missing in close quarters. From the size of the holes it punched through Marco and the shell, the gun was loaded with slugs. So I knew what Marco had hidden under his coat, but I had no idea how I'd gotten hold of it, or why I'd used it.

But of course, I did know how to use it. I had a history with violence. Mostly it was in the line of duty, and mostly I had made my peace with the things I'd seen and done. Only mostly though, and I didn't expect that to ever change. Like Marco said, I had trust issues, most of all with myself.

I wiped down the gun. Set it on the table. Wiped down everywhere I might have left prints. Took Marco's keys and wallet. The other guy didn't have a wallet on him. I didn't take time to search for it elsewhere. I turned off the lights. Walked to Marco's car. Drove away.

I was new to the department, new to the city. Only been there a few days. No one to call. I took back roads to a bad part of town. Left the car running, door open. Jogged ten blocks, took transit to my apartment. Kept my head down and my back to traffic cameras.

I showered. Changed. Tossed my old clothes in a garbage bag and poured in a gallon of bleach. Got into my own car. Dropped the bag in the river on my way back to the crime scene.

I called the department on my way. "I need to talk to Williams."

It was one of the only names I knew. Marco's immediate supervisor, head of Value Crimes. The guy who'd hired me despite my past, made me full detective even though I had some blemishes on my record. He must have had his own bosses too, but I'd never met them. I didn't know how the command structure worked exactly. From what I'd observed, it was a little disjointed.

It took a few minutes before they put me through. I ran a red light with an active traffic camera, to firm up my alibi.

"Williams."

"It's John Rojzik. We met last week. You just hired me."

"Flynn's partner. Johnny Boy. New guy, but not a rookie. Transferred in."

"Yeah. But I'm going to act like a rookie, Williams. I'm going to make a big deal out of nothing. You ready?"

"Shoot."

There was something in that particular choice of words that made me hesitate, hopefully not so long he noticed. "Marco was running down a lead on his own. He was supposed to meet me a couple hours ago. Coffee shop."

"Didn't?"

"No. Didn't answer my call either. Whatever. Some guys drink. Some guys have hobbies."

I would have kept going, but Williams interrupted. "Marco doesn't drink. No hobbies. Doesn't oversleep either. Doesn't miss appointments."

"Okay. Now I know."

"You sure you're okay, Rojzik? You sound a little jittery."

"I'm fine. Anyway, while I waited for Marco to show, I got to reading some articles he'd flagged for me. About polyamptery."

He had given me articles, but I hadn't gotten far on them. At first glance, I'd figured "polyamptery" was precinct jargon for some local flavor of organized crime. When the articles degraded into conspiracy theories about mystical forces and untraceable crimes… Well, it all seemed like a joke. Something to haze the new guy.

Two corpses later, I was willing to admit polyamptery was real. But I was still painfully ignorant of what that meant. I hoped Williams would take the bait and offer some real information. No such luck.

"Marco didn't ever show?"

"No. So, at first, I'm a little annoyed, but still willing to wait it out. Give it space. See what happens. I'm the new guy, right? Don't want to step on toes. Anyway, I'm leaving the shop and there's a note on my windshield. An address."

"Marco's handwriting?"

"I don't think so. I'm on my way there now."

"You want backup?"

"Should I? It's probably nothing. I'm ten minutes out."

"Drive slower. I'll run the address."

I drove slower. Williams came back a few moments later. "No backup. The house is connected to someone we've been watching. As it happens, neighbors reported possible gunshots an hour ago, but it's an abandoned house and a busy night and response hasn't gotten there yet. You'll be first. Keep me on the line and tell me what you see. Let's keep this in-house if we can."

I presumed he meant we weren't involving anyone outside our division. Perfect news. The fewer people investigating, the better. I stayed on the line all the way to the house. All the way to the door. I told Williams what I saw. Narrated the whole thing, step by step. For the most part, I kept it clinical. No emotion. Distances, positions, that sort of thing.

I didn't mention the smell of blood, of people turned inside out. Tried not to focus on it, even though it was everywhere and getting worse.

I heard Williams typing while I talked. When I finished, he said, "Wipe down any prints. Take their wallets. Make it look like a home invasion, and get out of there. I put in a few delays for our colleagues in the squad car. You've got ten minutes before anyone arrives, so do it right. Avoid the neighbors. Then come see me. Right away."

Not that I was in a position to judge, but he was suggesting I break the law, hinder an investigation. I wondered if he suspected what I'd done. If he was testing me. I tried to decide how to respond, and it took me long enough that he saved me the trouble.

"You have anxiety over those instructions, Johnny?"

"No." I shook off a wave of nausea. "Just processing."

"Act now, process later." He hung up.

I considered the possibility I was being watched. Decided it was low. That being the case, I didn't bother pretending to do all the things I'd already done. Just got out of there, quick. I was barely a block away before a police cruiser passed. Lights off but moving fast. In my rearview, I saw two more arrive. Officers got out. Surrounded the house.

Either Williams was innocently wrong about how much time I had, or horribly incompetent, or he'd meant for me to be there when they arrived. Meant for them to catch me in the act of covering things up. I considered all options, and decided Williams didn't seem all that incompetent. And not all that innocent.

I drove toward the precinct. Kept to the speed limit. Tried to think things through. I didn't know why someone wanted Marco and the vic dead, but it was pretty clear I'd been set to take a fall for their murders. Maybe I'd even done them—all signs pointed to it. I didn't know how I could have, but that didn't make the corpses less dead. Then Williams tells me to drive slower. Puts me on a timeline that would ensure I got caught.

But, like Williams said, I wasn't exactly a rookie. I'd smelled trouble back when they recruited me for the job. They recruited hard, despite my checkered past. So, I figured, maybe it was *because* of my checkered past. Maybe they needed a guy for the dirty work, for the stuff they didn't want to sign their names to. I smelled the trouble, but also a steady paycheck and a second chance, and a mystery I couldn't quite put together. At the time, I didn't have a lot of other options. So I took the job, but I also took precautions.

I parked a few blocks from the precinct, in a parking garage I'd scouted a week earlier. One with no security cameras and not a lot of traffic. I popped the upholstery on my seat and took out a gun I'd 3D printed, .357, rubber around the grip, no metal except for the tiny firing pin. Undetectable by scanners. Small enough to palm and only held one bullet at a time. But I'd practiced with it, and I was quick at reloading. Actually, I'd practiced with a previously printed version. Barrel only stayed true for five rounds before warping beyond usefulness, but five rounds is a whole lot better than nothing.

I tucked it into a pocket I'd sewn into my suit coat, under my left arm. One of those places where the coat pulls away from the body when somebody's patting you down. As for the five bullets, they were taped to the back of my badge, clipped to my belt. None of this would fool a close search, but most searches weren't that close. Not if you let them find the obvious things first.

I smiled at something a military buddy used to say. *It's only paranoia if you're wrong.*

Didn't look like I was wrong this time.

I walked fast. By now, Williams would know I hadn't been caught at the scene. He'd have people searching my apartment, looking for my car. Communications would be monitored. Maybe they'd arrest me at the front desk, but I didn't think so. Williams wanted to keep things in-house.

My ID got me through the front scanner, past the holding pen, down the hall. No one gave me a second look.

Value Crimes wasn't much more than a few rooms in the basement. At my old job, they called it racketeering and identity theft, but it had been clear from the outset that the new job would be…What had Marco called it? An "expansion" on my skill set.

I went down the stairs, past filing cabinets and storage rooms, past the place where you would have thought the basement ended. Then through bare cinder-block hallways lit by too-few bulbs. The temperature dropped a few notches. Goose bumps rose on my arms. At the end of the final hall, in front of a steel door, an enormous officer in a short-sleeved class B stood watch.

I tipped my head. "Krebbs."

"Far as you know," he said. His biceps looked like they'd split the fabric if he flexed. "What did you have for breakfast, Rojzik?"

I squinted at him. "Seriously?"

His hand moved to his gun. His stance widened on the concrete floor. "Not going to ask twice."

It was eerily quiet in the hallway. No traffic sounds filtering in from outside. No conversations from up above or beyond the steel door. Like being in a tomb.

"Didn't have breakfast, Krebbs. I don't keep regular hours. Ate a big lunch though. Burger and a double order of fries. Pickle on the side. Dill. Very tasty."

Krebbs nodded, but his hand didn't move off his gun. "Last time you were here, what questions did I ask Marco?"

"I'm supposed to remember?"

"Yeah."

"Williams is expecting me."

"He's expecting someone. Remains to be seen whether it's you."

Krebbs was a few inches taller than me and built like he was made of bowling balls. Maybe in my prime I could have taken him. But I was a few years past my prime.

I said, "Didn't you ask him if he watched the game?"

"What game?"

"I don't know. Hockey? Doesn't much matter, because he said he hadn't. Said he'd fallen asleep reading a western. 'Life of adventure,' you said. Something like that. Marco said it was better than singing in a choir. Do you sing in a choir, Krebbs?"

His hand relaxed. "Sing in two. Used to be three, but I'm pulling extra shifts lately. Look, Rojzik, questions are part of the drill. You got to answer them like everybody else. I know your file says you're immune, but I was hired to hold the line out here and I'm not taking chances. Nothing personal, you understand?"

"Sure." At least, I was starting to. As for what I was supposedly immune to, the explanation was probably in those papers Marco had given me, that I'd decided were a joke.

Krebbs patted me down. Did a good job of it too. Good, but not great. He took the gun at my side, and the one strapped to my calf. Took my phone. All of it went into a wall panel he opened and closed with a key hung round his neck. Same drill as last time. No phones or weapons allowed inside Value Crimes.

I lowered my hands, feeling the reassuring discomfort under my left arm.

He put his palm to a panel on the door, then punched a code into the keypad. The door clicked open. He nodded me through.

"Hey, Krebbs?"

"Yeah?"

"You know as much about polyamptery as Marco?"

He scoffed. "I've only been here eight weeks, Rojzik. Way I hear it though, no one knows more than Marco except maybe the boss."

"Williams."

He nodded. He was still talking about Marco in present tense, which told me what I'd hoped to learn. He didn't know, which likely meant nobody knew except Williams. Maybe I should have told Krebbs the bad news, but the door closed before I could.

It was probably smarter anyway, not to spread word, until I knew more about who I was spreading it to.

I walked down a short hallway to another steel door. Locked. No keypad. Before I had time to get properly claustrophobic, it buzzed open, and I stepped inside Value Crimes. Bad carpet and low ceiling tiles and a couple rows of plainclothes cops sitting in waist-high cubicles, about as lively as eggs in a carton.

There was something odd about the air. It was too still, and slightly warm, and I experienced a brief dizzy spell.

I shook it off.

The room was quieter than anywhere else I'd worked. There was the low electronic hum of computers, the whispering hiss of small speakers sitting idle. Fingers tapping lightly on keyboards. But no voices. No phones ringing.

I walked past rows of desks. Officers stared at screens, their faces washed out by the monitors. I hadn't expected so many people in the middle of the night.

Williams sat in his glass-enclosed corner office, watching me approach. If he was surprised to see me, it didn't show. He slouched in a worn-out office chair, behind a desk buried in newspapers and files. He looked back to his computer when I walked in.

"Sit."

I didn't. "Your officers arrived a few seconds after we finished talking, Williams. Not exactly keeping it 'in-house.'"

Williams motioned for me to close the door. I closed the door.

I took off my jacket and draped it over the chair in front of me. "Was it just easier to pin the murders on me? Nice and neat, not too many loose ends? Or was it personal?"

"Will you sit already, Rojzik? Or is that little gun in your jacket too distracting?" He moved some files from one side of his desk to the other.

I sat.

He said, "Got a wave scanner that sees through clothes, Johnny. Like at the airport. Can even do a cancer screening if we're in the mood."

"Krebbs knew?"

"Why else would he have done such a sloppy pat-down? The scanner's at the far end of the basement, before you even saw him. I told him to let you bring the gun in. It's a show of trust, Rojzik."

"Okay." I looked out at the officers in their cubicles. Did the rest of them know what had happened to Marco? Did they think I'd done it? "Pardon my paranoia. I might have mentioned my new partner getting killed and me getting framed for it. Wasn't sure I could trust you."

He pulled a file out of his desk drawer. "And now?"

"Still deciding."

"Smart."

"Did you kill my partner?"

"I wasn't there, Johnny Boy. You were."

"It wasn't me."

"No?"

"I was only there an hour later, remember?"

He gave a slow nod, then tapped the manila folder. "Look out there at the computer screens, Rojzik. Look closer."

I turned halfway, so I could still see Williams out the corner of my eye. Didn't take long to see what he was talking about. "No one's online."

"Bingo. Spreadsheets and maps and PDFs. Some of it we bring in on hard drives, but mostly it's internal documents. Other than the phone line, this office is a closed system."

"You can't even pull up somebody's bank records? Social media?"

"Not here."

"And what does that prove exactly? What does it have to do with my partner being dead and you telling me to cover it up?"

He slid me the file. "Primer on polyamptery. We were going to break you into this slow, but obviously things have changed. A few of the sections are Marco's work. Most are mine. Problem with the internet, Rojzik, it's a two-way street. Any time you're looking at someone they're looking at you too. At least, they are if they know what they're doing. And the people we track know what they're doing. I wasn't telling you to cover it up to protect the guilty. I was doing it so we'd have time to find the person truly responsible. The *being* responsible."

"Right. Well, that clears up everything."

"There's someone I want you to visit, Johnny. Goes by the name Adonis. Goes by a lot of others too, but Adonis is the one behind the faces."

"That supposed to make sense?"

"It will soon enough. Ask around in the morning, down by Ledgemet Pier. Maybe try the fortune teller."

"The fortune teller?"

"Lady Laura."

I glanced at the wall clock. "Already is morning."

"After ten or eleven, I mean. Adonis isn't an early riser."

I flipped through the file, both relieved and suspicious that I was going to have the opportunity to walk out of the building a free man.

The file contained some of the same articles Marco had given me, plus a lot more. Typewritten pages and photos. Handwritten notes in other languages. It looked like a dozen different things, rather than anything cohesive. "This Adonis got a last name?"

"Probably not."

"Male or female?"

Williams raised an eyebrow. "Depends on the hour, Rojzik."

I stared at the papers in my hand, trying to formulate an intelligent response. To learn more without revealing how dangerously ignorant I was.

"Look," Williams said, "we're not here to catch these guys. We're only here to track them. To remind them that things work better for all involved when they stay within bounds. Of course, if it does turn out they've killed one of our own, that calculus changes. We call in other resources. Domestic if possible, but even the most spineless of our foreign counterparts understands the need for an occasional dose of retribution. But it can't be done recklessly. We have to thread a very small needle, Rojzik. Spend some quality time with that file. And don't worry about Adonis. Completely harmless if you mind your manners. Been working with us a long time."

On my way out of Value Crimes, I tried to determine what the detectives behind the screens were actually doing. Far as I could tell, it wasn't much. Staring. Breathing in and out. There was nothing obviously wrong about it, but still... It made the hair on the back of my neck stand up.

It's only paranoia if you're wrong.

I walked out of the station and went the opposite direction from where I'd parked. Dropped my phone down a grate. Didn't trust it now that it had been in Krebbs's possession.

I made my way to the bus station. Emptied a locker I'd rented two weeks earlier, paid cash for a ticket. Rode the bus for half the ticket and slipped out while they were refueling. Walked in the starlight to a stand of trees at the edge of a field. Slept in my hammock.

I know. A hammock sounds like I'm on a beach somewhere, but you put your trust in a motel clerk, and you're asking for trouble. Put your trust in anyone, and you're asking for trouble.

Which is why I had the locker in the first place. Perfect place for a spare Sig Sauer P226 and ammo. Clothes, blanket, a little water and food. And the hammock. It's less conspicuous to sleep on the ground, but I'm getting old enough I don't *sleep* on the ground. Hammock's the best money I ever spent.

My first days in town, I got the locker and spent some time wandering, getting to know the homeless community, the parks and trails, how things fit together. You walk a neighborhood once and you learn more than driving through a hundred times. Then I let Marco drive me around to places I knew better than him. I noticed what he chose to tell me, and what he left out.

All things considered, Marco seemed like a stand-up guy. Until I killed him.

I looked at Williams's file before falling asleep, but my mind was too jumbled to make sense of it. Only thing that stuck was that every detective applying to work in the department submitted a blood sample, just like I had. A drug test. But drugs weren't all they were testing for.

Around noon the next day, I took a bus to the pier so I could read on the way. I'd had a dreamless night in my hammock. Slept like the dead as they say, which was rare for me. I'd have taken it as a good sign, but I was still having trouble keeping my eyes open. I did my best to focus. Even in the light of day, a lot of the stuff in Williams's file was way over my head, and the rest required me to accept realities that weren't real. Not to my way of thinking at the time.

Polyamptery. The word itself was a riddle, reportedly devised by the travelers themselves. *Poly* from the Greek, meaning *many*. *Ampt*, was a corrupted version of Latin, implying both *amp* and *apt*, *power* and *fit*. *Ery*, a *state of being* or *behavior*.

Beings of unknown origin, with the power to fit into many.

Ledgemet Pier wasn't somewhere I'd explored before. It turned out to be a couple of blocks full of tourist traps. Restaurants and trendy stores and souvenir shops.

I asked around. No one knew Adonis. The sun was blazing overhead. A breeze carried the smell of the ocean, but didn't cool things much.

Wasn't hard to find the fortune teller. She had a wooden sign hanging outside, between an ice cream shop and a tattoo parlor. Chimes jingled when I opened the door.

No receptionist inside the lobby. Good air conditioning though, and the place was thick with the smell of incense. Roses and cinnamon. A sign next to a thick curtain said: Ring bell and wait twenty seconds before entering. I peeked past the curtain but didn't see anybody. Just a little round table with a chintzy-looking crystal ball in the middle. Dim lights. Lava lamps at the corners of the room.

I rang the bell and waited twenty seconds.

Still no one there when I went in, but a note on the table said: Please sit. Madame Laura will be with you soon.

I didn't like it. Dark room. My back to the exit. Lots of things I couldn't see or know. But I had to start somewhere, so I sat.

Maybe a minute later, Madame Laura came in from a back door. Late thirties, early forties. Long hair, brown with a hint of red. Loose sweater, jeans. Big earrings, big eyelashes. Carried herself with easy confidence. Not worried about me in the least.

She sat. Looked me over. No accent when she spoke, no mystical affectation, just a nice voice. "You're a seeker of truth?"

I smiled. "Matter of fact, Madame Laura, I am."

"Do you believe I can show you the truth?"

I tilted my head. "We talking capital T, or small?"

Now she smiled. Nice teeth. "We should start with your palm."

I held it out. "That's where the truth lies? My palm?"

She took my hand in hers. Her hand was warm. "I start with the palm because it shows me more about you. Physical touch connects us. Then, the more I know about you, the more we're connected, the better I know how to see the things you don't see. How to know things you don't know. The small and the big."

I let her trace the lines, squint, do her thing. I figured she was waiting for me to ask what she saw, but I didn't particularly care what she saw. All the same, I didn't have all day.

"I'm here for Adonis."

She didn't look up. "Who?"

"Adonis. Someone told me you might know him. Or her."

Madame Laura continued tracing my palm, but a little slower than before. "I don't actually. I know *of* Adonis. But I am the least likely person you would ever meet who actually *knows* Adonis. I am never here when he is here."

"It is a 'he' then?"

"It varies."

"Polyamptery?"

She looked up with a sideways grin. "I take it you're Marco's new partner?"

"Yes and no. For now, let's say yes."

"You looked like you might be the long arm of the law."

"You let me in anyway?"

"Your money still spends. Speaking of which." She held out her other hand, without letting go of mine. I fished out my wallet. Handed her a twenty, then another when she kept her hand out.

"Cash," she said, tucking it into her back pocket. "You old-fashioned, or paranoid?"

"Both."

She gave a knowing nod. "Marco never did explain polyamptery, but he liked throwing the word around. I think he felt better about seeing a psychic if he could call it research. By the way, I charge by the minute. Did I mention?"

I gently pulled my hand away. "You did not. Am I paid up so far?"

"So far. But you're getting close. I'll keep you posted."

I'll admit, I liked her. But I didn't want her getting the wrong idea. "Name's John Rojzik. Marco's *former* partner, incidentally. He was murdered yesterday."

Up until then, she'd been cool as can be, but this rattled her. "Murdered?"

"Yeah. Sorry to be so blunt. I only served with him a few days, but I liked the guy. He seemed like a good one. I'd like to find out what happened, and that's why I'm here. That's why I'm asking for Adonis. Not a suspect, just someone I was told to talk to. He's supposed to help me get the lay of the land."

She leaned back in her chair. Took a few slow breaths with her eyes closed. "Adonis comes when Adonis comes. Good for business when he makes an appearance, but I don't have a great deal of control over it. So. If he shows up before your minutes are up you can talk to him. If not, you get me. You have any idea who killed Marco?"

"Yeah. In a way I know exactly who killed him."

She looked me straight in the eye, holding me there. I could tell she didn't want to ask, but when I didn't offer, she did anyway. "Who?"

"I can't tell you that."

She put the forty dollars back on the table. "John, right? You knew Marco for a couple of days. I knew him five years. We weren't involved like you're probably thinking, but he was a good guy. Maybe he only came around once in a while, and maybe it was only when he needed information, but he never tried to take advantage. He gave more than he took."

I pushed the forty back to her. "In other words, he was a friend."

"He was. So, I'm going to ask you again, Detective. Please. Who was it?"

It was stupid of me, what I said next. I'd like to say I did it without thinking, but really, I knew exactly what I was saying and why.

"Me. I pulled the trigger." I let the words sink in. "But, like I said, polyamptery. And, like I said, I really need to find Adonis."

I don't know what I expected to happen. Maybe I wanted her to scream at me, call me a murderer. Maybe I wanted her to tell me I wasn't. But Madame Laura had the fortune teller act down. Firstly, because I trusted her in a way I shouldn't have, and secondly because she didn't bat a beautiful eye when I told her I'd murdered her friend. She looked at me for five seconds. She nodded. She took out her phone and sent a text.

"Police?" I asked.

"You are the police."

"Might not be. Didn't show you a badge."

She scoffed. "I know the type. Besides, Marco came by last week. Said he was getting a new partner. Described you, told me you'd probably come to see me at some point."

Interesting. "And you didn't tell him I would kill him? Pretty big oversight for a fortune teller, Madame Laura." I regretted it as soon as I said it, but she actually smiled.

"Bigger oversight for a partner, John."

Ouch. But there wasn't any venom in her words.

"I never said I was a good partner."

"But you don't remember doing it, and you're here talking to me. Trying to make sense of things. Look, I know what it is to be jumped. That's what Marco called it. Only I get Adonis, and he's sweet as can be. Practically a guardian angel. Whoever got you? Sounds like you've got much bigger problems than confessing to a fortune teller."

She had me there. I said, "Only I'm supposed to be immune somehow. Something in my DNA. I'm not supposed to be *able* to get jumped."

"Otherwise, they wouldn't have given you the job."

I'd figured this was the case, that everyone at Value Crimes had to be immune, but it was good to finally have it confirmed. "So, either someone lied about my DNA indicating immunity…"

"Or you've got a traveler who can jump immunes. Wouldn't that be fun."

After what I'd read in Williams's file, the idea made me sick to my stomach. Seventeen shells killed, most of them female. And here I was sitting across from another female shell. "Who'd you text?"

"Adonis. Says he'll meet you outside."

"Really?"

"Really."

"You texted a near-immortal being without a permanent body? Does he have a near-immortal cell phone that he transfers with his soul?"

"You'd be surprised. Adonis is full of tricks. There're only a few shells he trusts with the number. Lucky you, you found one."

I nodded. "You know how to keep yourself safe in all of this?"

She shrugged. "Survived this long. You'd better get going, Detective. Like I said, Adonis will meet you outside."

I stood. "He meeting me with a gun, or without?"

She smiled. "Remains to be seen. If you do survive, Detective, don't be a stranger."

I put another twenty on the table.

"That's not necessary."

"Sure it is. It's for next time."

I waited in the lobby for a full minute, hanging back from the big window. I let my eyes adjust to the light outside, looked for anyone who might be an ancient, mystical being. No one stood out.

I almost loosened the strap on my holster, but thought better of it. If I was really stepping into this world, I had to accept the limits of what worked and what didn't. And, with that in mind, and considering what happened with Marco, I emptied my gun. Stashed the bullets behind one of Madame Laura's ferns.

I could have put the whole ammunition magazine somewhere, I guess, but my habit's to take the bullets one at a time. Count them, know what I've got and what I need. I don't even think about it, it's just automatic.

This time I was three bullets short.

I counted again.

Three bullets short. I didn't have the time or equipment to clean my gun in Madame Laura's lobby. But if I did...something told me that I'd discover it had been fired recently. Not by me. Not exactly.

Things were clicking into place, but mostly in a way that sealed my fate, rather than offering a way out.

So much for the safety of the hammock and starlit sky. I was in uncharted waters, getting deeper by the moment.

I stepped onto the boardwalk. Looked around. The place was constant movement. People walking past, looking in shop windows, sitting on benches, eating, talking, all the normal stuff.

"You gonna make me guess?" I asked.

An old guy on a bench a few shops down motioned me over. Bald. Thin, but with one of those old man bellies. Cane in one hand.

I sat next to him.

"Little tip," he said. "Lift your phone to your face before you do something like that. Otherwise, the locals think you're crazy."

"I'd have to agree with them. John Rojzik." I stuck out my hand to shake. "You're Adonis?"

He looked at my hand, then back at me. "Come on, Detective Rojzik. I know you're new to this, but surely you're not *that* new."

I put down my hand. "I am, actually. But I'm learning fast. So, you're the friendly type of body snatcher? Don't even want to touch me so you can have the option of jumping in later?"

"Frankly, Detective, you don't have much to offer. Besides, you officers are all immune, aren't you?"

"There's been a little doubt about that lately."

"There's always doubt when you're smart."

I acknowledged this with a nod. It was precisely my kind of thinking. "Kind of makes me wonder if you and I have met already. If I've already shaken your hand."

"We haven't."

I shrugged. "But how would I know?"

He got a slow smile. "I like you, Detective. You say what's on your mind, but only the part that might tell you what's on mine." He leaned both hands on his cane and breathed deeply. "Kind of wish I had met you before all this, but I didn't. Scout's honor."

"So, you're a scout now?"

"I'm a lot of things. Including busy, so let's get to why you're really here. Marco is dead?"

"He is. How'd you hear?"

"Mine is a small world, with many, many ears. Besides, it's all over the news. Doesn't take a genius."

"Williams told me to see you. Said you'd fill in the gaps for me so I had a sense of what's going on."

"Williams." The old man made a face.

"You know him?"

"Not my favorite human. Not lately anyway."

"Why not lately?"

"Sometimes power gets to a person."

I thought that over. "Makes them want things? Makes them take things, maybe?"

He wobbled his hand in the air, like I hadn't quite gotten it right. "Power mostly makes them want more power. It's like an addiction, and lately Williams is needy. So, I'm not too keen on doing his teaching for him. But I'll add a few things, for your sake."

I nodded. "Like what?"

The old man blinked a few times. He looked at me without recognition. When he spoke, he had a light accent. Southern. "What? You were talking to me?"

I raised my eyebrows at him.

He looked embarrassed. "Did you want something?"

"No—I...Sorry."

"You're looking for me anyway."

I looked to see who had spoken and saw a smiling blonde roller skating away.

"Or is it me?" This time it was a little kid with an ice cream cone, walking the opposite direction.

"Probably better be me." This was a guy in a gray business suit, talking into his phone. "Come on, Detective. I don't charge by the minute like Madame Laura, but if I did you couldn't afford it. Try to keep up."

I gave the old man another apology and hurried after the business suit guy. I was accepting new realities, but that's a far sight from knowing how to use them to your advantage.

We walked across a plaza filled with tourists and food stands. The pier and ocean came into view. Adonis put away the phone, but didn't say anything to me, even though we were side by side.

"You still you?" I asked. "As in *not* you, but Adonis?"

"For now."

"How long can you stay in one body?" I asked. "The stuff I read said if you stay too long you can't leave. Or you lose part of your soul or something."

"That's the rumor," Adonis said. "Depends on the person you're jumping, on how far you deviate from the way they usually think and act. Their sense of style and morality."

Not exactly a straight answer. "Style and morality?"

"Almost the same thing for most people. Personally, I'm a nice guest. I can stay a week or two without damaging myself or the host. Not that I'd generally be interested in staying that long." He brushed at his sleeves. "Sorry about the suit."

I looked him over. "What's wrong with it?"

"Cheap. This is Bill Arnold. Realtor. Bill owns his own office two blocks over, employs three junior agents and a secretary. Nice wife. Two kids. Slobbery dog. Bill comes down here for lunch twice a week, Tuesdays and Fridays, meets a buddy from college. Bill's a genuine All-American guy, and I simply adore him. But he has terrible taste in clothes."

"If you say so."

We reached the edge of the plaza and walked onto the pier. Our footsteps thumped in counterpoint to the sound of the surf. There were fewer people the farther we went, just a few fishermen to either side, some old ladies getting their steps for the day. I decided to ask the questions Marco had asked the last shell, the one I'd killed. I was starting to understand what the questions meant, and why they were important.

"How many shells do you run?"

"Three hundred forty-seven."

My surprise must have been obvious.

Adonis grinned. "All of them live or work within seven blocks of this place. The proximity makes it easier to tend a large flock. Men, women, young, old. Some travelers stick to one sort of person, but I prefer variety. I like people, Agent Rojzik. I've seen most everything the world has to offer. I've been everywhere, indulged and experienced, and all it made me was tired. Tired, and sick of living."

"A suicidal immortal?"

"Life doesn't get better for being longer, Detective. Not necessarily."

"So, you came here to die?"

We approached the end of the dock. A young couple sat on a bench, overlooking the water. Adonis motioned for me to pause. Then he startled, like someone had thrown water at him. His posture changed. He shrunk half an inch, just from slouching.

"Bill?" I asked. "It is Bill, isn't it?"

He looked at me without recognition.

"You sure this is a good neighborhood to rent in?"

That got me a smile. Teeth and gums. "Best anywhere."

He looked about to launch into his sales pitch when the girl from the bench shoved her guy friend and stormed away. She slid between Bill and me, winking as she passed. "Didn't come here to die, Rojzik. Came to live."

Her boyfriend hurried after her. "What did I say?" He gave me a sideways glance, not friendly, on his way past.

"Sorry about that." Bill's voice, but the posture was clearly Adonis again. "But, look—the bench is free. Let's sit."

I looked back at the couple. The girl was cocking her head at the guy. She had no idea why he was irritated at her, but she was irritated that he was irritated, and instead of trying to reclaim their seat, she stormed back toward the plaza. He followed.

"You did all that just to get their bench?"

"It's a nice bench."

"You could have just asked if you wanted it so bad."

"Don't worry." Adonis motioned for me to sit. "I'll smooth it over later. Got a million ways. Besides, they're not especially good for each other. So maybe I won't."

I sat. Looked out at the waves, the big ships near the horizon. It *was* a nice view. "You like people."

"You know, I do. I really do. I like knowing them and their spouses and the kiddos, and I like feeling how their clothes fit. How their lives fit. I like when the wife surprises the husband with some money she found on the sidewalk, just when they needed it most. I like leaving the money there so she can find it, taking it out of someone's pocket who was just going to hurt themselves with bad choices. I used to aspire to... I don't know. Greatness? Indulgence? Now I like quiet better. I like simple. I like taking a day off with people I've known for decades. People who look at me like they love me. Like they trust me."

I did my best to follow what he was saying. It shouldn't have been hard, but my mind kept wandering. "Even though it's not really you?"

"Isn't it? Oh, I don't know. It's as much me as I've ever been. Besides, don't we all put on a different persona for each new group of friends, for every new situation? Nothing strange about that. Are you feeling all right, Detective?"

I massaged my temples. "Headache coming on. Nothing to worry about."

"Mmm." He didn't look so sure.

"I'm fine," I insisted. "So, all this—this neighborhood—it's all your... your *nest*? Like, the whole town, instead of a few vacant mansions? It's your final resting place?"

"I don't know about final. No telling what the future holds. Maybe this is it. Or maybe it's only a rest stop between conquests. For now, I'm satisfied. In one form or another, I control eighty-three percent of the businesses and almost a hundred percent of the real estate in the area. I've been here a hundred years, and I've made so much money from and for these lovely humans that I could do absolutely anything I choose. And I—"

It was at this point I blacked out. Got jumped. I have no memory of it, not a blurring or pain. Just nothing. Next thing I knew, I saw my hand pointing my gun at Adonis.

"Back so soon?" he asked.

I nodded slowly. "Did I just try to kill you?"

"About eight times in a row. Lucky for me you took the bullets out of your gun. Or was that someone else?"

"That part was me." I noted the gun in my hand and put it away. "Was that why you were talking about how great your finances were? Trying to lure someone in?"

He touched a finger to his nose. "Had a suspicion. Wanted to see for myself. I was hoping there might be some conversation involved in my attempted murder, but it was just click, click, click. Not much information relayed."

I took a deep breath. "You should maybe walk away now. That shell you're in—Bill—I could kill him without a gun."

"Don't be so sure. I've been around a very long time, Detective. Learned a lot of things."

"Fair enough. But I'm guessing whoever's jumping me has, too."

Adonis nodded. "Fair enough."

"So, this other traveler, he can't just jump you directly?"

"What, like two of us battling it out in one brain?"

"Yeah."

He gave me a patient smile. "We're not human, Detective. We know how to seal the gate against an invader."

I thought this over. "Honestly though," I said, "the gun is still kind of a dumb choice."

"Public," Adonis said. "Messy."

"Almost like someone wants to be noticed."

Adonis took a deep breath. Let it out through his nose. "Sending a message, Rojzik. Poison, incidentally, is the way to kill someone like me. Poison or bombs, but poison doesn't have to be timed so precisely. If you find out who they are, you can attack all a traveler's shells at once. Do it right, and there's nowhere left to jump."

"You talk like you've experienced it before."

Adonis gave a small nod. "On both sides. We travelers are a territorial bunch, and none of us have what you'd call a clean slate. But I'm hoping it won't come to violence this time. I'm so tired of the noise. The conflict and constant paranoia. I realize you're just the messenger, Detective, but I really can't take much more of this. Not without responding in kind."

I didn't like where this was going. I stood. Took a step back. "How many of your shells has he killed?"

Adonis stared at the waves. "I lost some just last night. Unless my rival has already been at work this morning, that brings us to twenty shells lost. Twenty, Detective. But, as I say, I can't check everybody all the time, so maybe it's more by now."

"More women than men?"

"Does it matter?"

"Maybe."

Adonis nodded slowly. "Okay. It is more women. He's keeping me busy, Detective, but it's still a small number, relatively speaking. He probably thinks of it as just a warning. Probably thinks he's playing nice to focus on women first."

"What?"

Adonis frowned. "Some of my fellow travelers have antiquated notions about gender and the value of a life. I know that seems odd, since we inhabit both genders. But inhabiting isn't the same as *being*. You can inhabit to understand, or just to use. Big difference in the experience. Big difference in the outcome. As my realtor friend might observe, renting a place doesn't necessarily mean treating it with respect. Twelve women. Eight men. I guess that might mean something. But all of them pain me, Detective. As I may have mentioned, I like these people."

Twenty people killed. Marco had said seventeen. But that was yesterday, and today I had three bullets missing. Three bullets missing and I didn't remember my dreams and I'd seemingly slept all night but I was still tired. I looked back at the pier. "Think he's trying to scare you off?"

Adonis shrugged. "Most of my kind—the ones still alive—we've accepted that it's best to stick to our own territory. To leave each other's shells alone. It's a big world. Plenty of sheep to go around. Fighting the other wolves is a waste of resources. Especially when none of my sheep are all that impressive. These are run-of-the-mill, average folks. No world leaders, no supermodels. That's why I chose them. It's why I chose *here*. Nothing worth making a fuss about. Not to an outsider anyway."

"You make a pretty good case for not being the killer."

Adonis stood. He stretched. "I like you, Detective. It's my fatal flaw.

So, I'm going to give you a day to extricate yourself, before I kill you. Nothing personal. But you're my rival's shell. The only one I know so far, and I have to start somewhere. It will be a way to let him know I'm not backing down. To make him realize I'm not easy pickings. Might be enough to make him move on. To realize I do still know how to win a fight, even if I'm tired of starting them."

I puffed some air through my cheeks. "You're not into the whole 'don't kill the messenger' thing?"

"Not when it's the messenger of death. I enjoy pacifism as a lifestyle, but it's a lousy survival strategy. But you're a detective. Find out some of his other shells for me, and I may postpone your demise."

"Adonis. Not your true name is it?"

He gave me a patient look.

"No. I didn't think so. Look, all things being equal, I'd prefer not to die. But to be honest, I'm also not loving the idea of finding you other people to murder."

He nodded. "I sympathize. But not enough to matter."

"Suppose I could solve your problem? Really solve it, not just send a warning?"

"Well, I'd be most surprised by that, Detective. But I'm listening."

I only had half an idea, but it had potential, and the prospect of my inescapable death was a powerful motivation to pursue it.

"You did mention bombs, right?"

A familiar face stood guard at Value Crimes.

"Howdy, Krebbs. They don't have anyone else to guard this door? You taking every shift these days?"

Third time I'd met him, and Krebbs still folded his bowling ball arms and looked at me like I might be an intruder. At least now I knew why.

I kept my hands in my jacket pockets. "We talked about breakfast last time, Krebbs. And you singing in three choirs. How are the choirs going, Krebbs? You sing tenor or alto?"

Krebbs stared daggers at me. "You're not expected, Rojzik. You'll need to call and make arrangements."

"Got to talk with Williams. It's urgent."

"If it's urgent, then make a call."

"Have you been this obnoxious the whole ten weeks you worked here, Krebbs? Is Williams even in?"

Krebbs dropped his hand onto his holster. "Call and find out, Rojzik."

"All right, all right. Just seems like a waste of time. Since I'm already talking to him. How many choirs was it you sing in? How many weeks? You didn't even flinch when I said alto, and I know that would have gotten a reaction from someone in three choirs. Come on, Williams. You're getting all the details wrong. I might be a shell, but give me credit. Pretty clear I'm not the only one."

Krebbs smiled. "Think you're pretty smart, Johnny Boy?"

"Average. But I'm only keeping track of one person's memories. Seems like you got more than you can handle."

Krebbs pulled his gun. "About to be one less."

"Take your time, Krebbs. Williams. Whoever you are. I have a message from Adonis."

"Only message I'm interested in is him leaving. Unless—"

I slowly pulled out my left hand. The hand holding the detonator. "Dead-man switch. I figured you were bluffing about the scanner. Didn't see one, and didn't figure it would be in the department's budget. I think you only knew the gun was there because you jumped me for a second to find out. But this time you didn't. And since you didn't see what's strapped to me, I'll fill you in. Enough explosives to kill every shell in the building. You must have at least thirty shells here, right? Enough to set you back, even if you do jump clear of the blast. Pretty sure I'll lose my grip if you jump me, and you must be sure, too. Since I'm still talking."

Krebbs lifted his eyebrows. "So, talk."

"Like I said, I got a message from Adonis."

I shot him, two rounds through my other jacket pocket. One went wide, but one caught him in the leg. He shot back, but by then, I was moving, and his shot wasn't even close.

"Still Williams?" I asked, kicking his gun away. I could barely hear my own voice after the intensity of the gunshots. Hopefully we were far enough from the bustle of the precinct that no one upstairs had heard the shots.

Krebbs looked up at me in disbelief. “Did you just shoot me, Rojzik?”

He could have been acting, but I didn’t think so. I didn’t figure Williams would stick around for the pain. I skirted wide around Krebbs, but then thought better of it. I pushed his hand onto his wound. “Keep pressure on that. You might live.”

He said some rude things about my mother, but he did as instructed. I took his radio and threw it down the hall. Then I tried the door. It opened.

I entered the hall. The door latched behind me. The door at the far end buzzed open, and I walked through it.

“Do I just talk to any of these vegetables?” I asked. My hearing was almost back to normal. The second door closed. The people at the desks stared mindlessly at their screens. “Is this what happens when shells start to burn out?” With my free hand, I flicked a young officer in the ear. She didn’t even look up. “When you use them too much? Or make them do things they wouldn’t normally do?” I tried a few others.

“Hello? Williams? Whatever you call yourself? You want to pick a body so I can deliver my message?”

An older guy in a tan suit said, “Maybe I’ll just take your body again, Johnny Boy.”

I walked over. Slapped him. “You feel that one?”

A skinny guy started laughing, a row over.

“You jump quick,” I said. I walked toward him. “How many shells do you have anyway?” I flicked ears on my way over, counting out loud. “Seven. Eight. Must be twenty people just in the office. Was anyone ever immune around here?”

The skinny guy grinned at me. Big smile, crooked teeth. “Killed all the immunes, Rojzik. Marco was the last. Thanks for that. He didn’t trust Williams, so I had to give him an anonymous tip about that house. Are you really wrapped in explosives, Johnny? I find that fascinating.”

I lifted my shirt so he could see. "But that wasn't one of Adonis's shells you killed at the safe house. Marco asked him questions first, and I think he told us the truth. He wasn't running nearly as many shells as Adonis, and you knew his true name. You managed to kill the traveler along with the shell."

"Sure." The skinny guy leaned back in his chair. "You didn't think Adonis was the only campaign I had going, did you? It's a slow burn with most of the others, but Adonis is so cozy in his little slice of Americana that I hardly even have to work at it. His shells are everywhere. I'm practically tripping over them. Haven't found his true name yet, but it's only a matter of time.

"Thousands of years old, and you'd be amazed how dumb my fellow travelers are. How vain. They make one of their shells name a baby the true name, or tattoo themselves with it, or start a company with it. It's always around somewhere. We're all narcissists. Me included, but I'm a smart one. Kill enough shells and the traveler tends to get nervous. They reveal things by trying to hide them. And if they don't, I just steal their fortunes and kill as many shells as I can and move on. Try them again another time."

"You keep records," I said. "That's what you're really doing in that office of yours. You jump them a little at a time and gather information, try to fit it all together before you make your move."

I blacked out. When I came to, I was on the floor, gasping for air, a mouse cord around my neck, my own hands pulling it tight. My other hand was wrapped in tape, still holding the dead-man switch.

I pulled the cord loose. Tried to catch my breath. My heartbeat throbbed through my temples, but at least my heart was still beating.

The skinny guy wheeled his chair backward to look at me. He rubbed his neck like he'd been the one getting choked. I guess he had been, at first, when he'd been the one in my body. "You still trying to solve a case, Johnny Boy? Still think you're an ace detective?"

I shook my head. Slowly unwrapped some of the tape from my hand. "At first, I figured I was a fall guy. Someone you hired so you could frame them for the murders."

"You were. Still are."

I got to my feet. "And a way for you to send a message to Adonis?"

"Also true."

I steadied myself on a shell, a younger woman who might have been attractive if her eyes could focus. All this going on, me leaning on her shoulder, and she hadn't once looked up from the paper in her lap. "So why not just talk to Adonis yourself?"

"Oh, you know how it is. It's a dance. A little spying, a little murder. Testing each other's limits and skills. Corpses in place of greeting cards."

"Delightful."

"I can find another fall guy, Johnny. Tell me something useful if you're going to. You spoke to Adonis. What do you have?"

I took hold of a chair and dumped a shell out of it, onto the floor. A short guy, bald with a beard. He didn't so much as flinch. He didn't seem to mind either, when I sat in his chair. I propped my taped hand on the desk, where Williams could see. "I've got Adonis's true name."

The skinny guy leaned forward, half a smile on his face. "Who you trying to kid? No way he told you that."

"She did."

"She." Some color crept into the skinny guy's face. "Is that your attempt at creativity? Is it Cleopatra? Marilyn Monroe? Come on, don't make me hurt you."

Adonis had suggested using "she" to see if it got a rise out of him. Seemed to be working, so I leaned into it a little, each time I used the word. "Said she was tired of living. Wanted to make a deal with you. Something like that. It was more flowery, the way she said it, but that was the gist. She had a few conditions, and then she'd let you take over her shells. She has a list ready. Names, addresses, account numbers, and passwords. Said it would save you years of effort. I think she's been hoping for someone to take it all off her hands."

"Naturally. And why wouldn't *she* just kill all her shells, if she wants to die so much?"

I shrugged. "She likes them. Got a weakness for people or something. She's hoping you'll take over without causing them too much pain."

"Ha! You're slipping, Johnny. You had me going for a second. But do you have any idea the atrocities Adonis has committed? Believe me, causing people pain is not a worry. She probably didn't blink when she wired you up with explosives. Am I right?"

"You know her?"

"I know them all. I know their stupid superstitions and their meaningless exploits. I used to be like them. Superstitious. Then I got *religion*. I figured out how it *works*. They're not making more of us, Johnny. There were a thousand travelers when the world began."

"That's a big number, Williams. Just Adam and Eve and a thousand travelers hanging out in the garden?"

"The world didn't begin with you humans. Not properly. It began with us being born into your society. Into a world full of shells ready for the taking. We are what matters. There will be *one* of us when it all ends. That's *how* it ends, Rojzik. We are the purpose of life. The meaning and reason. All of existence is purely to choose which of us wins. The rest of you are just crops for us to harvest. Clothes for us to wear. At some point, I'll control the whole planet. I'll *be* the whole planet."

"You really are a narcissist."

"Everyone's a narcissist, Johnny Boy. I'm a god. You see these pitiful shells around you? I know tricks Adonis can only dream of. How to scoop out free will. How to control them without visiting so often. How to make someone a puppet even after I leave."

"Not very lively puppets."

"Lively enough for my purposes." The thin guy got a sideways grin, then his face went slack.

Williams walked out of his office. "You're still alive because I still think you might actually know something, Johnny Boy. Plus, I'm doing some creative thinking about those explosives of yours. Guy like me can have a lot of fun with something like that."

"Hair trigger," I said, nodding at my hand. "You'd have to be awfully fast to live through the fun."

"Oh, I am." Williams walked around back of me. He grabbed my shoulders and massaged them.

The skinny guy looked up. "See what I mean?"

Behind me, Williams's hands were still massaging. Made me want to crawl out of my own skin.

"It's an interesting thing," the skinny guy said, "how you can prime a muscle to do something, then slip away before the action is finished. Trick of the trade you learn early if you want to be one of the survivors."

"You've got more shells than Adonis has?"

"Tens of thousands more. About to be a few less, but so what? I'll regain what I've lost soon enough. Besides, right now the point isn't to have more shells. It's to edge out the competition. I've got shells I've forgotten about, forgotten to even kill. I get busy, you know? Doesn't matter. They served their purpose. Speaking of which. I think it's time you tell me the rest of Adonis's message. If she's sincere about me taking over, or even if she's not, I'd like to get things in motion. I have other places to go, people to be."

I took a USB drive from my pocket. "You mind?" I asked.

The skinny guy went slack. From behind me, Williams snatched the drive.

"Encrypted," I said. "In case you were going to kill me immediately, you should know there are three levels of information, and each requires a password. I've got it. You don't."

"Pity." He handed me the drive. Gestured at the computers scattered around the room. "Take your pick. None of them have internet, as you may recall." He pointed a gun at the back of my head. "Little advice. When you're a guy with only one life, you want to be real careful in situations like this."

I inserted the drive left-handed. Moved the mouse and pad to the left side of the keyboard. "I haven't been completely honest with you, Williams."

"Shocking. And you think this is a good time to confess?"

I got the file open. It prompted me for a password. "Good as any. I'm not really wrapped in explosives."

"No?" Williams backed up a step. "I'm not sure if that's good news or bad for me. Not sure whether I believe it either."

I pecked one-handed at the keyboard, making it look even more awkward than it was. "Turns out that Adonis has a lot of interesting supplies at her disposal. Unfortunately, plastic explosives and a dead-man switch were not among them. I asked, she said it would take a few days, and we came up with another idea. You want to know what's really in my hand?"

"Possibly."

I held my taped hand in the air. "Okay, then. Let's do it."

What happened next seemed all at once, but there was a sequence to it. There had to be. The attractive girl said, "Hezon." The skinny guy, "Luther." The heavy man, "Osiris." Every person I'd touched–or rather, every person Adonis had flitted into my body to touch–said a name, in blinding succession.

Meanwhile, I went slack, dropping my hand to the desk, letting my head bob.

Williams started shooting. Everyone who spoke got a bullet, until he ran out of bullets. Then I stopped playing limp. I turned and punched him in the throat. Right hand. Tape or not, the two-way radio made a fine core to my grip.

Williams fell back, gagging. I knew better than to give him a fighting chance. I shot him in the legs, six bullets. He dropped, bleeding profusely. He did not leap to another body.

An older woman got up from her desk and stood beside me. She reached for my gun. I let her take it.

"Would you be so kind," she asked Williams, "as to tell me which true name was correct before I kill you?"

"Adonis?" Williams pressed his hands against the holes in his legs. "You jumped Rojzik, too." He actually smiled. "Made a shell of my shell. Good trick. I've used it myself, a long time ago."

She stepped back to avoid the blood spreading across the floor. I did the same.

She said, "You won't again. Just because I prefer to keep to myself, doesn't mean I'm ignorant of our history. And our future. There are eighty-two of us left, after I kill you. It would be nice to know who's off the list."

"Why?" Williams asked. "You're too soft to be the last. Someone will kill you, too, before the end."

The old woman shrugged. "I don't particularly care. I've shared many good lives, and some bad ones too. But I never understood the impatience. The hurry to get it all over with. Let Armageddon wait a while, I say. Enjoy the ride. Get to know people."

Williams said some rather rude things, amidst his groans of agony.

I said, "Well, it sounds like a winning strategy to me."

"Thank you, Detective. What are the odds? A man with sense." She shot Williams in the chest, then in the head. She handed me the gun.

I looked around at the bodies. It brought back memories, some fresh, some old, all bad. In the moment, though, I felt calmer than I had in years. "How can you be sure he was still there at the end? How do you know he didn't jump?"

She clicked her tongue. "Because I said his name. Hezon would have jumped instantly if I'd gotten it wrong. He knew what I was doing as soon as he heard that list of names, but he was still here even after. Incidentally, you helped me narrow it down, Detective. More women killed than men. Practically a calling card for Hezon once you notice the pattern. So, no. He didn't jump anywhere. Like he said, we're a small club. Plus, he'd never endure the pain like that if he didn't have to. Not Hezon."

I nodded at Williams's corpse. "Not a pretty way to go."

"I don't mean that," Adonis said. "He's been shot before, more times than you might think. That's quick. I meant the pain of losing. And to a woman at that. A *pacifist*. Fate worse than death for Hezon. The pig. Serves him right."

"You're really a woman then?"

She shrugged. "Let's just say I sympathize with the cause."

I found the nearest chair and sat. "What now?"

She sat on a desk. "Quite a mess. Surprised no one's come in yet, guns blazing."

"Soundproof room. Locked doors."

"Well, they're going to want to blame someone eventually. You're the new guy?"

I let out a breath of air. "With a history of violence and insubordination. Some well-documented psychological issues too."

Adonis smiled. "I won't cast the first stone."

"Thanks."

"Let me tell you a secret, Rojzik. People want things to make sense. So, all you've got to do is tell them something that makes *this* make sense."

I snorted. "This? Nothing I could come up with is going to be good enough. Not for this, and especially not coming from me."

She went limp. Her eyes went vacant.

"Adonis?"

The woman's head swung slowly toward me, but there was no light in her eyes, nothing more than a burned-out shell.

I stood, closed the distance, took the gun from her hand.

A moment later, the light returned to her eyes. Adonis was back. She patted me on the cheek. "Let me worry about your credibility, Detective. There are six people alive in here, seven if you count Krebbs out in the hall. I jump fast, remember? I called an ambulance, and Krebbs already had police on the way down. You didn't throw that radio far enough, Rojzik. He's unconscious now, but he'll live.

"What I'm saying is that seven witnesses are going to corroborate whatever you come up with. That should take care of any questions. You want a medal while we're at it? Want an interview on the nightly news?"

"I desperately do not."

"Fine. Good answer. But you should stick around after this is all said and done. You may not be immune, but Value Crimes is suddenly short on personnel, and I'd like to see a certain kind of person promoted to lead it. A man who knows his limits. His limits, and ours."

I sat. Started to say something, but couldn't figure out what it should be. "I'm listening."

"See, that's another thing I like about you, Rojzik. Since you listen, I don't have to tell you again that this little town is a lovely place. I can help with an apartment if you'd like. I know these people. I think you'll fit right in." She raised her eyebrows. "Miss Laura might enjoy seeing you again too."

The responding officers must not have had the entrance code, because I heard something distinctly like a door being broken down. They'd be in the hall soon, and only a few moments after that, they'd be here with us.

"You should probably open the final door for them," Adonis prompted.

It was a solid suggestion. I stood. "You're a piece of work, Adonis. You just want me to stay so you can take over whenever you get the urge. Take my body for a spin."

She spread her hands. "I make no apologies for what I am. But I could do that anyway, no matter where you are. The thing is, I like people, Rojzik. I do believe I've mentioned that. But it would be nice to have someone I can talk to as myself. Maybe even someone I could trust. Someone who trusted me. That's a rare luxury."

She was right about that. "Rare enough you'd agree not to jump me again?"

"Well. I can't promise that." A smile played at the corner of her mouth. She straightened my collar. "I'm far too curious a person not to see how the world looks from behind your eyes. How about this? I'll only take over when you're having trouble sleeping."

"Who says I have trouble sleeping?"

She gave me a knowing look. "If I've got you pegged right, there's nothing you'd like more than to let your body rest without your mind replaying life's greatest hits."

She had me there. And, truth be told, it *was* a nice little town, when ancient forces of evil weren't battling for dominion.

I went to the door. I wasn't ready to sign up for anything permanent, but for a few months? Maybe. At least until I caught up on my sleep. Established a decent work history. Maybe paid Madame Laura another visit.

"You're almost smiling," Adonis said. "Does that mean we have a deal?"

I knew it wasn't necessary, but somehow I couldn't resist. I stuck out my hand. "We'll have to shake on it."

Canary

written by
Brenda Posey

illustrated by
RODDY TAYLOR

ABOUT THE AUTHOR

Brenda Posey was only three years old when she witnessed aliens invading from the sky. Even though the spaceships turned out to be spotlights on clouds, nothing could dampen her newfound excitement over the question: "What if?"

Growing up in Huntsville, Alabama, listening to the earth-rumbling sounds of rocket engines being tested, she dreamed up additional "what ifs." As she hiked the foothills of the Appalachians, foraging for blackberries and native plums, she continued to contemplate her "what ifs" while developing an abiding interest in woodlore. She dabbled with writing science fiction short stories before pursuing degrees in biology and computer science with an additional year of graduate study in entomology. After working as a researcher and programmer, she left the corporate world to raise two awesome kids with her also-awesome husband.

Reading her husband's extensive, slightly dusty library of classic science fiction novels reawakened her "what ifs." She returned to writing, drawing on her background in the sciences to create her own unique versions of reality. Her fascination with the contrast of high tech and the natural world inspired her story "Canary," where an Ozark recluse must defend her carefully crafted solitude against strange happenings.

When not plotting novels and short stories, Brenda can be found rehabbing a mid-century house, forging hot steel into almost-recognizable objects, tending a weedy garden for Gulf fritillary caterpillars, advocating for Oxford commas, and singing in choir. She continues to watch the sky at night because... well... what if aliens actually show up?

"Canary" is her first professional sale.

ABOUT THE ILLUSTRATOR

Roddy Taylor grew up in the forests of Washington State and the hills of San Diego. He was always a creator of drawings, movies, sculptures, and stories, but also found joy in math and science. After completing a degree in civil engineering, he went on to teach math and physics at boarding schools for ten years. He never stopped creating, though, and the choice to pursue a purely creative career evolved out of his side projects, especially role-playing games.

Nature, culture, and stories are Roddy's biggest inspirations for his work. The intricacies of the natural world and how humans have interpreted and interacted with it through the centuries are an infinite source of excitement and ideas. Whether it's imagining ancient life from fossils, reading the mythology of ancient Ireland, or watching birds move through the trees and sky, Roddy finds something to love in our world.

The biggest goal for Roddy's future work is to develop his skills, use his art to bring stories to life, and be a father to his six-year-old son. Living in the world provides endless opportunities to interpret and express it, and the great joy of illustration lies in the infinite options to solve the problem of making the imaginary visible.

Canary

Reverie Pearson woke into summer that day after a long dream of winter.

Yawning, she rolled to her side and scanned the gloomy interior of her single-room stone cabin. The windows were thickly curtained with doeskins against the Arkansas cold, but a shaft of sunlight gleamed past a place she'd missed. Dust motes danced in the light, and she watched, fascinated by the patterns until she remembered that winter sun never touched the north window.

Never.

With a groan, she sat up, kicking aside layers of quilts and coyote hides. Her gray-streaked hair, heavy with sweat, fell across her face, and she reached a shaking hand to brush the damp strands back into place.

The mushrooms, she thought, recalling the dried cluster she'd added to the squirrel stew simmering on her potbelly wood stove the previous night. *I must have misidentified the Lyophyllum.*

The mushrooms had plumped to a chewy texture, nutty and slightly spicy. But even a mushroom-induced stupor shouldn't have caused her to forget what season it was. Here in the Ozark Mountains, she lived and breathed by the years' chapters. Nothing—*nothing*—could change that!

She swung her legs over the edge of the cot. With another groan, she reached for her flannel shirt and jeans hung from a peg mortared into the wall. The effort made blood pound in her ears, but she managed to dress and lace her hiking boots before she stood and staggered to the nearest window. With one hand braced against the sill, she stripped the doeskins away, wincing as light knifed through the dusty glass. From the sun's angle, she judged the day to be pushing late afternoon. How had she slept so long?

"Frackin' 'shrooms!" she cursed, sliding the gauze curtains aside and heaving the window open. She repeated the process with the other two windows, and warm air breezed through the cabin, carrying the scents of damp earth and mimosa blossoms. Why had she thought that covering windows in summer would be a good idea?

But, no. She distinctly remembered banking the fire in the stove last night, shoveling a heap of ashes over a nearly fresh log. She knelt and pulled the stove door open to find that same heap of ashes inside, now cold and powder-gray with the log reduced to crumbles.

For a long moment, she stared into the stove. In summertime, she always cooked outside. Why had she left ashes in the stove?

"Maybe it's just a warm spell," she murmured, glancing out through the open window where green leaves shifted gently in the wind. A cicada's buzz began, rasping a steady rhythm of loud and soft.

This was no warm spell. This was full-blown summer. How had she lost track of nearly six months?

Without warning, pain lanced her temples, throbs beating in time with her heart. Blackness fogged her vision until she squeezed her eyes shut, breathing in shallow gasps.

I'm dehydrated. That's all. Her fuzzed tongue seemed to agree, but when she reached for her copper water pitcher, all she found inside was dust.

"Of course, it's empty," she muttered, tucking the pitcher under one arm and pulling the cabin door open. "I'm Ripley Van Winkle."

Scowling, she stepped outside into a grayish-yellow, air-quality alert day. Even here, as far away from civilization as she could get, there seemed to be no escape from the pall of humankind. She crossed the dirt yard and descended three stone steps to a spring that fed the stream running beside her cabin. Kneeling, she scooped up handfuls of cold, clear water and gulped greedily until she was breathless and dripping.

The effort made her head throb again, and she eased down to sit on the damp earth, thinking, *there must be something in the air that's making me sick. Giving me amnesia.*

She'd hoped that retiring to the Ozarks would help her detox from thirty boring years at a dead-end accounting job. Here, on four hundred acres of family land, she had everything she needed. A brain stuffed with her grandfather's survival lore. A cabinet packed with ammo. A thousand memories of camping and a deep loathing of anything urban. Nothing would ever shift her back to her apartment in the hellscape that Little Rock had become, with its smog and congested roads and addicts staggering around who'd shank you for the lint in your pocket.

She glanced up at the murky sky again. Pollution? Wildfire? She inhaled deeply, catching scents of damp earth and moss and the mineral leach from rock outcroppings. No hint of smoke. No stench of industry. No—

"Hey!"

The voice sent her scrambling to her feet. A little kid stood at the edge of the woods, a boy about six years old, maybe seven. Maybe ten. She wasn't an expert on children.

"What're you doing here?" she said, blinking against a stabbing headache. "Where're your parents?"

"I'm lost," the boy said in a whiny voice.

She took a cautious step closer, studying him. He didn't sound lost, and he certainly didn't look lost. He looked like some prep school brat, dressed in crisply pressed olive shorts and an orange polo. Dark hair mussed artfully. Not a scratch or a bruise or a smear of dirt anywhere on his pale skin. How had he simply appeared without the slightest warning from the solar-powered proximity alarms?

"You're on private property," she said, jabbing a finger at him. "Didn't you see the Keep Out signs? The purple paint on the tree trunks? That means no trespassing, you know."

The boy shrugged. "I didn't. Can you prepare food? I want to watch."

She snorted at his pretentious tone. "You haven't missed a meal in years. Now, where're your parents?"

He gestured vaguely toward the woods. "Out there. We were hiking, and I got separated."

"That was stupid. Really stupid. There're things here that'll kill you. Razorbacks. Mountain lions. Rattlers."

He shrugged again. "You can protect me, right?" He managed to look both contrite and hopeful.

She glared at the kid in a stare down until her eyes began to water. How long could this brat go without blinking?

"Okay," she finally said. "I'll get you something to eat."

She filled her water pitcher and glanced around the clearing. If summer was full-on, then the chickens should be…

"Oh, no," she whispered. "Chickens!" She took off in a half-stumble, half-sprint toward the coop behind the cabin. If she'd forgotten winter and spring, then the chickens would be…would be…

To her relief, she found five of her hens and the rooster scratching about in the yard, with two more hens ranging out in the woods. She checked the nesting boxes and found nothing.

"I don't understand," she murmured.

"Understand what?" the boy asked, watching the chickens with wide eyes.

"If it's summer, there should be eggs." She wiped away a sheen of sweat from her forehead. "There should be chicks running around, or a lot of fat snakes."

"Snakes?"

"Egg eaters, chick eaters…you know. Rat snakes, corn snakes. Maybe a possum." She brushed past the boy and climbed the steps to the cabin. "I'll check the coop after we eat."

Another wave of dizziness passed over her, and she nearly dropped her water pitcher as she bumped against the wood-planked food-prep table.

"Are you okay?" the boy asked from the doorway, though his voice carried no hint of concern.

"I'm fine," she snapped, glaring at him. *He* wasn't staggering around. Whatever was causing her dizziness couldn't be bad air. "Just…low blood sugar."

"Low blood sugar," the boy repeated, stepping into the cabin and squatting to gaze intently at her boots.

She snorted and reached up for a strip of venison jerky hung from the ceiling. "Here," she said, folding the jerky back and forth until it tore. "You can have some of this."

The boy stood and backed away. "I... I don't eat that."

"What? You're vegetarian?"

The boy blinked rapidly, and for an instant, she thought she saw something blue spark behind the dark brown of his left eye. Before she could process what she'd seen, he repeated, "I don't eat that."

"Fine," she said, gnawing off a hunk of jerky. The salty meat seemed tougher than she'd expected, and her molars put up an aching protest as she chewed. "If you won't eat meat, I'll do pancakes."

But when she uncovered her crock of sourdough starter, she found a layer of brown goo fouling the deflated sponge.

"Great," she muttered.

The boy stood on tiptoes to peer past her. "What's wrong?"

She covered the crock. "My starter's gone bad."

"Is this something you need?"

"Not desperately," she said, rummaging inside her potato bin for two of the least sprouted. Still chewing the wad of jerky, she quickly peeled and sliced the potatoes into rounds and arranged them in a greased iron skillet.

"Here," she said, handing her fire kit to the boy. "Follow me." She picked up the skillet and a spatula and went outside to a tarped camp stove.

"You're cooking?" the boy asked, watching as she uncovered the stove and knelt to insert sticks into the fuel chute.

"That's how food happens," she said, lighting a match.

"Food," the boy said in an awed whisper, watching her intently.

Moments later, smoke rose from the chimney, then cleared as flames emerged. She set the skillet on top of the chimney and eyed the potatoes.

"Forgot the salt," she said with a sigh. "Hey, kid, make yourself useful and go get the salt. It's on a shelf where the skillet was."

The boy looked up at her blankly. "I... what...?"

Useless brat! For an instant, she thought she saw another glimmer of blue light spark inside his left eye. She blinked, and the blue disappeared.

"What's that in your eye?" she asked.

But the boy ignored her, turning instead to stare at the potatoes which had started to sizzle gently.

I'm definitely hypoglycemic, she decided. *Hallucinating.*

"Never mind," she said irritably. "I'll get the salt."

She climbed the steps into the cabin and returned with the salt and some plates and spoons. The boy watched her sprinkle a flurry of salt into the skillet.

"Is this something you need?"

"Need? You ever eaten unsalted potatoes?"

The boy seemed to consider her question, staring off into the woods. He blinked a few times and then met her gaze. "I don't think I have."

"Figures," she muttered, settling onto a rickety bench beside the stove. "You probably have a private chef."

She flipped the potatoes a few times until they'd cooked through.

"As soon as we've cleaned up," she said, sliding a portion onto the boy's plate, "I'm taking you to the corner store. You can call someone to pick you up from there." She served herself the remaining potatoes, chopped off a section and wolfed it down.

The boy bounced his spoon on his plate like he wasn't sure what to do.

"Let me guess," she said, taking another bite. "You've never had plain potatoes before."

The boy watched for a moment and then copied her. But as soon as the spoon touched his lips, he shuddered.

"I'm not as hungry as I thought," he said, nose wrinkling as though he'd smelled something foul.

She shrugged. "I can't eat mine and all yours, too. Go chuck it to the chickens." At his uncomprehending expression, she explained, "Dump your food on the ground near the chickens."

She watched him trot away. What a useless little spawn! She halfway hoped she'd meet his parents so she could give them a piece of her mind about taking helpless kids out into the wilderness and losing them!

Her stomach burbled a reproach, and she slowed to chew more thoroughly. No sense in bringing cramps from swallowing too much air with her food.

The boy came sauntering back with a pleased expression on his face. "They liked it!"

She tried to nod an acknowledgment, but the effort, combined with an escalating twist in her gut, froze her. Had she not cooked the potatoes enough? Had the stream water picked up some sort of contaminant?

"Are you okay?" the boy asked.

She looked up at him, and his features blurred. "I'm not—" She slumped to one side, caught herself, and then fell completely off the bench.

"Help," she managed to croak. "I can't..."

The boy knelt beside her. "Just breathe," he said in a strangely calm voice. "Keep breathing."

She couldn't... pull... enough air. He spoke in some kind of high-pitched gibberish and then pressed something to her forehead—a metal sheet, thin, like foil. The metal adhered to her skin, and a paralyzing coolness spread from her face to her torso to her legs.

The boy looked taller, somehow. Then there were more boys. Where were their parents? They gathered around her, gazing down with eyes that blazed with bright blue fire. How had they gotten here? What was...?

Why...?

But summer faded away from her into night like a dream forgotten.

She was floating, buoyed in some kind of thick liquid. A glass tank surrounded her, outlined with tiny white bulbs.

In the shadowed reaches beyond the tank, black amorphous shapes drifted. Glowing blue flecks, like welding sparks, arced from shape to shape. Some of the dark shapes pressed against one another. Where they touched, the sparks disappeared, replaced by blue glows that moved in shifting patterns, reinforcing, fading, and reforming.

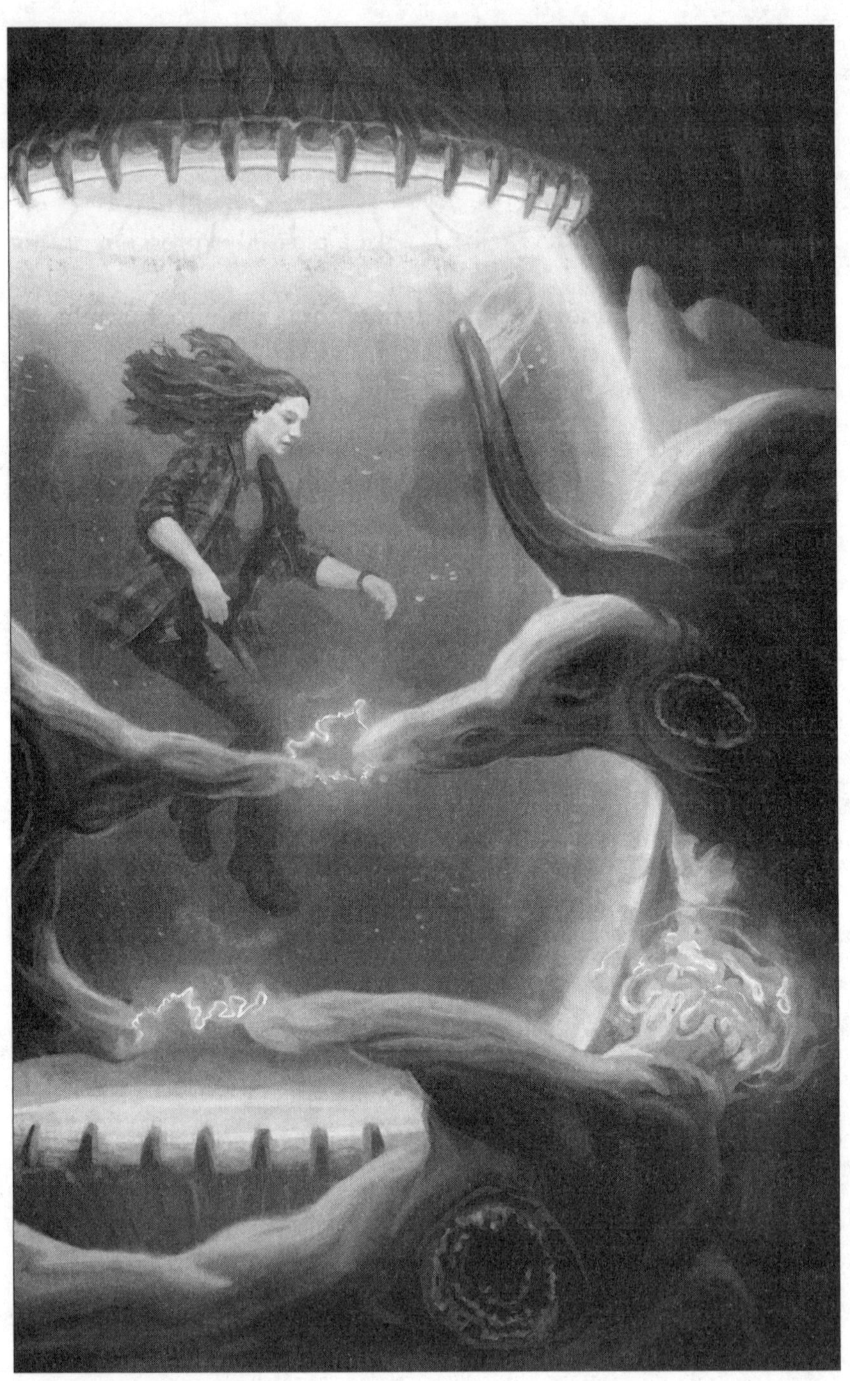

RODDY TAYLOR

She blinked slowly and tried to turn her head.

At her movement, the blue glows brightened. The shapes moved closer and flowed over her tank, spreading, joining together like giant black amoebas, blocking any view of the room behind. Blue-rimmed mouths sucked onto the glass, smearing tracks of mucous as they contracted and relaxed, contracted and relaxed, like the rhythm of her heartbeat. Whispers echoed through her prison.

"Is this something?"

"Something you need?"

She opened her mouth to scream, but she had no breath.

She closed her eyes to stop the nightmares, but her eyelids couldn't block the blue-lit rings gulping against the glass.

Her fingers curled through goo so thick it felt like gelatin.

She was trapped, unable to speak or move. But she could hear with perfect clarity, and all she was able to do was scream silently and listen as the voices whispered over and over, "Is this something you need?"

She woke again into a dew-fresh morning.

A cool breeze wafted through the open windows of her cabin, shifting the curtains in a mesmerizing dance. Her fern-stuffed mattress crackled as she snuggled deeper under her quilt. She took a deep breath, savoring the wind-blown honey scent of wild plum blossoms slightly tainted with an overlay of dust.

Wait... Plum trees? Blooming in summer?

She pushed herself upright and leaned over the edge of her cot to look through the window. Outside, the bare-limbed oak trees were fuzzed with dangling, greenish-brown catkins shifting in the morning breeze. This had to be springtime. March. Maybe April.

The door creaked open. Intruder! Before she could panic, a soft voice asked, "Are you better?"

That dratted boy again! Still dressed in his olive-green shorts and orange shirt, not a smudge or wrinkle marring the fabric.

"What... how... did you...?" She glanced down, realized she was naked, and clutched the quilt to her neck. "You little pervert!"

"You wanted to sleep," he said, stepping inside the cabin. "I helped you."

She glared at him. "Go outside!" She lifted a hand free of the quilt and pointed. "Go…check on the chickens."

The boy grinned. "Already did. I saw eggs. Got bitten." He held up a bloody finger.

She grimaced. "I'll take care of that after…" She paused to self-assess. Was she hungry? Her stomach pinched in answer. "After breakfast. Now, go! Scram!"

He turned obediently and left, pulling the door closed behind him. She stood, keeping the quilt wrapped around her, and watched through the window as the boy strolled to the stream. He paused at the edge of the bank, lost his footing, and slid down to the water on his backside. What a freaking stupid kid! She couldn't wait to be rid of him!

Her clothes were piled on the floor with no attempt at folding. Had she left them there, or had the boy dumped them? She shook each piece vigorously, dislodging spiders that scuttled away and disappeared into cracks between the floorboards.

"Yeah, I'll feed you," she muttered, dressing quickly, "and then it's straight back to your mommy." She tugged a dresser drawer open, took out a fresh pair of brown wool socks, pulled them on, and stepped into her well-oiled hiking boots.

As she adjusted the laces, she glanced around the cabin. Dust lay thick on all the horizontal surfaces. As soon as she'd ditched the boy, she'd come back and do a little spring cleaning.

Spring?

She finished tying double knots and stood up, shifting to settle her feet inside the boots. If yesterday had been summer, how had she gone straight to spring in a single night? What had happened to fall? To winter?

Must still be coming off that mushroom high, she reasoned, though the explanation wasn't reassuring. She reached for the skillet hanging on the wall. The *wall*?

She opened the door and searched for the brat. He was climbing up the stream bank, dripping wet.

"Hey, kid!"

He waved at her. "Hey!"

"Did you clean up last night? Put the skillet and the dishes back?"

For a moment, he frowned. "Yes?"

She stepped down to the yard. "That's a question, not an answer. You're not sure? Or you don't remember? And where did you sleep?"

"I, uh..." He waved vaguely toward the woods. "Over there."

"On rocks and tree roots?"

"Oh, I didn't sleep. I watched the stars."

She snorted. Didn't sleep? Could this kid get any weirder?

She collected three eggs and got the camp stove and skillet heating, but when she went to add grease to the skillet, she hesitated. Vegetarians could eat eggs, but what about the grease? Did the boy know where grease came from? Was that why he'd refused his food yesterday?

She cracked the eggs into the skillet, and when the boy raised no objections, she sprinkled on a liberal dusting of salt.

"Is this something you need?" he asked.

"The eggs? Or the—"

She broke off as memories of whispers echoed in her thoughts. *Is this something? Something you need?*

She glared at the boy. "What did you say?"

The boy cringed back. "Sorry. I just wanted to know."

She took a deep breath and closed her eyes. In a few hours, she'd be rid of him. She could go back to normal. Back to the solitude and peace she craved. Back to keeping her little corner of the world on track. Had it been only yesterday she'd met the kid?

She gathered plates and spoons and served up the slightly overdone eggs. The boy, as expected, wouldn't accept anything.

"You have to be starving," she said before cramming a spoonful of egg into her mouth.

"Oh, I ate some berries out there."

She chewed and blinked. Blinked and chewed.

"There're no berries this time of year. At least, nothing that won't make you sick. Do I need to make you vomit?"

He shrugged and stared past her at the chickens. "I forgot. I didn't find any."

She put her plate down. "At least let me look at your finger."

The boy shrank back. "My finger's fine." He held up his hand. "See? All better."

She picked up her plate again. "Well, someone can look at it when you get home."

As she ate, she studied his clothes. They'd already dried back to a perfectly clean, crisply pressed state.

Must be some new fabric technology, she reasoned. *Insta-clean clothes. Maybe I should barter for some.*

As soon as she swallowed her last bite, she stood up. "Time to get you home. You ever ride double on an ATV?" When he shook his head, she sighed. "You'll learn quick enough. Now, go say goodbye to the chickens. And watch your fingers!"

He grinned and trotted away down the well-worn path to the coop.

As she burned off the skillet residue and wiped a film of fresh grease back in, she pondered what she was going to say when she handed the kid over. Would she have to wait for police to collect him, or could she just dump him at the store and leave? What if they took her in for questioning? Would they fine her for riding double on the trails? How could she corroborate her story that he just showed up at her cabin?

It all came down to the boy. If he backed her up, this would all go smoothly. If not…

She checked to see he was still busy with the chickens before she went inside and tugged a wooden panel off the wall. She reached inside the hidden cavity and pulled out a loaded revolver and a handful of twenty-dollar bills.

"Hey kid!" she called when she'd secured the revolver in a holster clipped to her belt. "Let's go!"

The boy followed her to the ATV she kept parked under a lean-to.

"Put this on," she said, handing her helmet to him. "Over your head."

While he inspected the helmet, trying to figure out which opening to use, she checked the tires on her dusty Trail4, climbed aboard, and tried to start it.

Nothing.

She tried a few more times with no success. Not even a whine.

"What are you doing?" the boy asked.

"I'm trying to get you home," she said irritably. "But the battery must be dead."

"Is this something you need?"

"Quit asking that!" she snarled. "Of course, I need it!" She gestured around. "I need all of it!"

The boy seemed to freeze with a blank expression on his face. "All of it," he repeated in a monotone, a flash of blue sparking behind one eye.

She stared at him for a moment. Maybe he had some kind of implant, the latest in body mods. Whatever. She had more important things to worry about, like how she was going to get rid of this glowy-eyed, obnoxious kid. The solar charger would take at least a day to juice the battery, and she was *not* going to babysit for that long!

Well, if she couldn't pack him out, she'd have to walk him out. Judging by the boy's scrawny legs, he'd never hiked more than a kilometer in his life, and it was a good eight to the store. Would he be up for it?

"Hey," she said, snapping her fingers in front of his face. "We'll have to walk. It's less than two hours, but you'll have to keep up. Think you can do that?"

He nodded and followed her to the cabin, watching as she collected venison jerky and dried apple slices. She stashed everything inside her nylon backpack and handed two stainless steel water bottles to him.

"Go fill these." At his blank expression, she added, "With water. From the spring."

He took the bottles cheerfully and headed to the stream, but when she checked on him, he was still trying to get the lids unscrewed.

"You're useless," she growled, opening a bottle and filling it. "Here… This one's yours."

He took a step back, shaking his head. "I don't need it."

"Not optional!" She shoved it into his hands and dipped her own bottle into the water. "Now, follow me. Stay close and don't wander."

They hiked single file along the winding path leading to the main trail. The boy kept stopping to look at everything. Flowers. Vines. A herd of wild goats. A pile of boar scat. Everything!

At this rate, she thought, dragging him away from a copperhead puddled in a sunny spot at the edge of the path, *it'll be 3:00 before we get there!*

At last, they reached the metal farm gate barricading the overgrown entrance to her property.

"We'll be heading south," she said, like the boy would know or care what that meant. She led him around the gate and stepped out onto the trail, keeping a wary eye out for other hikers and ATV riders. Most of them would be the lean, tanned, loaded-with-superfluous-gear types, but there were always a few who wanted trouble.

But she saw no one.

Must be a weekday, and everyone's at work. She grinned. *Except for me!*

An hour passed. They'd been climbing steadily uphill, and the boy was lagging. She was breathing hard, but he wasn't. At least, she couldn't hear his breaths.

"I never asked your name," she said, glancing back. He'd dropped his water bottle somewhere.

"Mason," he said, trotting along dutifully.

"Mason," she repeated as the trail topped the ridge and began winding down toward a valley. This was the easy part of the—

She stopped suddenly. Somehow, the trail had dead-ended at a boulder field of rounded limestone outcroppings. Thorny bushes and shortleaf pines sprouted between the stones, completely obliterating any trace of a path. She glanced behind her. The trail they'd just hiked curved away, through oaks and hickories and loblolly pines, disappearing over the ridge.

"This isn't right," she muttered, studying the ground. She was standing on packed dirt, a path perfectly plain to see. And two meters farther on, the dirt stopped at stone, like a line had been drawn. Had the earth cracked along a fault line, exposing bedrock during a quake? If so, how had the stone eroded away so quickly? And why hadn't she felt the tremors?

She shaded her eyes and squinted, trying to focus on the valley beyond the trees. Instead of mountains in the distance, she saw a vast flatland stretching to the horizon. A river cut through the plain in a lazy S-shape, glinting under the late morning sun. River?

She backed up, trying to get her bearings. Had she missed a turn? How was that possible? She'd grown up here, hiking and riding the trails since she'd been able to walk. There was no way she'd gotten turned around.

And where were the rest of the freaking mountains?

Mason stopped beside her.

"What's wrong?"

She glanced down at him, panicked breaths coming fast. "We'll have to…backtrack. I've…lost the trail."

"Is this something you—"

"Stop asking that!" Her shriek rang out across the valley. "What's going on?" She jabbed a finger toward the flatland. "That's not possible! Where's the rest of the wilderness? The rest of Arkansas? And the whole winter-summer-spring thing. It doesn't make sense. And you, showing up out of thin air!" Her legs grew weak, and she collapsed to the ground, thoughts spinning in confused circles. "And your eye! What's with the blue glow in your eye? Are you…are you…?"

Mason held up his hands. "One question at a time, please!" He took a deep breath and continued in an uncharacteristically baritone voice. "I'm going to tell you something that may seem rather confusing."

"Confusing?" She choked on the word and fought back a coughing fit. "Like what's been happening lately *hasn't* been?"

He knelt beside her, gazing intently into her eyes like he was searching for something. "I think you're ready to hear this." He leaned closer. "Arkansas no longer exists because your world no longer exists."

Of all the explanations she'd expected, this wouldn't have made the top hundred. For a moment, she sat with her mouth half-open, trying to parse the meaning of his words. "But…I'm still in the world. How could it not exist? Are you saying this is some kind of simulation?"

"This"—he gestured around—"isn't a simulation. It's a transplanted fragment of Earth."

His words made no sense. She stared at him in disbelief. "A fragment?"

He nodded. "Before... the end, we took a sample." He made a circle with his hands, fingertips touching, and swept them downward. "A sample of Earth cored all the way from the atmosphere through bedrock. We preserved it, embedded it here, and isolated it for study."

She tried to picture what a cored sample of Earth would look like. Failed.

"Why?"

"We needed a reference to terraform this planet to match yours."

"*This* planet?" she said, shaking her head. "That's... crazy! Wait! You said my world doesn't exist anymore. Do you mean *Earth* doesn't exist anymore?"

He took another deep breath. "The planet itself still exists, but not as a viable world. We observed your planet for millennia as your civilizations developed. And, predictably for your type, unthinkable weapons became more valued than life."

"Weapons?" Her voice cracked. "You mean...?"

"The worst kind."

Nausea burbled inside her gut. Had some monumentally stupid person triggered apocalypse while she'd been tucked away in cozy isolation, oblivious to the escalating tensions? Who had started it? Finished it?

She squeezed her eyes shut, imagining horrible, horrible visions.

Mushroom clouds coiling up and up into the atmosphere. First one, then another and another. Flashes over Los Angeles. New York. Dubai. London. Shock waves. Things worse than nuclear bombs. Destruction on a global scale. Dust clouds swirling over fires that burned and burned across continents...

She opened her eyes. No. This couldn't be real. Her gaze wandered back to the unfamiliar valley and river. They didn't belong here. Or maybe *she* didn't belong here.

"As for your own role in this endeavor," Mason continued, as though he'd read her thoughts, "we needed a test subject. Someone who was attuned to the nuances of the environment. Someone to demonstrate how sustenance was procured and—"

She'd heard enough. "Stop!" she cried out, struggling to her feet. "I don't care what *you* need!"

She stood there, breathing hard, trying to process. Earth on fire?

Everything gone? A different planet? She'd always craved isolation, but this, *this*, was too much to comprehend!

She took off in a stumbling run, but after a few steps, she fell and couldn't get up again.

"Why?" she said as she pushed herself into a sitting position. She could barely catch her breath. "Why?"

She knew she should be crying, but somehow, tears seemed inadequate to the magnitude of the loss. Who'd started it? And why? Had it all been a misunderstanding? Some kind of political tipping point? Religious frenzy?

"Stupid people!" she said, pounding the ground with a clenched fist. "Stupid…stupid…"

Mason approached and sat down beside her. She leaned away, arms hugged around her knees.

"How long?" she whispered after a few minutes. "How long has it been?"

He gazed at her, one eye glowing with familiar blue. "Just over twenty thousand of your original years. Enough time to establish stable gas cycles here and introduce Earth flora and fauna."

"Twenty thousand years?" She shook her head. "I…I can't be that old!"

He picked up a pine needle and rubbed it between his fingers. "Chronologically, you are over twenty thousand years old. But you've been in stasis, so in actuality, you've aged only a few days."

"Stasis," she murmured as memories of blue mouths whispering, *"Is this something you need?"* made her shiver.

He dropped the pine needle and gestured toward the woods. "All living things within this sample, except you and the chickens, have been preserved using geostasis. At periodic intervals, we unlocked the geostasis to mix atmospheres and test whether or not we'd succeeded in duplicating seasonal changes."

She dug her fingers into the ground, wincing as dirt wedged under her fingernails. "That's why I lost track of time." She fixed an accusing glare on him. "You engineered it."

He shrugged. "We had to. Your own stasis was more sophisticated, as you've observed. Even more so than the other humans."

For a moment, she gaped at him. "You've done this to other people, too?"

"Oh, yes," he said proudly. "We've preserved about a hundred thousand humans for a stable breeding population. When you prove that conditions are adequate, we'll settle them here, close by—"

"What? Here? You mean here in...?" Words escaped her. A hundred thousand disoriented, clueless people tromping around her woods? Her home?

"No!" she cried out, attempting to stand. Dizziness crushed her, and she dropped back to her knees, breathing hard. "You can't! All I've ever wanted was to be left alone, and you... you can't take..."

Her breaths tightened. Mason reached to her, holding a thin metal sheet, but she slapped his hands away. His skin felt slimy. Cold. His eyes burned bright and blue.

"Who are you, really?" she asked in a choked whisper, trying to dodge as he darted his hands toward her face. She wasn't fast enough, though, and something wet adhered to her forehead. A familiar paralysis gripped her as blackness squeezed her vision. Buzzing noises filled her thoughts. She felt herself slipping away... slipping...

But, before she lost consciousness, she whispered, "*What* are you?"

She was floating again in the same tank of thick liquid.

The glass around her sparkled with lights like stars in a black expanse. Blue glows approached and adhered to the glass. This time, though, she could understand their whispers.

"She refused to accept the reality."

"Will the others?"

"Is she ready?"

"The world is ready."

"The people must be ready."

She opened her mouth. No sound came out, but she willed her thoughts to ask, "Why me?"

"We have learned from you."

"You are the test."

"You will lead the people."

She closed her eyes. "No. I don't want people here."

The whispers grew urgent.

"Is this something?"

"Something you want?"

She listened to the whispers. Need. Want. What was the difference? And she smiled as the whispers flowed over her, asking again and again, "Is this something you want?"

She woke next into the chill of an autumn night.

Shivering, she dragged out of bed, pulling the quilt around her. She struck a match and lit a candle against the darkness. In the dim light, she examined her cabin. A thick layer of dust flocked every surface like gray snow. Only her clothes, hung from their pegs, were clean. For a panicked moment, she thought her revolver and holster had gone missing, but when she checked the hidden cavity in the wall, she found them stashed inside.

"They're getting better at resetting me," she murmured as she slid the windows closed and then cupped her hands around the candle flame for warmth. She stared into the light, trying to remember all that had happened yesterday. Or whatever had passed for yesterday.

It's that kid, she thought. *Mason. He's the key to all this.*

She strained to recall what the blue whispers had said. Something about learning from her, about leading people. Nope. She'd had enough of people in her old life. She would *not* be responsible for watching over a whole planet's worth!

But where could she go? How could she pack up and leave everything she'd spent a lifetime building?

She shucked off the quilt and dressed quickly. Still shivering, she knelt to start a fire in the potbelly stove, arranging kindling and logs. The tinder caught sullenly from the candle, but she persisted, blowing on the infant smolder until she had a decent start. The kindling burned more enthusiastically, and she kept feeding in sticks until the logs began to smoke.

By the time the stove was radiating, the sky had brightened to the pale gray blue of dawn. She pulled a rocking chair close to the stove and sat, basking in warmth as the rising sun lit the fall leaves outside into riots of orange and yellow and more subtle hues of caramel and butterscotch. Were the other people out of stasis, too, watching the sunrise like she was?

She sighed and glanced up at her mugs and glass jars all in a row on the shelf above the prep table. Maybe some hot tea would help her make sense of her questions. She opened her wardrobe and pulled out a jacket, shrugged into it, and grabbed the predictably empty water pitcher. Just as she reached for the door, it opened on its own.

Mason stood there. He was still dressed in his absurd little shorts and polo shirt, but she knew that everything, from his clothes to his knobby knees, was just an illusion. She'd already guessed what he really looked like behind his human facade: a black shifting shape covered with flowing blue glows.

She gestured him inside. "You can drop the pretense, *Mason.* I know you're not a little kid. You're not even human."

He turned to study her, eyebrows quirked in a perfect imitation of consideration. "I think I'll maintain this appearance for a little while longer."

She marched out the door and crossed the yard to the stream. *He needs something from me*, she thought, kneeling to fill her pitcher. *Or wants something.*

Need. Want. Who was she kidding? There was a *big* difference!

She returned to the cabin and poured water into a kettle on the stove. While the water heated, she chose a mug and spooned in a serving of dried mint. "I'm guessing you don't want any tea."

"I don't, thank you."

She eased down into her chair. "So, how did you learn English? You speak it well enough to pass as human, no question."

"I learn by using"—he made a noise like crows cawing—"to touch minds directly. As a result, I can vocalize over seven hundred languages and dialects."

She stifled a laugh. No matter what he really was, he still looked like a little kid. Her amusement passed quickly, though, and she fixed him with a glare. "Then tell me, oh speaker of many languages, what is it you've been learning from me?"

Mason sat on the floor beside her chair. "Each time we broke your stasis, we thought we'd succeeded in making this planet habitable, but you quickly showed us how we'd failed. You are so much more sensitive to the environment than our own testing equipment."

She stared down at him for a long moment. "So, you've been using me like some kind of... what... a canary in a coal mine?"

He frowned at the reference.

"Never mind," she said quickly before he could ask for clarification. "So, when we were hiking yesterday—"

"Forty-three years ago," Mason corrected.

She waved a hand impatiently. "Whatever. The trail ended where my world stopped... at the edge of the transplant?"

He nodded. "We've smoothed the transition between transplant and terraformed planet to make it less of an obstacle. And we bumped the atmospheric oxygen up just a bit. You seemed too breathless on our hike."

A low whistling note began. Mason glanced around for the source of the sound.

"Water's ready," she said, standing to take the kettle from the stove. She poured hot water over the mint in her mug. The familiar ritual soothed her as she stirred honey in and took a sip, careful to cool the tea with her breath.

Mason studied her as she returned to her chair, cradling the hot mug. She tried to ignore him, but his persistent gaze unnerved her, like he was examining a bug under a magnifying glass, and she was the bug.

"So, why do you do this?" she asked, taking another sip of tea. "What do you get in return?"

He leaned forward, the perfect imitation of an eager child. "My people were rescued, too. We learned from our mistakes and we... return the favor by saving others. Now that you've proven this planet is ready, we've prepared compounds with houses like this one and food sources based on your garden and—"

Her thoughts drifted as he rambled on and on about how his people had turned the planet into a giant human habitat with shelters and gardens and water and recreational opportunities. *Like we're hamsters or livestock or something. It's disgusting!*

"Then we'll take the seeding population out of stasis and start them here where the biodiversity is—"

"Stop!" She jumped up, dropping her mug. It hit the floor and cracked in half, spilling tea out. "I've already told you! I don't want them here! What part of 'no' don't you understand?"

Mason stared at her. "I thought you'd want to mentor them. To lead them."

She snorted. "You definitely picked the wrong person for your little experiment if you think I'd want that!"

He glanced down at the tea spreading across the floor. "Just so there's no misunderstanding," he said, scooting away from the liquid, "you're saying you *don't* want the companionship of your fellow humans?"

"Of course not! I moved away to escape them!"

"But these people aren't the ones responsible for destroying your world. You wouldn't be in danger."

She shook her head. "They're human, aren't they? That means they'll just do it all again. There's no way to break the cycle."

"You'd have a chance to try," he said, a puzzled expression on his face. "You could set the example!"

She grabbed a towel and knelt to wipe up the tea. "Doesn't matter how hard I'd try. Destruction is what humans do. They always ruin everything. I don't want to be a part of that."

He watched her scrub until the floorboards were nearly dry. "I suppose we could delay the repopulation seeding for a time."

She glanced at him. "Delay? Really? You could do that? How long?"

He hesitated. "I don't know why I suggested that. Maybe... maybe... No, there's an ever-increasing risk with delay. We've already lost over two-thirds of the original collected population due to effects from prolonged stasis. Not unexpected. We made allowances for attrition, but with every passing year, we lose ten or so more."

She shivered at the callous way he referred to human deaths, like he was talking about strawberries going moldy in the refrigerator. "No," she said, "I won't be responsible for anyone else dying."

His eyes glowed cold blue for a moment. She guessed he was communing with his fellow... what were they called? And surely his name wasn't actually *Mason*.

"We can't delay anyway," he finally said. "They tell me we're too far into the reanimation process. There's not a safe way to pause it."

She groaned. A hundred thousand people? Seriously? Crawling all over the perfect solitude she'd carved out for herself? Knocking on her door twenty times a day? Ugggh! There had to be another solution!

She swiped her towel across the floor one last time. Maybe…

"Could you put them on the other side of the planet? Somewhere I'd never see them?"

He shook his head. "We crafted this region to support the initial population."

"So, that's it." She stood up with an exasperated grunt and walked to the door. "You've ruined my life. For what? So I can watch the same suicidal cycle start all over again?" She opened the door, wadded the towel into a soggy ball, and flung it as hard as she could into the yard.

Then she bolted for the woods. She could hear Mason calling after her, but she cleared the stream with a practiced leap and kept running, leaves crunching underfoot as she sprinted between the trees.

This is my land, she thought as she ran. *My world! No one else has a right to it!*

She ducked under the branch of an oak tree and tripped over a root, stumbling off balance until she landed on her hands and knees. She sat, breathing hard, and examined the scrapes on her left palm.

Careless. *Stupid!*

She wasn't bleeding, but a wound like this, minor as it was, could turn septic.

A rustling sound startled her. She turned, and…there he was again, so close she could almost touch him.

She scowled. "Can't you leave me alone? It's bad enough I'll never have a moment's peace again. I don't need you tracking me wherever I go, dropping in out of thin air!"

He stood there, gazing down at her. "Were you serious about the 'suicidal cycle' part?"

"You don't get it, do you? It won't take long for people to destroy this planet, too!"

"Are you sure of this?"

She glared up at him. How could she make him understand the darkness hidden in people's hearts? Even in her own? Here, at least, in her tiny slice of the world, she only had to deal with her own faults…her own limitations…her own…

"Wait," she whispered, sitting up straighter. "Tiny slice…" She blinked a few times.

Would it be possible?

She shifted in place, dry leaves crackling under her. "I'm very certain this planet won't last long," she said slowly and carefully. "You might as well start getting another one ready."

Mason looked confused. "Another planet?"

For an instant, she glimpsed his black, ever-changing true form behind the wavering appearance of a child. "Another planet," she repeated. "One that would be easy to prepare because it used to have humans living on it and could be ready again if…*when*…humans need it."

A squirrel began its scolding chatter on a branch above, and Mason glanced up before meeting her gaze. "You mean Earth, right?"

She nodded. "Has it been long enough?"

He froze in place, both eyes lit with an ice-blue glow. She watched intently, hardly daring to breathe for fear of interrupting a consultation that could change her life forever.

After a few moments, his eyes darkened back to their normal brown. "The planet has been self-healing long enough to make the terraforming process worthwhile."

"Good," she said, standing up to brush off dirt and leaves. "You can use this same area as a sample, yes? And you'll need a test subject for the transplant. Someone who wouldn't mind being the only human on the planet."

He tilted his head, studying her. "I think I know someone who fulfills that requirement."

"You'd do that for me?" She took a quick breath. "I can go home?"

He continued to study her, one eye sparking blue. "We've never been proactive like this before, but yes, you'd be back on your native planet. And we could delay any resettlement, if things here go as you say, until the end of your natural life cycle."

For a moment, she tried to imagine spending the rest of her life completely alone. Would she regret her decision? She'd have no hope of rescue if she got injured. There'd be no one to call if she needed a tooth pulled or help moving something heavy. No one to talk to but herself and the chickens and whatever animals she could eventually capture and pen.

And what about supplies? She'd never be able to replenish her flour or salt or canning jars. Not that she'd be able to do that here, either. Whatever she had now would have to last the rest of her life.

She grinned. Not a problem. There'd be no regrets. This would be the grandest test of her homesteading skills she could ever imagine!

Leaning over, she held out a hand. "Then we have a deal."

He stared at her, one eye glowing blue.

"This is a handshake," she explained, reaching to clasp his tiny hand, trying not to flinch from his cold touch. "It shows we're in agreement."

He looked up at her and returned her grin as he squeezed her fingers. "It's a deal!"

She pulled away and wiped her hand on her jeans. "So, when do we start? Now?"

"Now is good," he said, gesturing her toward the cabin. "Just don't expect Earth to be exactly what you remember. There'll be some changes."

She started back through the trees, retracing her flight. "But no one else would be there, right? No mutants or zombies or anything like that?"

"No one."

She glanced over her shoulder. "Not even you?"

"I'm not ephemeral like you," he said, stumbling and landing on hands and knees in a mud puddle. He stood up, skin and clothing completely clean. "I'll be overseeing your settlement, but after that, I'll be moving on."

They reached her cabin. She darted inside to dress the scrapes on her hand and smear on some salve. Then she shut the vents on her stove, set the broken halves of her mug on the shelf, and went back outside.

Mason was waiting for her just beyond the steps. "We're ready if you are."

"I'm ready," she said, watching as he pulled a familiar, thin sheet of metal from his pocket.

"Are you sure? Last chance to change your mind. Is this something you want?"

Blue whispers filled her memories. *Is this something? Something you want?* But the voices in her thoughts no longer terrified her. The next time she heard them, she'd be back on Earth in total, exquisite solitude.

She laughed and knelt down, pulling her hair back. "It's something I *need*!"

He met her gaze for a moment. Then he smiled and leaned forward.

"See you in summer," he said, and pressed the metal sheet to her forehead.

The Triceratops Effect

written by
S.J. Stevenson

illustrated by
ART IKUTA

ABOUT THE AUTHOR

Shaun Stevenson (who monkeys with words under the moniker S.J. Stevenson) is a professional nonprofessional in the UK civil service, based near Liverpool. He spends most of his working life wondering how he's mucked up the spreadsheets this time.

Shaun lives with his wife's kindly suppressed frustration at his growing collections of broken laptops, Star Wars LEGOs, and barely clinging-on zombie houseplants. Sometimes his almost-tame teenagers find him amusing, but generally only by accident.

Shaun has scribbled stories for as long as he's been able to clutch a pencil. He likes to think his writing has improved over the years but admits it's difficult to tell since his handwriting hasn't. The stories he enjoys writing most tend to be science fantasy, with a pinch of humor for added flavor.

"The Triceratops Effect" hatched from ambling thoughts about time-tourists building an observatory to follow the trajectory of the asteroid that wiped the dinosaurs out—and then wondering what would happen if they couldn't find it. From there, it was a short hop to riffing off Ray Bradbury's classic "A Sound of Thunder" to tell a goofy tale about overcoming logistical challenges, toxic bosses, and the risk of being squished to death by an irritable ten-ton lizard.

Any resemblance to social commentary about humanity's avaricious talent for screwing up entire ecologies is purely coincidental and should probably be written off for tax purposes.

ABOUT THE ILLUSTRATOR

Art Ikuta also illustrated "Form 14B: Application for Certification of Consciousness Transfer (Post-Mortem)." For more information about him, *please see page 23.*

The Triceratops Effect

I knew I was in trouble when the Suit shambled over, with a scowl that could have hacked through diamonds.

I sighed. I was already having a bad morning with the endless whine of bugs that seemed to think the back of my neck was Michelin-starred cuisine. Unfortunately, the squished smudges of their slapped kin didn't seem to be any sort of deterrent.

"Take five, everyone," I said to my crew. Because the only thing worse than a bollocking is a bollocking with an audience.

I watched emotions tumble across faces. Natural nosiness versus the opportunity to squeeze an extra brew into the sweltering morning. Supposedly the sun was dimmer than it was going to be in another sixty-six million years, but it sure didn't feel like it.

In the end, the brew won through. But possibly only because everyone realised that bollockings have a nasty habit of turning contagious.

"Good luck," Sam mouthed as she hopped on a helter-scooter. She was a petite woman in her mid-thirties, who kept a bob of silver hair tidied away beneath her hardhat. She also had a ninja smile that appeared without warning and always seemed to go for the heart.

Although I suspected it was too late for luck, I gave her a resigned wave as the crew skimmed away, following a curve of ferns towards base camp.

I was right. When it comes to guessing bad news, I usually am. As talents go, I'd rather this one went.

The Suit reached me. The rainbow-sheened gown he was wearing wasn't really a suit, of course. Nothing with that much *glint* would ever cut it as office-wear when I come from. But then, being a Suit seems to be a frame of mind rather than an article of clothing.

He loomed above me, a bulky, scowling man with blond hair scraped back in a failed attempt at a ponytail. He had scrunched-up eyes almost lost among hanging jowls, which made him look like the back of his face had been attacked by a vacuum cleaner. "What's this?" he said, poking me in the chest.

It was adolescent of me, but I couldn't resist. "A finger," I answered promptly.

Perhaps the Suit reddened with anger. Beneath the blotches of sunburn, it was difficult to tell. "You being funny?"

Various responses slid through my head. Most were so far over the polite horizon, you could have looked back and no longer seen its curve with a telescope. But getting pissy with Suits only ever ends with learning they have fuller bladders, and they're not shy about spraying.

I wiped sweat from my forehead. "How can I help, sir?" I asked instead.

He squinted at me, trying to find something wrong with my answer. He smelled of some awful honey-spice perfume. It was almost enough to blot out the more fragrant aromas of a nearby hillock of triceratops turd. Then he tugged me around.

"Tell me what you see!" he demanded.

But hordes of drone-diggers milling in a vast crater beside a sun-glittered ocean were a lot of "seeing" to choose from. Certainly, too much to squeeze into, however long the fuse on the Suit's patience was going to last.

So, I kept to the fundamentals.

"Bloody big hole in the ground, sir," I said.

At which point the Suit's temper exploded.

Not that anyone believes me, but I never intended to get into the time-travelling business. After all, it wasn't an available career choice when I was growing up, what with not being invented for another two hundred years or so.

But that's life. One day you're in a dead-end job in Public Immigration Screening Services, the day after there's an unexpected influx of refugees from the future (due to Temporal Integrity collapsing like a house of cards in a washing machine). Next thing you know, *kablam!* Your life ricochets and you find yourself in a dead-end job in Chronological Recovery and Protection instead.

(Note: Organisation names may have been changed to protect my sanity.)

Sometimes, I think I should find it all the more remarkable. After all, what's more bonkers than getting to watch dinosaurs bonking, for God's sake? But it's hard to stay awestruck when someone's yelling at you for being over budget, three months behind schedule, and (irony of irony in my trade) running out of time.

If there are any deities willing to do me a favour, may you bless Sam and all her descendants, because she brought me a brew back. And one of the better biscuits too.

Since Sam had been born around three hundred years after me, from a strictly linear chronological perspective, I was old enough to be her multiple great-grandparent or something. However, from a more sensible, actual years-alive count, we were roughly the same age.

"Bad?" the visor on Sam's hardhat whirred back, as she handed me a mug. It was filled with a steaming liquid, blacker than Satan's nightmares.

"Pretty bad," I said. Actually, if the verbal I'd been given had been administered with a baseball bat, I'd be in a full-body plaster cast between now and the oceans boiling away. Although, to be fair, that was currently scheduled to take place in just five months' time.

"What's the outcome?" Sam sat beside me in the control-pod, lit with various blinking lights and data-dials.

"The usual." I sipped my brew too fast, burning my tongue. "Time Management is saying we need to do more and do it faster, but with fewer people, less equipment, and a reduction in perks."

"That's the usual, all right." Sam nodded. "I expect officially they're grateful for our hard work, but unofficially, we're incompetent slackers."

"Bingo!" I said. "So, we get it done or they leave us behind when they drop the nukes. Though, I think the Suit was joking about that. Hopefully, anyway."

"What can they do, though? Really, I mean. Given what'll happen if we don't get the job done?"

I paused before answering. "Who knows?" I said. "But you know what Time Management is like when things go over budget. I'm not sure I trust them not to get into a sulk and just cancel the whole thing."

"You're joking!"

"Yeah," I mumbled. "Probably." But I wasn't sure I was. It would be spectacularly stupid to disband the project. But then, spectacular stupidity has always been one of humanity's most reliably renewable resources.

We stared out across the glittering tropical sea. Somewhere nearby, deep among the jungle of ferns, boomed the bellow of the very last triceratops. The crew had nicknamed him Gary. I didn't think it suited him. He looked more like a "Keith" to me.

Gary had been bellowing a lot recently. I figured the poor bastard was starting to realise he really was alone.

"I don't suppose there actually could be an asteroid?" Sam asked, after Gary finally fell quiet.

I recognised the pleading look on her face. Hell knows I'd seen it in the mirror often enough. I shrugged. But I meant "no," and Sam knew it.

"Best get digging then." Sam's ninja smile flashed, then the visor whirred back down on her drone-control hardhat.

The main problem with time travel, it turned out, wasn't the laws of physics, or paradoxes, or even getting the math right and the technology reasonably cost-effective. The main problem was the same as it was in a hundred other fields. The problem was people. Or more specifically, the ability of people to screw up just about anything. And the problem with screw-ups is they defy gravity, as they rarely get screwed-down again afterwards.

It's a gift of our species or something. Though as gifts go, I wasn't alone in wishing this one would.

"Remember, they're all idiots," Billie Tourist had spat, when training my class to join Chronological Recovery and Protection. "If you scraped around the insides of their skulls with a butter knife, you wouldn't find enough brain cells to fill a spoon."

Billie Tourist wasn't her real name of course. We just called her that because of how much she hated Travellers. She was a deceptively short woman with bright orange hair, skin like polished coal, and a chronic dose of volcanic misanthropy. Everyone assumed either the hair or the skin colour was false, but no one dared ask which. Knowing Billie, it was probably both. But then, knowing Billie, it could easily have been neither.

"I ever tell you how many times I had to go to damn Woodstock?"

"Twenty-two," I mouthed silently, which was probably numerically less than Billie had told the story.

"Twenty-two times," Billie said, her voice dripping with enough disgust to overflow a bucket. "It got so I could barely take a leak without running into myself leaving the bathroom on my way in. And trust me, you do *not* want to get too close to another iteration of yourself. Not if you don't want your quasitrons to gang up behind your eyeballs and take a dump right through your synapses." Billie scratched an armpit. "You know why I had to keep going back?"

We knew; of course we knew. Billie couldn't have hammered it into our skulls any harder if she'd used a trained brontosaurus on a pogo stick. *Anachronisms*. Tourists leaving something behind that shouldn't have been there.

Of course, there was sometimes the opposite problem too—tourists pinching unapproved "souvenirs" from the past. We'd had Neanderthal spears, an incubating dodo egg, a selfie with Genghis Khan, and at least three extinct venereal diseases.

For a long time, the consensus had been that it didn't seem to matter too much; until quite suddenly it had.

And that was how I'd ended up with a long line of twenty-fourth century refugees, queuing at my Immigration desk; stranded with only hand luggage, bewildered faces, and a faint tracery of tachyon-radiation scars.

Around the campfire that evening, the crew were playing "guess the constellation," trying to work out how the jumbled stars might unfurl to a more familiar configuration. It was pointless since we were in a completely different part of the galaxy. But it was a pastime.

As the night progressed, we moved to an altogether stronger brew, but even this wasn't softening my anger.

The Suit had a job to do, too, I told myself. Perhaps he was getting it in the neck from the shinier Suits, higher up in Time Management. Perhaps it was only because he was under pressure. Perhaps he wasn't always such an *utter* asshole.

At some point the conversation shifted to more serious topics.

Big Benji, a tiny tattooed guy with flowing blue dreadlocks, was burning a sausage on a stick in the flames. "What I don't understand, now," he murmured, "is why there aren't more of us? A crew of forty? Digging a crater this size? It don't make no sense, see?"

"Yeah," Mad Swifty sniffed, scratching her tightly cropped dark hair and squinting through the fire smoke. "Job this size? Should be eighty of us. A hundred even. Cost's all in the hardware, ain't it? Our wages?" She spat into the fire. "Nickels, dimes, and change from peanut crumbs."

"Yeah, but what happens if we screw it up though?" Benji asked, just as he always did. Every. Single. Night. "How'm I supposed to look my mam in the face?"

"We've just got to accept it, kid." Mad Swifty leaned back, her hands laced behind her head. "This job's the pits."

It was an old joke that I usually found funny. But after today's bollocking, I suddenly couldn't bear to listen any longer. And I couldn't make myself explain—yet again—that we'd been picked because we were discreet. That there just weren't enough of us with the right skills. That the window for funding had been tight and, idiocy of idiocy, apparently almost hadn't been approved at all.

"Where you off to?" Sam's voice startled me. I hadn't even realised I was standing.

A plausible lie came to me: "Some of my insides need to be on the outside!" I faked an embarrassed grin then headed towards the toilet block beyond the circle of firelight.

The late hour was cool, and the chirrup of glowing insects gave a texture to the night. I found myself walking towards the perimeter fence. Not that it was really needed anymore. Only eighty years ago, relatively speaking, venturing beyond the fence was a shortcut to "splatteration," with several incautious Travellers heading to their final destination as a red-smudged footprint.

Even so, the gate in the fence still had more locks than hell. Oh, eyes, too, as it was covered with cameras.

I disabled locks and cameras both, then stepped beyond the camp. I told myself I only planned to gather my thoughts, but somehow, I was grabbing a helter-scooter and skimming off into the darkness.

The moon was rising, coating the night in a silver glaze. It was noticeably bigger than my Home Time and splattered with an unfamiliar pattern of meteorite strikes. "Genuine meteorite strikes," I muttered.

I'd thought I had no destination in mind, but soon I was back at the crater's edge.

The drone-diggers were barely shadows, hanging bat-like from the sides of the pit, following another heavy day of microwaving rock into vapour. They looked tiny, though I knew each was the size of a jet plane.

The crater was, indeed, a bloody big hole in the ground. It was also, as the Suit had screamed back, nowhere near big enough.

Stats I'd been vaguely aware of since childhood staccatoed through my mind. At the end of the Cretaceous, an iron-nickel asteroid the size of Mount Everest had blasted into Chicxulub with the force of a billion atomic bombs. It had left behind a crater sixty miles wide and twelve miles deep, along with a convenient apocalypse that had cleared the way for humans to evolve.

I turned and kicked a tree. *If only*, I thought, *that had actually been true*.

There was a sky-shaking roar and something huge crashed through the ferns. *Gary the Triceratops!*

I stood frozen as the dinosaur thundered towards me, immense in the primeval moonlight, and bellowing as its pounding feet pockmarked the trembling earth. Gary was seven metric tons of granite flesh, cabled sinews and overwhelming pulverising death. And then, there were those *horns*. Three feet of solid bone that made elephant tusks look like tickle sticks.

Rather than sensibly leaping out of the way, I found myself

thinking that being impaled on one of those horns would hurt like hell. Just not for a very long time.

But Gary was slowing, *slowing*, until he came to rest on the crater's lip, barely an arm's width away. He raised his huge head and bellowed again.

It felt like the world was shaking to the sound of the last, lonely triceratops. And, afterwards, it felt like the world still shook in the aftermath of a silence louder than the sound of thunder.

My heart hammered, thinking about the tourist hordes, and the many, *many* hunters among them. So many, I was sometimes bitterly amazed we'd left anything behind to damn well fossilise. "I'm sorry," I whispered. "I'm so sorry."

If this had been a neuro-flick, Gary's huge head would have turned towards me. There would have been a moment of connection with his golden eyes meeting mine. Perhaps, I'd even have rested my hand gently on his beak.

But it wasn't, so I didn't. Instead, we stood beside an earth ripped open to cover up a lie so huge, even Gary looked tiny beside it.

The moment ended. Gary crashed away. Still shaking, I grabbed the helter-scooter and skimmed back towards base.

Not that I ever believe me, but I sometimes tell myself that I had a choice. That I didn't have to get involved with faking the space rock that didn't *actually* wipe out the dinosaurs. But then, maybe all choice is illusory, since we're always condemned to be ourselves.

I'd just got back from retrieving water-repellent clothing from tenth-century England, when Billie Tourist had tracked me down. Not that I'd have been difficult to find, since I was in the Departure Lounge canteen, tucking into my usual protein-slab burger that I'd nicknamed the McFly. Beyond the canteen's glass, the various Departure Gates had fizzed as they spat stray quasitrons.

"No way," I'd said, after Billie had explained the position, via every obscenity ever invented.

Billie's response had been so softly spoken, I almost missed it over the crackling hum of the Departure Gates. "You don't believe me, pal? Or you won't do it?"

I'd put my burger down, no longer hungry. "Both, I guess," I'd said.

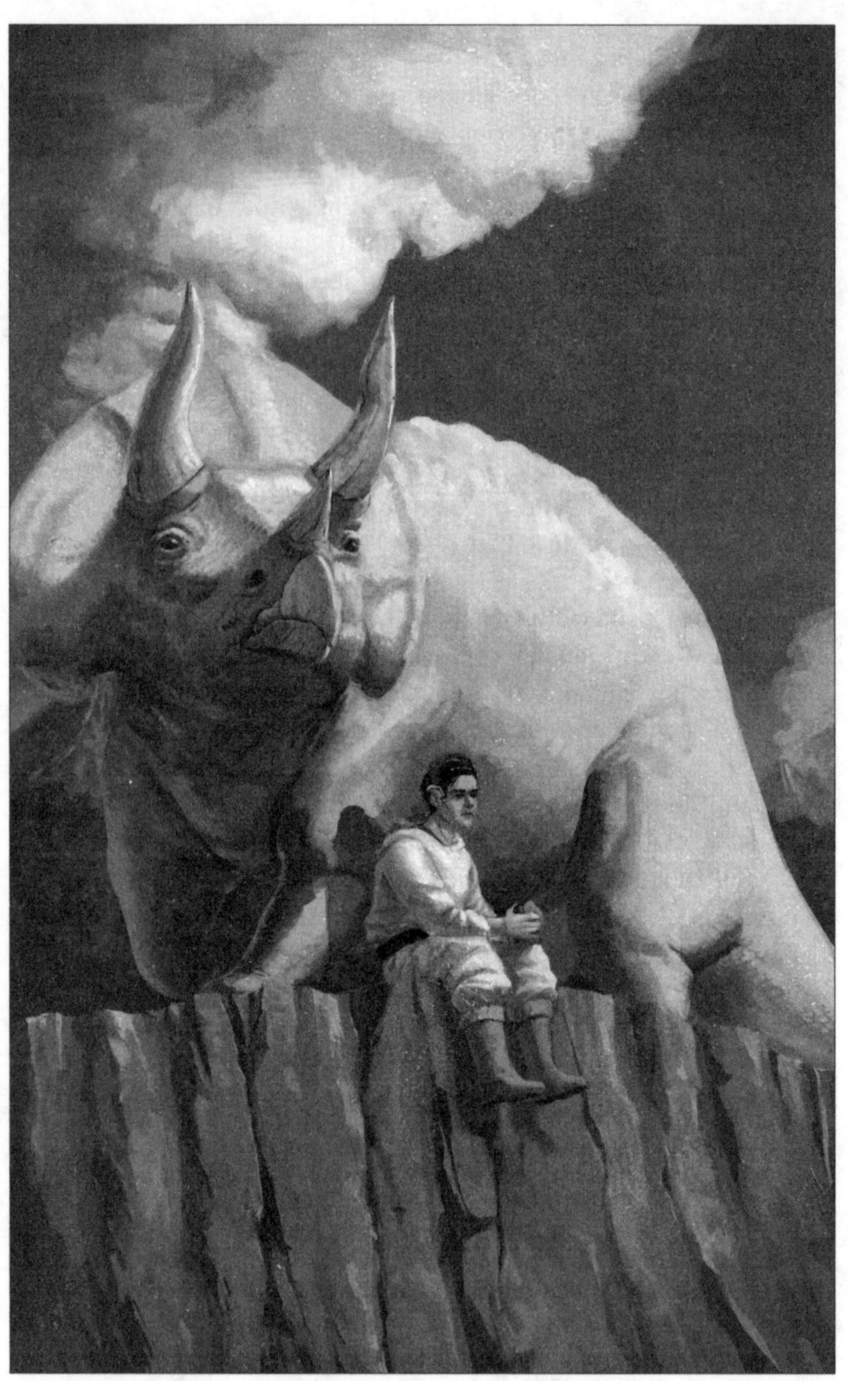

ART IKUTA

Billie had scraped her chair closer. The spasms of purple light from the Departure Gates had given her a demonic look, and somehow helped her loom over me, even though she'd practically been staring up my nostrils.

"Listen, pal," Billie had said. "I don't blame you. This isn't just the mother of all cock-ups, it's the 'killing your grandmother before she does the sideways tango with grandad' of all cock-ups. But if any job we do is necessary, this one is."

My mind had reeled. The Late Cretaceous was supposed to be the ultimate guilt-free tourist destination. Following all the work that had gone into rebuilding the collapsed timelines, there was still some eccentric twenty-fourth century villionaire who'd built an entire mansion out of sauropod bones. He'd even had security cameras fitted into a hundred robotized T. rex skulls.

Because how could hunting to extinction really matter, when an asteroid was going to pulverise the entire ecosystem anyway?

Except...except now, according to Billie, it wasn't.

"Think it over." Billie had clapped the top of my arm as she stood. "First Gate Drop to sixty-six million years ago is three days and eighteen hours from now."

It took three days and fifteen hours before I caved in and packed my bag.

"Knew I could count on you," Billie had said, when I showed up at the Departure Lounge clutching my hand luggage, my work boots, and an expression of bleak despair. Billie's expression had twisted into something that was almost a grin. "Like a piece of advice?" she'd added.

"Sure."

"The Time Management Suits are crawling over this like maggots hatching from a turd. Watch your back."

Thing was, long-range time travel was a bastard. Tunnelling through the quantum strata of twenty-four billion yesterdays took colossal amounts of energy. And then there was the need to cross the six hundred light-years the earth had travelled, during the intervening years.

Consequently, the Gate was only opened once each month for supplies. And because of Entangled Tachyonic Causation, for each hour spent in your destination time, an equivalent hour had to pass in your departure time before you could travel back.

I'd once asked how Entangled Tachyonic Causation worked. A scientist had answered using sentences containing more syllables than I had IQ points. A technician had mentioned semi-synchronic-chronological alignment. And the Suit had muttered about leveraging operational cost-effectiveness, while exploiting synergistic windows in Regulatory Framework meta-space. Which I understood meant he had less of a clue than I did.

In the end, Billie Tourist had told me it was about not making the laws of causality throw a tantrum so huge, that it'd crap a hole right through the diaper of the universe.

I gave up asking after that.

Ultimately, it meant the budget for our Cretaceous cover-up stretched to only a finite number of Gate drops. Time may be money, but time travel took finance which, ironically, was money transported from the future to pay for the now.

Perhaps it should have balanced out somehow. Except it never quite did.

Which was why the thunderstorm that hit two weeks after my bollocking was, in Suit lingo, *"A mis-opportune development vis-à-vis meeting challenging delivery targets, with pace and grip."* Or, in the less ridiculous lingo of everyone else, *"Oh-crappity-crap, what the hell do we do now?"*

Sam and I bunkered in a shelter as the sky fissured with lightning. In the night beyond the glass, winds shredded clumps of trees, and thunder detonated like an orchestra, with the musical talent of a wounded cement mixer.

I stared at the carnage and worried about Gary. Mind you, I was often worried about Gary these days. He'd been bellowing less and less. I pushed him out of my mind.

"What d'you think?" I asked Sam, gesturing at the storm with my glass of red wine.

"Three weeks delay. Probably four." She slumped into a rocking chair and slugged her own wine.

She was right. There'd be floods and equipment damage. There'd be an unholy hell of a mess.

"Maybe more if we lose power," Sam added.

At which point, because fate's humour is not just sick, but terminal, the lights went out.

"Screwed, aren't we?" Sam said, and I realised she was right behind me, her ninja smile flashing, as her arm snaked about my waist.

I lifted her hand and kissed it. "Totally," I said.

And so, we screwed to feel better about being screwed. We'd done it a couple of times before, so even though the sky certainly shook, we both knew that was only the storm.

Afterwards, as the torrential rains faded to a mere downpour, we talked through the problem. Get more people with the right training, discretion, and willingness to help? You can't hire people who don't exist. Additional digger-drones? You'd still need those nonexistent people to shepherd them. Extra nukes? Too much radiation which might affect evolution in even the most resilient timelines.

Sam had thrown away the next comment as a joke. "If only we could just clone ourselves!"

And then, we'd stared at each other. "Oh!" we both said together. And then, a little later, we said "Oh" again. Only this time with more enthusiasm and extra exclamation marks.

If long-range time travel was a bastard, hopping back a few hours was a kindly Great Aunt with a kitten and a passion for knitting. The quantum foam separating *now* from *then* would be cobweb thin.

And this, of course, was reflected in the cost. To get to the Cretaceous, we'd needed a vastly expensive Supremely Long Operational Gate. However, if this was the Golden Gate Bridge of Time Travel, a Short/Near Operational Gate, like the one that we'd just had delivered, was a plank of half-rotted wood wobbling over a stream. It had a temporal bore capacity of barely one hundred and fifty years, although this was more than enough for our needs.

Sam and I stood beside it, admiring the twin curved pillars of silvery darkness, like shadows woven with threads of moonlight. Above us, a flock of rainbow-coloured birdysaurs flitted about the tree line, ca-cawing as they crunched up dragonflies with tiny needle teeth.

The Suit was sweating profusely, so despite having air-con built into his robes, damp patches were soaking through in darkening stains. "This had better work!" he hissed. "Or I'll have your daddy's balls for earrings."

"It'll work, sir," I said, biting down a reply about how they'd match the giant scrotum that passed for his face.

The Gate's crackling fuzz snapped into focus with an elevator ping and a blast of air that smelled of ozone and frozen mornings. Through the Gate we could see a familiar Cretaceous landscape, except the sky beyond was a star-swept vastness. On the other side, it was eight hours ago and twenty miles to the south. Beyond, my crew and I were still sleeping in our cabins, in the newly built Base B.

I felt a fluttering apprehension. It's rarely a great idea to be too close to an earlier version of yourself. Apart from the causality risks, the atoms you're made of get pissy. It amplifies quasitron interference or something.

This was why we'd set the Gate to drop us off a small distance away from where we were sleeping. After all, the perimeter of the crater was three hundred and fifty miles, so it was simple enough to control the diggers from slightly farther away.

We trooped through the Gate into the cool morning darkness beyond. As soon as the last woman was through, the Gate crackled shut.

I clapped my hands. "Okay, everyone," I shouted over the murmuring conversations of my forty-strong crew, "Mr. Suit may be the biggest pain in the arse since rectal cancer was poked with a cattle prod, but let's get going anyway."

And moments later, with a roar of ion turbines, the first of the drone-diggers thundered into the air and *thrum-thrummed* deeper into the crater.

The weeks oozed past. Long weeks, because twice each day, my crew and I marched through the Short/Near Operational Gate, risking migraines and mild tachyonic scarring, to work thrice around the clock.

It worked like this: One session of work, one session of leisure, and one session of sleep. But always with one iteration of us at work somewhere around the deepening crater, shepherding the digger-drones.

Naturally, I'd negotiated treble pay for the crew and a much bigger bonus as an incentive, but I couldn't shake the suspicion that they'd have done it anyway. Especially when I overheard Big Benji and Mad Swifty discussing it over a tuna sandwich one evening.

"I mean, I thought about us asking for more, see," Benji had said, brushing a blue dreadlock away from his eyes. "But then I thought humanity might be completely evaporated from the chronology of the universe. So, then I thought my mam'd give me such a walloping if she found out!"

"I expect you'd never hear the last of it neither," Mad Swifty added archly, with a sideways glance.

"Exactly!" Benji said, then looked hurt as Mad Swifty fell about laughing.

Inexorably, the crater grew deeper. Progress was so good that sometimes the Suit forgot to be an asshole. Once, he even tapped me on the shoulder and told me I wasn't doing quite as crap a job as he'd expected.

"Think we might actually pull this off," Sam said, as we stood perspiring, sipping brews, and staring into the abyss. Thankfully, it didn't seem to be trying to stare back. Instead, it seemed perfectly happy belching gouts of incandescence from the churning magma sea in its distant guts.

I sometimes thought how strange it was that even an actual vision of hell could become mundane. Especially when viewed with a brew in one hand and a chocolate biscuit in the other.

"What's the depth now?" Sam asked, slurping the last of her brew.

"Six miles in the basin, eleven in the throats," I answered, referring to the honeycombed pattern of extra-deep pits we'd dug throughout the main crater.

"The nukes?"

"Next Gate Drop."

Sam shivered, and her face was pale as she glanced away. "Close," she said.

"Close," I echoed, feeling the familiar wash of despair.

Soon, they'd drop those nukes down the throats and blast the crater into a hurricane of flame and ashes. This would usher in a decades-long nuclear winter, along with catastrophic environmental collapse.

The result? The mass extinction of seventy-five percent of the planet's plant and animal species.

The nukes had even been layered with iridium, an element common in asteroids. The final piece of false evidence planted to frame "natural causes" as the murderer of the dinosaurs, rather than time-travelling human bastardry.

I wasn't even sure anymore why we were going to so much trouble to hide the reason. The Suit seemed to think it was mainly to protect the past, and to keep causality on track so the right people would be born at the right time. By which he most likely meant himself.

Typically, Billie Tourist had a different view:

We tell the world what we're really doing, and someone, somewhere will have a moan. Then all the money needed to get the job done will be spent explaining why we need to get the job done. And then there won't be any money left to get the job done. And then, we're not even shafted. We're just... not.

So, there it was. We had to dig the crater so we could evolve to be able to dig it in the first place. But it still didn't make me feel any better. It was like pulling a lever in the trolley problem, so a train hit a crowd of other people rather than you and your family. You needed to do it to survive, but that didn't stop it being a crime.

Gary's sudden bellow scattered a flock of birdysaurs from a distant tree line.

Sometimes whole days would pass now without hearing Gary. Stupidly, each time I thought he might have migrated far away, and some miracle could mean he'd even survive. But then I'd hear his lonely call again, and I'd feel the sauropod-kick of shame followed by the velociraptor bite of my own ridiculous hopes.

Sam rested her hand against my arm. "It'll be all right," she told me.

But she knew she was lying and so did I.

It was the Last Day. The last day of our mission and the last day in every conceivable sense.

A grimy sunlight filmed the sky with a haze the colour of picked-clean bones. But still, the scent of the world was rich with the soil's perfume, teeming with every kind of life.

"You okay, Boss?" Big Benji asked as he passed me.

"Yeah." I rubbed an eye. "Just my allergies."

"I didn't know you had any allergies, Boss," Benji said, even as Mad Swifty elbowed him in the ribs.

"Give a little bit of space here, kid!" she said, ushering him away.

I got back to supervising packing everything from Base B that wasn't going to be left for Obliteration. But it was strangely quiet, all of us lost in our thoughts. Last night, we'd tried to have a party, drinking cold beer and barbecuing under the stars. A couple of the crew had sung bad karaoke as smoke trailed upwards from beds of spreading ashes.

The Suit had even given a speech that said "thank you" without patronising us too badly.

Billie Tourist, who'd come through the last Gate Drop to oversee delivery of the nukes, had done a convincing impression of projectile vomiting. It was almost enough to make me laugh.

But no one's heart had really been into partying, and we'd all gone to bed early. Sam and I had sloped off together, though we might as well have sloped off apart, for we lay in silence staring at our cabin's roof and listening to the orchestral animal susurrus of the world beyond.

Tomorrow, we knew, there'd be a thousand thunders, followed by a screaming empty silence.

"Boss," Mad Swifty shouted, "we got a problem with the Short/Now Gate!"

I sighed and went to take a look. I found Mad Swifty and Sam squatted beside the Gate's Quasitron Decompression Accelerator, with half a dozen of the rest of the crew milling around pointing. Violent pink sparks sizzled from the quasitronics, and the space between the Gate's curved pillars was shimmering silver.

I bit down my irritation. I needed this now like I needed a kick in the head from a titanosaurus. "Did someone spill something on it?" I asked.

"No." Sam looked baffled. "It just erupted into life."

"Looks like interference from an attempted G2G analogue transmission loop," Big Benji murmured. Then, seeing our startled faces added: "I read about it once, see? You can send messages via the Gate, if you know how to invert the accelerator and..."

Suddenly the shimmering silver space slipped into focus, revealing a small, forested clearing, inhabited by a single person, whom I recognised at once. Which wasn't actually that difficult, since I'd grown up with exactly the same reflection.

A bout of nausea hit me so suddenly that I staggered forward. *Damn it!* Pissy quasitron syndrome.

"Get yourself to Base A, now." The other me, looked equally unwell, with sweat cascading down his forehead. "And when you're there, come quiet and stay out of sight."

"What? *Why?*"

"You really want to stand here nattering, when it feels like someone's knotted an electric fence through your guts?" the other me snapped. And then the Gate blinked shut.

"Woah, Boss," Mad Swifty said. "You're mean when you're cross!"

I stood stunned, even as Sam rested a hand on my shoulder. "You have to go," she told me. "You know you do."

I cursed. It wasn't like I had time to go jaunting. We had barely three more hours before our final departure.

"You can't break the causality," she said. "Well actually, you can but..."

Sam didn't need to finish the sentence, because no one needs a Paradox Aneurysm, especially since sixty-six million years is a hell of a long time to wait for an ambulance.

Cursing again, I grabbed a scooter, kicked the rotors into life and purred off into the jungle, wondering what could be so important that I'd put myself through this.

I soon had my answer.

As I approached Base A, there was the faint sound of a triceratops bellowing. My first thought was Gary. And then I realised the call was different somehow, it contained an edge of *trilling*. The call was... *female*...

A coldness hit as I guessed.

I leapt off the helter-scooter and skirted the shadows of Base A's warehouses. Shaded beneath a towering red flower, the Suit was hunched over a speaker system. The speaker system shuddered as it looped through the recorded roaring trill.

Beside him stood Billie Tourist, wearing a face of such radiant fury, that I half-expected the Suit's head to explode in its incandescent glare.

And, cradled in the Suit's arms, chrome barrel glittering, was a plasma-rifle.

Billie and the Suit were lost in an argument, which made it easier to slip closer without being seen.

"In four hours, the damn animal's going to be a scattering of atoms." The Suit ran fingers through sweat-greased hair. "Who cares if I take a souvenir?"

"Souvenir, my warty butt!" Billie spat. "How much d'you expect to coin for it?"

When it comes to guessing what utter bastards people can be, I'm rarely disappointed. It's a gift I have. Though as gifts go, etc., etc.

"None of your damn business," the Suit was saying. "Now get outta my way before I have to remind you who's in charge here."

I thought about the satisfaction of strolling from the shadows and smacking the Suit in the mouth. But I wrestled with myself and won. Though, as I slunk back to Base B, I felt like I'd lost, too.

And suddenly, the idea was in my head, wholly formed, as if it had been simmering away in my subconscious already.

I radioed Sam. "I need you to see if Big Benji can do me a favour," I told her.

Some hours later, I eventually made it back to Base A again, my skin lightly streaked with tachyon burns from urgently transporting the Short/Near Operational Gate. I was so exhausted that I just wanted to curl up and sleep. It hadn't helped that the shell of Base B had been filled with an unsettling stillness, its emptied cabins silenced, where I'd been used to a richness of noise.

For some reason, I'd felt compelled to lock its gates before we left. I expected the crew to tease me about it as we helter-scooted through the undergrowth, but nobody did. In fact, nobody teased anybody about anything, and even Mad Swifty seemed lost in her own thoughts. Instead, as we skimmed for the last time through that Cretaceous wilderness, with sunlight slanting through the towering ferns, we did so in unjoking silence.

Sam found me a little bit later, her expression solemn.

"Gary's taken the bait," I said, and only after I'd finished speaking did I realise it wasn't a question.

And, from the jungle, the roar of the last lonely triceratops boomed.

"The Suit?"

Sam took my arm and led me to Base A's main square. The Suit was on a souped-up helter-scooter, hovering high above the camp. He was stroking the plasma-rifle as the pulses of recorded triceratops trilling boomed from the speaker below.

"You sure about this?" Sam asked.

"No," I said. But I was lying, and we both knew it.

I clambered on my own helter-scooter and purred towards the Suit.

"Hey, asshole!" I shouted. "Mind if I borrow this?" And I grabbed the speaker of triceratops noises and accelerated back towards the jungle.

The Suit's outraged shriek crackled over my radio: "*You idiot!* What d'you think you're saving the damn dinosaur from?"

"Mostly your ego, sir," I muttered.

If the Suit felt the barb, he didn't react. "It'll all be the same in another sixty-six million years anyway!" he screamed. "Or hadn't you noticed, that's the sodding point of us being here?"

He was right, I knew, as I thought of the various other extinctions humanity had absent-mindedly inflicted on the planet. In another sixty-six million years, it would indeed all be the same.

Some things never change, even though they should.

But there was something else as well that the Suits in Time Management never, ever seemed to really get: just because nothing really matters in the end, that doesn't mean nothing matters now.

But the thundering ground shook my thoughts away. *Gary was closer*. I swerved among the trees, skimming through swirling clouds of biting bugs...

And then I broke into the clearing, where the Short/Near Operational Gate loomed.

It spat sparks of crimson-tinged light as its cloying icy stink filled the clearing. Trails of smoke drifted from its machinery, and the Gate was juddering, having bored through the quantum strata to its maximum range. I hurled the speaker through it, even as Gary thundered into the clearing.

Once again, I was struck by his size; the vast muscles churning beneath those downy, iron scales. He paused, then bellowed with such ferocity that a cloud of birdysaurs squawked into the air.

Then he plunged on, one hundred and forty-seven years into the past.

"He's through!" I radioed Sam.

"On it."

I caught one last glimpse of Gary pounding into the lusher world beyond, then with a final crackle of quasitronic darkness, the Gate guillotined shut.

"Good luck, buddy," I murmured. There was just one last thing to do. I gave my stomach an anticipatory rub at what I was about to put it through.

It was time to give myself a call.

When I got back to Base A, the Suit was nowhere to be seen. Since I'd been expecting the bollocking of my life, this was worrying. But Billie Tourist was there, wearing a grin almost wider than her face.

"Suit's gone ahead to Home Time. He went after you but managed to fall off his scooter."

"Hope he had a hard landing," I muttered.

"The opposite actually," Billie said. And for the first time, I realised laughter seemed to be spilling out of her. "He fell right into one of the leftover piles of your pal Gary's turds!"

For a moment, I stood stunned, then I started to laugh too. We'd always known the Suit was full of crap, so just for once it seemed right to have a crap full of Suit.

A friend of mine once said there're no such things as happy endings, because all endings are a kind of death, and how can any death, even a kind one, be happy?

Those words were in my mind as we trooped, exhausted and despondent, through the main Gate, and back into our Home Time.

The sterile clatter of the vast Departure Lounge thrummed, lit by purple fizz. It was crowded with jumbles of equipment, including a towering stack of blue tesseract-crated drone-diggers. It was also full of scurrying technicians, while illuminated readouts hovered in the air. The largest showed a countdown.

"Asshole Suit's going to get you sacked, you know?" Billie appeared at my shoulder, crunching through a bag of chili-chocolate pretzels. "Price tag for the skull of the last ever triceratops? Cost him gazillions."

I shrugged. I'd had my fill of time travel anyway. Besides, there were two meanings to career, and mine had always had more of the zigzag than the ladder. So it doesn't go.

"Staying to watch?" Billie's tone was light, yet her dark eyes bored into me more fiercely than the microwave lasers that had vapourised the Chicxulub crater.

I wanted to say "no." I wanted a proper shower. I wanted to sleep for a week. Besides, I didn't want to. Oh, hell. I really didn't want to, *so, so much*.

But my head was nodding. Somehow, I owed this, though I didn't know whom or what I owed it to.

Shortly after, I found myself on a mezzanine with Billie and Sam. A panoramic sweep of screens curved around the silenced lounge, showing the expiring light of the last day of the Cretaceous. The speakers were on so we could witness the whoops and cries of its doomed jungle life.

I gripped Sam's hand as the countdown closed in on zero and felt the despair in her answering squeeze.

Then every screen blazed incandescent as we watched the ending of a world.

Collaboration

BY LARRY NIVEN

Larry Niven began working toward a master's degree in mathematics at the California Institute of Technology—until discovering a bookstore jammed with used science fiction magazines convinced him to drop out and write SF instead.

He broke into professional science fiction in 1964 and has been going strong ever since. Niven is best known for his Known Space future history, a still-growing series of more than thirty novels and stories, including Ringworld, *which won the Hugo, Nebula, and Locus Awards. He also coauthored several celebrated national bestsellers with fellow Writers of the Future judge Dr. Jerry Pournelle, including* The Mote in God's Eye, Lucifer's Hammer, *and* Footfall.

Beyond fiction, Niven's interests include saving civilization—particularly defending Earth from meteoroid impacts—and advancing human expansion into space by making space endeavors commercially viable. He has played an active role in space policy discussions, helping convene the Citizens Advisory Council for a National Space Policy during the Reagan administration.

Niven received the L. Ron Hubbard Lifetime Achievement Award in 2006, was named a Grand Master by the Science Fiction & Fantasy Writers of America in 2015, and has served as a Writers of the Future judge since 1985.

Collaboration

In the years since I last wrote about collaborating, I've learned a little more.

I'll suggest rules here—but I've broken each of them...

1. **You don't have to collaborate.** Don't worry about it. Many of our best writers have followed visions that were too individual to share. I include Robert Heinlein, who never collaborated, and Piers Anthony, who did. Jack Williamson wrote with everybody.

2. Collaborating generates feuds, friendships, love matches, and marriages. No guarantees. **But it's always less lonely than what you've been doing: sitting alone at what was once a typewriter.**

3. **Before you suggest a collaboration, you should already have a body of published work.** It helps if you have established your own voice and your own reputation before trying to work with someone else.

4. I broke that rule. I won't name the writer I tried to work with. She claimed to have published a lot, all academic work. She's legitimately a bear for work and for research. We wrote two stories and turned one into a script before she revealed her real motive for working with me. We've sold nothing. So— **Learn when to get out.**

5. **Be open to the idea.** I don't know how to seek out a collaborator. They've all come to me, except the first:

My wife Marilyn and I had invited David Gerrold to dinner. We got to talking...ran to a dictionary to look up "mauve"...generated a novelette, then expanded it into a novel. Later: David became story honcho for the *Land of the Lost* television show (1974), and invited all his friends in. I wrote one episode and two halves. **Socialize! There are side effects.**

6. **You'll learn from one another.** Jerry Pournelle suggested writing together after the first Bouchercon; we knew each other from LASFS (Los Angeles Science Fantasy Society). Steven Barnes tracked me down: already his habit was to find an expert to give him some training. Brenda Cooper had tried writing; I looked at some stories, found logical flaws and phrasings to correct, and some ideas; wound up writing with her, teaching—and learning. Matthew Harrington wrote several stories for my The Man-Kzin Wars anthologies, then suggested collaborating: "I need the money." That's not a tempting approach! But he *was* a genius, and—

7. **The easiest answer to "shall we collaborate?" is to give the writer something you've stalled on.** I tried Steven Barnes out with a stalled "The Locusts": the premise was too pessimistic for me. Steven warped it into an optimistic ending. We've done nearly a dozen novels since. I sent Harrington a beginning for *The Goliath Stone,* and he ran with it, blitzed it! Don't forget this option. (Of course you need some stalled stories first.) So: maybe not yet for you.

8. Steven Barnes stuck with me during several years of writer's block after Jerry's death and while my wife Marilyn was growing gradually sicker. I wasn't getting new ideas. After Jerry's death, we finished *Starborn and Godsons,* our third triple collaboration, using Jerry's notes for the ending. Then we took some of my old ideas, and one of Steven's suggestions brought to me 40–50 years ago while we were still fumbling, and we built around them. My writing skill oozed back...and *Ice Vegas* will be out in August 2026. **Don't lose old ideas, or old friends.**

9. Terry Pratchett was in Los Angeles autographing. He was stalled six hours by a delayed flight home. I took him back to my house to kill time. We talked collaboration (of course!), and I rapidly realized: my beanstalk wouldn't fit the Discworld, and Terry wrote too fast for me. I'd never catch up. But we had a wonderful time! I've done this more than once: talked collaboration with a friend through an afternoon.

Remember: Talking about writing is recreation. You're not working until you're producing text.

10. One author found my email address (many such offers come by email) and asked about writing some novels set in Known Space. **This is intrusive; you shouldn't. Or so I felt.**

I had already rejected opening Known Space, as offered by Jim Baen of Baen Books. However, I...changed my mind and took the writer up on it. What the hell. I did forbid him to use the Ringworld. I used our five Fleet of Worlds novels to correct some flaws in the Known Space universe, particularly for my character Beowulf Shaeffer—

11. Math students had determined that Beowulf would die in the original story. And the Ringworld is unstable. And integral trees would be too small, and spinning. **Read your mail. Readers' critiques are your key to new stories.**

12. **If you can stand seeing your work extended by another writer... you may be entertained.** Give TV a try? There's no way to avoid collaborators in TV; you don't even get to choose them. But there's money, and a hugely expanded audience.

13. Gregory Benford and I were friends for over half a century before thinking of writing together. A plasma physicist and honored writer, he has credentials up to his nostrils. The Ringworld and other megastructures fascinated him. I introduced him to a fifty-year-old fannish cartoon: "Cupworld," by a fan named Hollander, a LASFS member, published half a century ago at the back of the fanzine called APA-L. I'd played with Dyson spheres; Greg had too. We wrote *The Bowl of Heaven*...and I got exhausted in the middle. Cobbled a "shrug" ending. A cliffhanger! I hate when a writer does that to his readers.

Greg and I resumed, wrote *Shipstar* to make a proper ending. Then, *Glorious*, the half-expected sequel, to wrap up loose ends. **Remember: you're doing this for fun!**

14. **Editing can lead to almost-collaboration.** Scores of writers older than me got hundreds of ideas from John W. Campbell. Those ideas made them famous. I have his twelve-page rejection letter detailing flaws in what became "Death by Ecstasy."

And once, Marilyn and I were driving to a Westercon with James Patrick Baen in the back seat. Jim and I had already published two books of other writers' takes on my Magic Goes Away universe, with moderate success. I knew it was fun. Now he said, "Let's open Known Space to other writers!"

I said, "No way. Known Space is mine, too individual for other minds."

15. A little later, I said, "We could open up The Man-Kzin War period. I don't do war stories; I've never been in the military." **Maybe other writers could re-inspire me.**

The Man-Kzin Wars now numbers fifteen volumes. Some of the stories are mine. Mostly the other writers do fine without me. There were a few—

In "Galley Slave" by Jean Lamb, a soldier captured by a Kzinti warship found that her captors couldn't adjust to the reality of a sapient female. They made her run the kitchen. Poison? A Kzin catches on, jumps her, and she kills him. I couldn't buy that—unless Jean gave her heroine a better weapon, a war blade hidden in the cutlery. A minor change was all it needed, and she gave it that.

16. **And your collaborators will learn from you.** Hal Colebatch's first story had a wonderful middle: a fur rug from Rudyard Kipling's time is recognized in the early ARM (Amalgamated Regional Militia) era as a Kzinti pelt. The middle, set in the 1800s, was wonderful! However, I made him rewrite the opening and closing, over and over. His second story shows, perfectly, a basic mistake. He kept the original idea after the story went elsewhere. I worked up the nerve to ask him to chop off the back half, and I outlined an alternative. Instead of telling me off, he did it my way.

Several later stories he wrote perfectly.

To me, that rewrite led to a world settled by humans and Kzinti telepaths, and I wrote that into "Fly By Night" and added a Jotok, a wonderful alien I borrowed from Donald Kingsbury's Man-Kzin War story.

17. **Editing is as much work as you want it to be.**

Your editor is your most important collaborator, and someone else is paying him! And—pros tell each other: When your editor tells you something needs fixing, believe him! When he tells you *how* to fix it, ignore him.

Skinny-Shins

written by
Orson Scott Card

illustrated by
CIRUELO CABRAL

ABOUT THE AUTHOR

Orson Scott Card is the author of the novels Ender's Game, Ender's Shadow, *and* Speaker for the Dead, *which are widely read by adults and younger readers. His most recent series, the young adult Pathfinder series (*Pathfinder, Ruins, Visitors*), the fantasy Mithermages series (*Lost Gate, Gate Thief, Gatefather*), and the Side Step series (*Wakers, Reawakening*) are taking readers in new directions.*

*Besides these and other science fiction novels, Card writes contemporary fantasy (*Magic Street, Enchantment, Lost Boys*), biblical novels (*Stone Tables, Sarah*), the American frontier fantasy series The Tales of Alvin Maker (beginning with* Seventh Son*), poetry (*An Open Book*), and many plays and scripts, including his "freshened" Shakespeare scripts for* Romeo and Juliet, The Taming of the Shrew, *and* The Merchant of Venice.

Card was born in Washington and grew up in California, Arizona, and Utah. He served a mission for the LDS Church in Brazil in the early 1970s. Card currently lives in Greensboro, North Carolina, with his wife, Kristine Allen Card.

"When Writers of the Future *offered me a chance to write a story to support (or be supported by) a preexisting illustration, I jumped at the chance—mostly because I had never written a story that way.*

"With song lyrics, I usually welcome a preexisting melody, because otherwise I fall into iambic or anapestic tetrameter quatrains. With 'Skinny-Shins,' however, the need to explain/justify the features of the illustration required a far greater degree of attention to the artist's work than illustrators have ever paid to my fiction when the shoe was on the other foot!

"In other words, I wanted to make sure that close examination of the illustration, after reading my story, would leave nothing unexplained. And that, dear readers, is hard.

"I can't even explain all the details in my own writing. And I'm the one who made all that stuff up.

"However, for my entire career I have found that my stories grow best when they spring from at least two completely unrelated ideas. The effort to invent reconciliations is what provokes the story to grow and become thick and rich (not unlike a good ketchup). I won't belabor the point by telling you the two ideas that spawned this or that book of mine.

"Suffice it to say that the illustration provided me with the revelatory scene that ends the story, but all the material from the Antarctic Scan was from a completely different idea I've been toying with for decades, ever since I became fascinated by the intelligence of octopods.

"It gave me my alien invaders who have long meddled in human genetics, and why shouldn't they also have meddled with dinosaurs?

"It also allowed me to completely ignore so-called UFOs, which myriad persons have declared to be identified flying objects. If millions of people already think those aerial phenomena are adequately explained, to me it no longer belongs in the realm of science fiction. The TBS series People of Earth *was a brilliant attempt to realize the UFOs of contemporary mythology, but I wouldn't have the heart to attempt anything similar myself. (Though I do find myself muttering the mantra 'don't get weird' while picturing the head of a stag saying it.)*

"The only thing that wasn't all that challenging was to create the story as if it had been written by a mostly untalented nonwriter. It was disheartening to me how easily I brought that off. And when sharp-witted editors pointed out expository holes in the story, my true (and lame) reply was, 'That's what I was trying for.'"

"Still, I hope that you enjoy (or enjoyed) the story as much as I enjoyed writing it. Since I'm fairly close to becoming a Writer of the Past, I'm glad to have a chance to take part in Writers of the Future. *Who knows? Maybe I do have a future in the sci-fi business."*

ABOUT THE ILLUSTRATOR

Best known for his paintings of dragons, fantasy artist Ciruelo Cabral has collaborated with some of the most prestigious publishing houses in Europe and the United States for the past forty years. He has created book covers for George Lucas, album covers for Steve Vai, Spinetta and The Flower Kings, a comic story with Alejandro Jodorowsky, The Official Eragon Coloring Book *for Christopher Paolini, and cards for Magic: The Gathering. As both author and illustrator, his work has been published in nine personal books, including the long-selling* The Book of the Dragon. *Ciruelo also pioneered a painting technique on rocks, which he calls Petropictos, creating pieces that have the appearance of sculptures. Exhibitions of his art are held in different countries, which require him to travel frequently each year. He lives with his family near Barcelona, Spain. Ciruelo has been a judge for the Illustrators of the Future Contest since 2016.*

Ciruelo created the cover art for this anthology. About it, he said, "It is a true pleasure for me to create a painting for the cover of the Writers of the Future *anthology. I greatly admire the dedication and talent behind the making of such a book, and I am especially pleased to support the emerging authors and illustrators featured within.*

"In my cover painting, titled The Fire Tribe, *I portray two warriors and a mighty dragon defending an island dominated by an erupting volcano. They appear to belong to a people who have lived in the presence of volcanic fire for generations, developing a deep understanding of its dangers and power. Fire is central to their world—an ever-present force that must be respected, harnessed, and endured."*

Skinny-Shins

They've been here all along. That's what Rho and I found out when we were working together on the ADS—Antarctica Deep Scan. Rho was the scan; I was the pilot. The programmer and the pilot. A team. A superb team. So superb that we realized the flaw in the program, the assignment, the technology. We were, in essence, spending a lot of time confirming that there was nothing under the ice of Antarctica except dead mountains and, between the islands, water. We were never going to see anything under the water, and that's where everything was.

"Why are you reading that?" she asked me one morning—a morning when I was still awake from the night before.

"It helps me sleep," I said.

"Which means it keeps you awake. You're going to be worthless today."

"You're the one who scans. I'm a mere facilitator. Your assistant for transportation."

"What's going on?" Rho has very little patience with bullshit.

I held up the book so she could really see the cover.

"Oh, please, Sigmund," she said. "Why would you read that? You're a scientist."

"A pilot. Your chauffeur. *You're* the scientist."

"Where did this inferiority complex come from?"

"Because you're going to feel free to—you know, feel *obliged* to ridicule everything I'm about to say, so it will accomplish nothing except perhaps getting me taken off the project."

"If anybody should know that there was *no* ancient high civilization on Antarctica, it's you."

"That's why this guy is so close but so tragically far," I said. I like being melodramatic. Rho loves that about me.

"I hate how you talk like everything is some big portentous ominous apocalyptic deal," said Rho, which was her way of telling me how much she respects and leans on me.

"This guy—a scientist, by the way, a zoologist—"

"Not anymore. Not if he wrote that book."

"You're judging a book by its cover," I said.

"I'm judging the book by its title. 'Paleo Civilizations Under the Antarctic Ice.' Please."

"The one thing wrong with this book is he didn't believe his own title."

"The worst thing about that book is if he *does* believe his own title. Do you want breakfast before you go to work with me and fall asleep on your desk?"

"You drive. Then I can fall asleep on the ride in."

"We work right here. There's no commute," said Rho.

"In my heart, we leave home and then go flying over the ice."

Rho thought for a moment. "So, you're taking this guy seriously."

"He marshals his evidence. He doesn't claim that any *non*-evidence is significant. He seems to try to disprove his own hypotheses, like a good scientist."

"You said he was a zoologist. So, he knows how to make his bullshit look like science."

I got up and washed my face, ran my wet hands through my hair, and took my caffeine pills because the doctor said that it's all the other stuff in coffee that was going to kill me.

"Do you think washing your face and wetting your hair is a sufficient shower substitute?"

"Absolutely not," I said. "But to people who don't live in the same flat with me, it'll look like my hair dried naturally after my shower."

"You barely have any—"

"That's why I keep my hair so short. So that it air-dries without any effort on my part. Are you going to listen to my great insight of this morning?"

"Am I going to have a choice?"

"Are you going to *listen*, and at least entertain the possibility that I might be onto something?"

Rho sat down on a chair beside the table where we kept our two laptops. She folded her hands and rested them on the table. Then she looked unblinkingly in my eyes.

She was at least going to pretend to listen, and that's the best I was going to get. But without her help, I couldn't carry out my little test that morning. Or, probably, ever.

"He has a lot of evidence that Earth's biota started getting messed with about three hundred million years ago."

"Messed with?"

"Zoology stuff, but I don't think he's faking. Genetic codes that don't make sense. Sudden jumps that don't look like punctuated evolution; they look like—"

"Intelligent design?" she said scornfully.

"He's not a creationist."

"What *is* he, then?"

"A scientist trying to figure out biological history."

"Cut to the chase, Sigmund."

"He believes that the evidence, scanty as it is, tracks back to pre-Pangaean time, when the supercontinent drifted across the South Pole and got shrouded in ice. One of the snowball Earth phases, he speculated. Lowered sea levels wiped out a few coastal species; but it wasn't bad enough to be an extinction event. He tried to track through the fossil record on Pangaean tectonic plates that were near enough to Antarctica to give us an idea of what the flora and fauna were."

"From the fossils on the adjacent lands," said Rho.

"And he couldn't find anything except a possible reconstruction of the biome of that time."

"All guesswork."

"'All science is guesswork that we haven't disproven yet,'" I recited. It's a mantra I use to win arguments because it's, like, true. I went on. "He figured that there could have been a major civilization on Earth between the Ordovician and Silurian."

"The Late Ordovician Mass Extinction?" she asked.

"Definitely not. He ruled that out because Earth had nothing to sustain a major civilization at that time."

"He could have ruled it out *also* because there is no *trace* of a civilization in that era before there were flowering plants."

"See, that's his point. Not the flowers, not the seeds—the traces. We keep thinking that civilizations have to build huge stone monuments. They have to have agriculture that generates surpluses. They need to show housing, tools, *artifacts.*"

"They don't *need* to do those things," said Rho, "but they *did* them and we study them."

"It's like the Australians," I said. "Thirty thousand years before the Mesopotamians or Egyptians supposedly invented agriculture, the Australians had completely mastered the water supply in order to nurture all the necessary flora and fauna in that dry, dry land. They weren't wrestling with nature, they were tending crops where they naturally grew. It was all water management, which is the main reason civilizations grow. To manage water."

"They were not a civilization."

"They invented agriculture," I insisted. "Just because they didn't also invent writing doesn't mean they weren't a civilization. One so in harmony with the land that it endured for millennia. Unchanged. Supporting a population that was far larger than anything along the Nile."

"Based on what artifacts?" asked Rho.

"They didn't *need* metal tools. They made baskets and—"

"No spears, no bows and arrows—"

"You don't know that—"

"Will you get to your point?" asked Rho.

"My point is that civilizations don't always have to have the kind of artifacts that the post–Ice Age civilizations used," I said.

"In other words, this fine zoologist has accounted for the *complete* lack of evidence by explaining that the absence of evidence is not definitive."

"You stated it very well, Rho."

"Why does this matter?" she asked.

"Because I want to refocus the Scan for the next few days. Weeks, maybe."

"Refocus?"

"This guy recognizes that the proofs of this paleo civilization can only be under the ice, and he expects that scans like ours will produce the evidence."

"Evidence of an alien civilization back when grass hadn't been invented yet," said Rho.

"But we haven't found it the way we've been scanning, have we?"

"That author is going to be so disappointed," said Rho. "Or he'll come up with reasons why the civilization actually existed, but left no litter behind."

"Maybe," I said. "I don't know him, I can't vouch for him. But I can vouch for *me.*"

"So what's your brilliant idea?" asked Rho.

"My barely possible, barely adequate idea, but worth a week of our time, maybe."

"And your idea is?" Rho asked.

"Under the ice. He kept saying that the civilization's detritus will be under the ice."

"Well, if it was *over* the ice we would have found it from the air."

"For some ridiculous reason, he assumed that if it was under the ice, that meant it would be *on land.*"

Rho stopped looking at me. She fiddled with her laptop, opening it, waking up the mouse, but really just thinking. I knew she had to keep her hands busy, when she was thinking.

"OK," said Rho. "You've got my attention. You're saying that this semi-possible paleo civilization was aquatic."

"Deep in the water under the ice."

"So the civilization would exist only where the Antarctic land masses were *not.* In the water between them and around them."

Rho thought again, fiddling again, calling up a couple of nuisance games on the computer but playing none of them, just opening them and shutting them down. "If they went to the trouble to cross light-years of space, why would they choose to live underwater?" asked Rho.

"Well," said I, "I thought of that and for a while I imagined that instead of maintaining an atmosphere inside their starship, they maintained an aquasphere."

"Whole different engineering problem," said Rho. "And if they could build spaceships—"

"But there were never any spaceships," I said.

"Well, it couldn't have been an Earth-born civilization because we didn't have any animals that achieved sentience."

"That's just wrong and you know it."

"Do I?"

"Smartest invertebrate," I said. "Quick, off the top of your head."

"Cockroaches are tough, but they're not exactly smart," said Rho.

"Under the water," I reminded her.

She thought again. "Octopuses," she said.

"Non-chordates," I said, "but with dexterity and strength to build things or move things around to their benefit."

She laughed. "Octopuses working as crab-boys, herding the crabs into the corral and feeding them till they're big enough to be worth eating."

"It's funny, sure," I said. "But octopuses are by far the smartest invertebrates."

"Or dekapuses—the oldest octopus fossils had ten arms."

"Why would you even know that?" I asked.

"So I could win on Jeopardy," Rho answered.

Still deflecting by joking, but I knew right then that I had her. A home-grown "alien" civilization, three hundred million years ago. Maybe three hundred and *thirty* million years ago. But soft-bodied boneless creatures like octopoda weren't going to leave any fossils. They could survive out of the water for half an hour at a time, breathing through their skin, and zip along from tidepool to tidepool to eat everything edible in each one while their occupants were stranded.

Smart creatures. Can't be domesticated. Escape through slots too narrow for a credit card. Oh, Rho and I used our laptops for a couple of hours, finding out everything we could. Until she said, "I'll bite. There might have been a species of octopus that achieved sentience, made tools and habitations. But suppose we refocus the Scan to look under the water. What will we see that we can recognize as signs of civilized octopuses?"

And I gave her the only answer I had. "We'll know it when we see it."

"That's your *plan*?" Rho asked.

"This is science. First you gather all the data you can. Then you analyze it, categorize it, recategorize it, and only *then* do you start making hypotheses and trying to test them."

"But how will we know *data* when we see it?"

I shrugged. "Maybe we won't. That's why we only spend a week, tops. I'll design the paths we fly so that we only scan down into water, not on land. And you record it so maybe we can study it and find something if we don't spot it right away."

Sounds like a great plan, doesn't it? But we gave it a week and a half and scanned the whole ocean floor underwater within the boundaries of Antarctica. And if there were any signs or artifacts or anything, we never saw them. Failure? Well, no. If you eliminate a possibility, that's good science.

We both put copies of the entire Scan, eleven days' worth, on our laptops and in the cloud, not even bothering to password-protect it because *nobody* had the slightest interest in what we were doing in pursuit of a ridiculous theory. Then every few weeks, I'd pull it up on a high-res screen and try to figure out some pattern in the crap we scanned. Wouldn't civilized octopoda leave us *some* kind of breadcrumb trail?

Unless they didn't want to be found.

And when did that old civilization fall? How did it fail? The Australians lasted till the Europeans came and destroyed their whole system by turning sheep loose on the grasslands and completely wrecking their water distribution system. Their system *worked* and it had to be destroyed by outsiders—it wasn't going to be destroyed by stupid choices of the citizens.

Found nothing. A few years of this, pretending we were only pursuing a hobby. We kept on finding nothing. I was about to delete it from my laptop because it took up a *lot* of room—I still had a complete copy in the cloud—when the world changed.

When *they* came.

Everybody remembers it. Various manifestations in the nighttime sky. Lots of lights. But absolutely nothing on radar. Nobody sees their ship land. Nobody can find it by aerial or satellite reconnaissance. They show up as pedestrians in Plano, Texas. The first three, they looked just like regular people. One Hispanic, one African American, one pale as the moon. Albino, people speculated, but his eyes weren't pink. When we asked him later, he said, "They just told me white."

They just told me white. Which suggested to a lot of people, like me for one, that what we were seeing was not the real appearance of these alien tourists. And it suggested to almost as many that they weren't tourists, either. Yes, they said that Earth was famous throughout the galaxy for its beautiful flowers. Flowering plants had not developed on any other worlds. They all relied on wind pollination instead of insects, which do not pervade the exoplanets the way they do Earth. Our insects look for nectar; on other worlds, they're all looking for blood. "Aren't you riddled by insect-borne diseases," asked some curious scientist or journalist or cop.

"No," said White Guy. "We fixed ourselves so that we're impervious to invasive microbes. Every cell has mechanisms to destroy any virus or bacterium or fungus that initiates a dangerous sequence of actions. Like trying to enter the cell. Or introducing toxins between the cells. Disease germs exploit any vector, but we've been fighting that war so long that we have no viable vectors left," he said.

They would not submit to examination, and to return the favor, they did not investigate our bodies. When we reminded them of all the alien abductions and probes aboard spaceships, they said they had never heard of such things. "Why would we do that?" they asked. And nobody could think of an answer. Because they already knew about us. "We watched you grow up," said the Hispanic woman. "You have no anatomical or biological secrets from us."

So, did that mean that they were able to scan us from the outside, with no physical penetration of instruments? Everybody speculated about everything. Now and then they'd suggest that a particular line of inquiry would lead nowhere—that's why we assigned a lot of good scientists to try to figure out how they could see inside us with no detectable ray or sensor or tool of any kind.

Oh, and they refused to shake hands. They *had* hands, they *used* their hands. But they wouldn't shake ours. Some taboo in their culture? A religious restriction? Maybe they couldn't shake hands without inadvertently getting impregnated by any cell in our bodies? That was my favorite for a while.

But we *never* had the chutzpah to ask them about *that.*

We were too busy accepting their help curing diseases. Their methods of keeping climate changes under control. Their survey of the entire ocean around the planet. Their deadening of all radioactivity within every nuclear weapon, owned by any country. Their putting volcanoes on a limited schedule with a boundary set on their ferocity. We didn't know how they did any of that. They didn't want us to know. They basically said, "We already did everything that *can* be done. If it needed to be done again, we'd have told you how to do it. But you wouldn't have actually understood, you'd just be repeating our actions blindly. We did what we set out to do, and it will never have to be done again. You're welcome."

Yes, they used sarcasm. They really did understand our language. But also any other language they were exposed to. These weren't your garden-variety sci-fi aliens. No giant heads, no spindly bodies, just ordinary-looking humans whose X-rays, surreptitiously taken, revealed absolutely nothing. Not "nothing useful." I mean *nothing.* No shadow on the film. No digital impression. For all we knew, they could be bags of empty flesh that wouldn't shake hands.

Rho and I read the written reports, we talked to government contacts to see if they could tell us more, and we speculated like crazy. Or *I* did, anyway, and Rho tried to find substantive reasons to shoot down every idea I had. The usual. She always won, but I also kept thinking about some of the shot-down ideas in case they weren't completely dead. A lot of the ideas, to tell the truth.

In that whole time, the first few years the aliens were with us, while we still hadn't settled on one non-insulting, non-demeaning, but also non-worshipful name for their species, Rho and I spent a lot of time using high-penetration satellites to continue the Antarctica Scan, though with no intention of finding artifacts left over from a three-hundred-million-year-old civilization. Well, Rho had no such intention.

Anyway, it was in the colder part of autumn, which isn't a very clear distinction in Antarctica, but we were no longer *in* Antarctica, because our satellites could be operated from anywhere that had electricity and a clear shot at the sky. So, we chose a place with zero mountains, trees that weren't very tall, and now that the water was receding, dry land to live on and work on, and good grocery stores and restaurants: the Outer Banks of North Carolina. All of the drowned houses from the global warming phase were demolished and removed, and new, sturdier, much better-looking ones were springing up in their place.

Rho and I were ensconced in one of the older beach houses that had survived by having a great foundation. It had been a luxury place with room for all your cousins to visit, back in its vacation-rental days. Therefore, the Geological Survey had us sharing a not overly large bedroom with all of our equipment. And every other bedroom was occupied by similar teams doing similar jobs that could be remotely operated. From that house, stuff was being controlled all over the world. Our assignment was Antarctica.

There was no security anymore. What with the aliens kind of intimidating everybody—well, candidly, *daring* any nation to start a war—and you know all the other beneficial things, this was a Pax Extraterrestrialis, or Paxet, and people wanted better jobs than standing around waiting for danger that never came and was never going to come. Why waste budget on featherbedding some security-guard union?

The colder part of autumn, like early November. Before Thanksgiving. You'd think I'd have the date memorized but you'd be wrong because I have a lifelong block against remembering dates or even what day of the week it is today. So, the *season* is all I can remember, and *that* sticks in my mind because I had the privilege of standing outside in that cold, with wind coming off the Sound, waiting for our visitors to return our house to us.

"We have permission from your…" the alien looked at his paperwork, "Secretary of the Interior."

The shorter one looked over his shoulder. "The Interior of what?" she asked, a hint of derision in her voice. I knew immediately that she *wanted* me or maybe all of us to overhear. Kind of a snotty move, but they all had their own personalities and as long as nobody got violent, you had to kind of put up with whatever they said. Usually, they were nice.

Rho started to explain the idea of national borders and the United States, but the aliens knew all this stuff. The woman (or whatever she was) pretended not to know because it gave her an excuse to be rude without causing an interplanetary ruckus.

"America, I know," she said.

The other alien looked at her very calmly, and she shut her mouth and went outside. So it was just us and the taller guy with short hair. "We're not top secret or anything," said Rho.

"And neither are we," said the tall guy. "You can tell anyone at all that we came to call on you."

"Just us?" asked Rho.

To me, the answer was obvious—none of the other teams were being rousted while the aliens did some sort of inspection.

"If you would kindly wait outside while we examine your papers and computers and communication equipment."

"It's cold out there," said I. "We're going to need our jackets."

"I promise that you won't die of exposure before we finish up and leave."

"Then why can't we just sit in the kitchen and maybe heat up some sausages for lunch?"

"Do you know the names of the animals who contributed their flesh and fat and, probably, random glands and organs to be ground up for your sausage?"

I had long since gotten used to the fact that they always condemned meat-eating but never punished us for going ahead and eating it. So my answer wasn't respectful. "I only know that one of the cattle was a bull named Frank, and his penis and testicles are in the sausage."

The alien just looked at me. I didn't apologize. I didn't care if he got the joke or not.

"Some of my best friends are bulls," I explained. "We tease."

"Please go outside. Now. Now."

The repetition implied that he actually meant *now.* So Rho caught me by the arm and pretty much dragged me outside. There was a breeze, not gale-force winds, and it was steady, not gusty. It was also chilly. Don't ask me a temperature number. I was born when America used Fahrenheit, and the Celsius numbers still sound ludicrous to me.

I immediately wrapped my arms around Rho to stay warm and she shoved me and stepped back. "That's not one of the benefits," she said. "I was smart enough to wear a sweater today."

"I was smart enough to remember that I have an indoor job," I answered.

Within five minutes, Rho came and put her arms around *me.* But she made sure to turn me so that my back was to the breeze so I was more a windbreak than a biological heating unit.

The alien came out of the house.

"Did you leave any sausages for *us*?" I asked him.

"No," he said. "Sausages should not be your food."

"Did you find out anything about our work that we should know about?" asked Rho.

"Your supervisors will explain," he said. Then he got in their skiff and started motoring back over the Sound. Probably heading for Roanoke Island. Though nobody had the faintest idea where the aliens hung out when they weren't walking into your home office with a letter from the Secretary of the Interior.

Rho started back inside and I followed her. It was warm in there. No breeze. Of course, I went to the fridge and all the sausages were still there. So the alien had been making a joke? Who knew?

Then, because I really was hungry, I started frying up a couple of them, and the stink was amazing. Rho rushed into the kitchen fanning the air in front of her face. "What in the name of Zeus and Aphrodite are you doing to make that smell?"

"Cooking things that I thought were sausages," I said.

"So, he really did take all our meat," said Rho.

"Not joking after all," I said. "Unless it was a prank, in which case he's laughing all the way back to Jupiter."

"I don't want to eat any of that," said Rho.

"Wasn't offering," said I. I used a fork to probe the stinking junk in the frying pan and determined that whatever it was, it was all organic material. I ran it down the garbage disposer.

"You sure it won't harm the pipes?" asked Rho.

"The sausage didn't come with instructions."

"But, still."

"Rho, if I had put that stuff in the garbage, it would've stunk up the house for a week."

"Much better for it to eat a hole in the PVC pipe and leak this stink all over the basement?"

"I'll bag up a sample of it, and you can find some chemist willing to analyze it," I offered.

"I already know what it is," said Rho. "They *really* had a sense of humor."

"And yet you're not laughing."

"Because the look on your face is so sad," said Rho.

"Do you think this mess could be their feces?" I asked.

"Thanks for making it worse," she said, as she left the kitchen.

I poured some anti-odor chemical down the drain and let the disposer soak in it for about a minute, then ran it again with hot water and yes, the odor was gone. Except for my clothing. I had to get rid of that because even after five washings, if I put it in my dresser drawer with something else, the stink would be passed along.

But that's all later. That day, Rho and I only figured anything out when we settled down and got back to work. Our computers were off—very helpful, our boot-up sequences always took forever. What could we do? We started the computers, the boot-up sequence started, and then our computers froze. No key brought any result.

"I've never had any trouble with this computer," said Rho.

"That's because it has never been visited by aliens before," I said.

"Oh, you can't blame *them*."

"Post hoc, ergo propter hoc."

"That is a fallacy and always has been." Rho sounded so superior.

"They came in, they left, the computers won't work. *Both* of them failing at once. Rho, there's a causal connection in there."

"Paranoids always think so," said Rho.

"Rho," I said, chidingly. Showing her my sad face. "I'm booting from a flash drive."

"It's working?"

"You've got eyes," I said.

Up came the screen from the flash drive. I got a file manager program and looked for whatever was on the laptop's main drive.

Nothing.

Not a single file, not a directory, nothing.

So, I got another flash drive, this one with software that could read the hard drive bit by bit. I still got the nothings. Not a surviving bit, let alone a byte.

I tried again with forensic software and this time there were a few widely separated regions where there was a flicker of a palimpsest. Nothing that formed an actual letter, let alone a word. Just the default I-can't-read-this-character symbol.

"Check mine too," said Rho. "Maybe they thought mine wouldn't have any serious information because I'm just the girl assistant."

"You're the analyst," I said, "so don't give me any feminist victim stuff."

"Why do you even have software that can do this?"

"I worked in cyberforensics before I qualified for drone and missile work."

"Were you any good at it?" asked Rho.

"I got too good," I said. "They wanted to move me to Cardiff, and I would have done it, if I hadn't found out in time that you and I were going to be teamed up."

"You didn't know me," said Rho.

"I still don't," I said. "But I'm glad to see you every day."

"How did they wipe us out clear down to zero on every file and directory?" asked Rho.

"Before we raise a stink about this," I said, "let's see if our bosses know what's going on."

"Have they ever?" asked Rho.

"They probably always knew everything, we just weren't cleared to be told."

We did check with our management and they said they'd get back to us. They came back to explain that everything was above their security clearance. I asked, "How come all the other teams in the house were left alone?" None of my business, they told me.

How are we supposed to continue our scanning with all our data wiped out?

They told us, "Give us a day to find out." And the next day they sent a lady from Human Resources to say, "Here's your final check and your severance pay. The Scan has been called off completely. Your team has been terminated. But we'll give you great recommendations for your next job."

"This is what I'm qualified for," I said. "There's no other employer."

"Look at your severance check," said the HR lady.

It was a very, very nice amount of money.

"Take that as an offer to go back to school and get trained in something else," she said.

"You think I can go to university for a year on this amount of money?" I asked.

"I know you can go to college for four years and get a new bachelor's, or three to get another doctorate."

I wanted to say, this is ridiculous, this is offensive even, to think that money can make up for my career getting snatched out of my hands.

But I didn't say anything more. I just nodded, gave a wan smile, and said, "I assume we're not taking anything home with us."

"If you have a change of clothes or something, or a picture of your great-grandparents, by all means, show it to me and I'll tell you if you can take it home."

Rho chimed in. "But it's not a change of clothes, it's *all* our clothes," she said. "We live here! You're throwing us out of our home!"

The HR lady looked at her blankly. "Let me look at your boxes and suitcases before you close them up."

"And go where?"

"Your family's house?" HR lady asked.

"They're living in about six different places," said Rho, "none of them big enough to take me in."

"Sucks to be you," said the HR lady. She turned to me questioningly.

"I'm wearing everything I own that matters."

"Well, your phone matters, I'm afraid. Please hand it over."

"I bought it myself," I said.

"You signed an employment agreement that said you could only buy phones from the Interior Department and grant your supervisors total access to the contents."

"Why am I being treated like a spy?" I demanded.

The HR lady nodded with apparent sympathy. "It's out of my hands," she said. "You know that."

So, we packed up—I helped Rho with everything except folding underwear, and in a tote bag I carried a change of underwear, my beach flip-flops, and the flash drives with my special software.

The HR lady confiscated the flash drives.

I didn't argue. I had all the software duplicated in several different cloud locations under different names. And if they found those too, I could always reacquire anything I needed.

After about a week, Rho called me. Since I had texted her about twenty times since we were shut down, I wasn't surprised that she still knew my number. We were going to go to a park together, she told me. Bring a lunch but no software or computer stuff and definitely not my new company phone. I toyed with the idea of hiding flash drives in a peanut butter sandwich. Then I remembered that I had a peanut allergy and even if I could hide the drives in a club sandwich, why bother?

She spread a blanket on the grass because it was damp and there were grass cuttings that would stick to our clothes. "If you wanted to jump my bones for old times' sake," I said, "I bet we both have apartments with beds."

She said nothing. Chewed on her bagel and cream cheese.

"I always thought," I said, "that we had something that could outlast our being teamed together."

"Thought so, too," she said. "And now look at us, being together even though we're no longer on a team."

"Oh," I said. "Still friends." She nodded. "With benefits?" I asked.

"Don't push it, Sigmund," she said. "That remains to be seen."

"Why are we meeting here?"

"I'm hoping they don't have listening devices scattered throughout the woods," she said.

"These aren't woods, it's a park."

"Sigmund," she said, "you know and I know that it was the aliens that shut us down."

I nodded.

"Why?" she asked.

I shrugged. "The Inscrutable Alien," I said. "I should write a book."

"They are not inscrutable. In fact, they're not even subtle. They come without warning and without any plausible excuse, they push us outside while they 'examine' our computers. We come back in and our computers have been completely zeroed out. No data. And then, we're sent home with a warning that we can *never* work in surveillance or satellite scanning again."

"Nobody told me *that,"* I said.

"If you try to get a job in those fields, *then* they'll tell you. Or the HR people at the new job will inform you that your application is not acceptable because you have no qualifications in the field."

"What were they so afraid that we might find?" I asked.

"What did we already find without realizing it?" she said.

"If we could get our hands on the data," I said.

"What an interesting idea," she said. "What did *you* store in the cloud?"

I shook my head. "They locked me out of both of my cloud storage areas."

"Sigmund, I *know* you. Only *two* cloud storage units?"

"Yeah, I have all my data backed up," I said, "in storage units they didn't search."

"Sigmund, you have just confessed to three felonies. Backing up top-secret material in the cloud with insufficient security."

"Rho, if they haven't found them, it means the security *is* sufficient."

"If we download any of it and use private computers to analyze it, it'll be ten times easier for them to catch us."

"Rho," I said, "I'm with you on this. We just need to be careful. Download in small batches, alternate so we're never studying the data at the same time. No emails or phone calls about what we're doing."

"What *are* we doing?" Rho asked.

"We're looking for what they don't want us to find," I said.

"And *where* are we looking?" she asked.

"In the Antarctica Scan," I said.

"You still think there might be ancient civilizations there? Because three hundred million years sounds ridiculously ancient to me."

"I don't know *what* is there. But the only scan we did that was ours alone was Antarctica. And they must believe that we already had data they didn't want us to find."

"Sigmund, it isn't just *us* they wanted to shut out. It's the whole human race."

"They're keeping secrets," I said. "But everybody has stuff they don't want to tell."

"It's one thing," said Rho, "if we *know* what kind of thing they're concealing. But with Antarctica, we have no idea what they were afraid we'd find under the ice, under the water."

"So, now it's time to really analyze our data."

"But not by computer," said Rho, holding up a finger.

"It would take me a thousand years to copy out all our data by hand," I said.

"Nobody can read your handwriting to begin with. We'll read printouts of the data and analyze them with our brains."

"And we meet in the park?" I asked.

"Or wherever," she said. "We never say the name of the southern continent out loud again."

"They're not going to be interested in us," I said, "not unless we give them reason. They think the threat we represented has been neutralized. They don't care about us anymore."

"We don't know what the aliens care about or don't care about."

"They cared about the southern scan," I said.

"So it seems."

"Do you think they pose a danger to us?" I asked.

"I think anything we don't know is a danger to us," said Rho.

All the sneaking around was putting a strain on us. Officially, Rho and I were *not* in a relationship, but of course that's always a lie—not necessarily both people but at least one of them is lying. And, since it was me, I never really imagined that Rho might actually, like, love me. I was just Sigmund, the nickname she gave me in psych class at Stanford. I was always analyzing people, she said. I didn't think I did it any more than anyone else, but hey, the smartest woman in the class (and maybe in the whole university, I just didn't have enough data to be sure) had given me a nickname that could be taken as affectionate, or at least not hostile.

Then, when we saw each other in our summer internships after sophomore year, she remembered to call me Sigmund, and I started calling her Rho, because that's the letter before Sigma in the Greek alphabet. She only complained once. "Calling me 'Rho' makes me think you're going to try to start up the round 'Row, Row, Row Your Boat.'"

"I'm not," I said. "I have too much respect for the letter R."

We dated off and on. I was head over heels, and she stabbed me all the time by saying, "I think it's so cool that we can be friends and do everything together, without anybody doing something as silly as falling in love."

"Yup, yup, yup," I said. "Wouldn't that be silly?"

She started fast-forwarding the movie we were watching. "This is the icky part. The alien comes out of his chest. Just right out, it's horrifying and scary and disgusting."

"So…pretty much like P.E.," I said.

"You're in good shape."

"I do pushups in my room, morning and night. Hundreds of them. That's not P.E. because nobody's making me do them."

She laughed.

"You think I'm in good shape?"

"You know you are."

"You said we can be friends and do everything together," I said.

"You listened!"

"I always listen," I said. Which was only true when she was the one talking.

"Of course," she said, "we don't do *everything* together."

One thing led to another, and I thought that once a physical connection had been made, we'd be off that "friend" train. But, no, all we really did was add a pleasant little caboose to the train: "With benefits."

What benefits? I wondered. Of course I enjoyed making love to her, except that I was pretty sure she was just having sex with me. I did my best to show her how careful I was, how attentive, even worshipful, but she just said things like, "You must have been popular with nerd girls in high school." Like she thought I had had a lot of romantic experience. I never went on a date or had anything *like* a girlfriend in high school.

So I was in heaven, I was in hell, but I was pretty useless in our project because instead of trying to decode the meaning of the aliens' behavior, I could only think of, look at, dream about Rho. Was Einstein thinking about some babe when he was doing thought experiments about trains going at the speed of light? When the apple fell by or on Newton, did he think, "I bet *she* would like this apple"? No, he thought about the apple as a tiny moon being attracted to the earth, and then imagined the earth also being attracted to that apple. That was an amazing leap. Only possible because Newton was not scheming to try to get with some Edinburgh lassie.

Rho kept suggesting lots of ideas, and—give me credit—I never said "stupid idea" or even "that couldn't be it" because I didn't know anything, so how could I contradict her? She said stuff like, "Have they been charting all the nearby wormholes and accidentally took one that dropped them off in our Solar System?" or, "We don't know enough about them. What's their lifespan? What if they don't *need* a generation ship because they don't die?"

That got me thinking. How many of the aliens were known? They were never anywhere in great numbers, but I knew I had seen several dozen different faces in both genders. And I had never heard any of these individuals mentioned by name. Did they *have* names? Perhaps they didn't need them.

Rho and I talked about our questions and speculations. We never fooled ourselves into thinking that just because we liked an idea, it must be true or at least plausible. And I came to rely on Rho for her fine-tuned hypotheses. She was never just throwing mud at the wall to see if anything stuck. Everything was reasoned out, at least enough to imagine the consequences of each hypothesis.

After a long period of speculation, Rho came up with my favorite. "What *are* they?" said Rho. "They present as human, but they don't claim to *be* human. Have you ever heard of one of them, like, mating with a human?"

"And I don't *want* to hear about it," I said. "I also don't want to see it or do it."

"Good. I don't have to worry about catching interstellar gonorrhea."

"Why is the idea so repulsive to me?" I asked. "Some of the alien women I've seen are kind of fine."

She shook her head. "You don't know what they really look like. Do you think that on some faraway planet, it just *happened* that a sentient race came along that happened to look just like us?"

This had such a ring of truth about it that I allowed myself to think that she was right. It was unreasonable for them to be humaniform. "So, what are they really," I asked, "amoebas?"

"I don't think a single cell could ever be that big. And where do they get their human suits from? Hollywood?"

That's how our sessions went. We had to assume the aliens were smarter than us, and that they were able to make better machines, or how did they get here? Ask, guess, guess again, ask again with a different slant. We were both smart, though I'd say she had two-thirds of our collective intelligence, which often filled me with the standard question that a guy who is dating above his station has to ask: What does she see in me?

Yeah, that's what I'm thinking about when she interrupts my reverie to say, "Sigmund, what if they don't come from another planet?"

I tried to wrap my head around that. "Then where are they from? Nebraska? New Guinea?"

"No. The place you started us looking all those years ago. In the ocean. No artifacts on land because they spent little time there. And under the ocean, what kind of shelter do you need to build? You don't need to keep the rain off. You can't possibly install a furnace to keep warm."

"If they evolved on Earth, Rho, why haven't we run into them before?"

"Who says we haven't? Look, when they 'arrived,' did anybody see them come out of a spaceship? For that matter, did they really come from space? Remember that radar detected nothing when their spaceships did their fly-bys. What if they have a technology that can make us see lights in the sky, even shapes, but they don't actually get into vehicles and fly. All illusion. Smoke and mirrors. CGI."

"But *why* go to all that trouble instead of walking up and saying, 'We're here, too. Think we can get along?'"

"I don't know. What if they're more vulnerable than we think. Most of their advanced technology is really biotech. Healing. Immunizing. Have we ever seen them in *any* vehicle?"

"They ride in cars all the time," I said. "Usually limos, because we're so hospitable."

"Do they ever *drive* a car?" she asked.

That was not the kind of thing I would have noticed. "They do have unusually big eyes," I said. "So, they're not *completely* like us."

"Drive. I asked if they drive."

"I never noticed one way or the other," I said, knowing how she admired obliviousness.

"I think never. I think they might not be comfortable with our technology."

"Rho, are you saying they developed under the ocean?" I tried not to allow triumph to be detectable in my voice.

"Probably near a coastline, because that's where the nutrients and the sunlight are."

"And we never saw them."

Rho closed her eyes. "I'm trying to think of the underwater creatures that are smartest. Dolphins and whales, of course, but those are mammals."

"Rays?" I suggested. "I don't know how smart they are."

"Smart as sharks," she said. "And they have cartilage instead of bone, so maybe they're more flexible."

"You want flexible, you gotta go with the jellyfish," I said, joking.

Of course she took me seriously. "How and why would jellyfish evolve brains?"

So, I took *her* seriously. "We already know what the smartest non-mammal is, in the ocean. Doesn't have any skeleton. Just a beak, the only hard part of their body. Two big eyes."

She nodded now. "Eight arms. Or legs. Suckers on them."

"We see octopuses all the time," I said. "But way back before the aliens came, I think we talked about octopuses as the possible source of—"

"Source of an impossible thing, a civilization millions of years ago that left no artifacts on land."

"But they don't *make* solid, durable things. They work with biology. Genes. Directed evolution."

"You're saying they're *God*?" she asked.

"I'm saying, what if, instead of toolmaking, they learned how to do what CRISPR does, tweak the genome of a creature right after fertilization, to make it grow as you want."

"So, where are the creatures they designed?" asked Rho.

"How do we know that we don't see thousands of them every time we go under the water?" I asked.

"They've been running the oceans for thousands of years?" She sounded skeptical.

"Why not millions?" I asked. "What about the speculation of that zoologist. A civilization born during and after a great extinction. Three hundred and thirty million years ago. Or three hundred million."

"How do octopuses live on land, breathing air?" said Rho. "Pretending to be human."

"If they've been guiding their *own* evolution, maybe it took since humans first ventured out on the oceans for these hyperpoda to decide they might want to look like us and come to the surface and act like us?"

"Ten thousand years is an awfully short time for them to evolve human appearance. From an eight-legged mollusk to humankind just since the last Ice Age?"

"Rho, we're not talking about evolution here, are we? It's deliberate *self*-modification by the best genetic scientists in world history. They're certainly better than us."

"My family was cancer-prone. I expected to have to have my breasts removed sometime, because that's the only way women in my family survive to be grandmas. Now that's not even a worry."

"Geneticists. And they've turned their attention to helping us. As for that ten-thousand-year problem, humans have been on the water since *long* before that. We talked about the Australians before. They've been on Australia for forty thousand years, and they *had* to come on boats, because there has never been a land bridge connecting Australia and New Guinea on one side, and the rest of Indonesia and then Asia on the other."

Rho said nothing yet. Hadn't thought of a devastating put-down. She could wait.

"The Australians had *dogs,"* I went on. "The canids got away and went feral, but dingoes are dogs, and those evolved along with humans. Forty thousand years ago, there were no dogs."

"Are you proving yourself wrong?" Rho asked. "Falsifying your hypothesis because Australians couldn't have started on Australia forty thousand years ago?"

"They still had boats. Maybe there was plenty of contact between Australia and the Asian mainland, where they could have acquired dogs twenty thousand years later."

"So what? Why did you bring up dogs?"

"The way humans and dogs augmented each other and grew mutually dependent, why not suppose the hyperpods had some other creature they needed."

"Pilotfish?" suggested Rho.

"Something like that. Or maybe a symbiote, a parasite inside the hyperpods that contributed to the growth of their intelligence."

"All of this is crazy stuff, Sigmund," she said. "The aliens are *nothing* like octopuses."

"Just because we still look like apes with a shave and a nose job doesn't mean that *they* had to choose to continue to look like octopoda," I said.

And there we were, coming up with a weird consensus with no evidence at all.

"What difference does it make, Sigmund?" she asked one night after our speculations. "They haven't done us any harm. They've made our lives vastly better. I haven't had a cold since they distributed that immunization virus, that prepped all our bodies to destroy the common cold every time it pops up in our bodies. They're peaceful, they're kind."

"But they stopped us from scanning Antarctica. *Stopped* us, erased us. They're afraid of *something.*"

Rho couldn't argue with that.

Until one day, when we were living together again as a research team, figuring out how to mine oxygen and water on the moon, she got up in the morning and dressed warmly, which meant that she wasn't going to stay indoors all day. "Come on, hurry up," she said.

"Easier to hurry when I know what I'm hurrying *for.*"

"If I tell you then you won't go."

"Thanks. I also won't go *unless* you tell me."

"The aliens teach classes, don't they?"

"Sure. Yeah."

"At most of the major universities, there's at least *one* alien instructor."

"Probably true. Stanford, Harvard, MIT definitely."

"Where's the closest university with an alien teaching at it?" Rho asked.

"Johns Hopkins, probably," I said.

"Do you know how to get there?"

"I know how to enter Johns Hopkins into the GPS."

"And while you're driving, Sigmund, I'll try to look up the alien's teaching schedule."

It's not like we had worked out a script, because everything depended on how the alien answered us. But I'm pretty sure Rho had not thought that her idiot boyfriend would ask a question like the one I blurted out, after he took us to his office and closed the door.

"People call you the aliens, but that's descriptive," I said. "And there are other names that are kind of crude and disparaging."

"I'm aware," said the alien. "We all are."

"I don't think there's any harm intended. Nobody hates you."

"Nobody shows it if they do," said the alien.

"What *should* we call you?" I asked.

"My name is Lytton Strachey," he said.

"Come on, really?" asked Rho. "A nineteenth-century biographer and snob?"

"Nobody was using the name, I learned a lot from his book, and it was easy for humans to pronounce."

"Well, Professor Strachey," said Rho, "we have a couple of—"

"I know who you are," Strachey said. "Rho and Sigmund, you call each other. Good name choices. I also chose my own name. It's not a bad custom."

"What do you mean, you know who we are?"

"You were doing the Antarctica Scan. Our governing council voted to stop you."

"Yes, and they succeeded, though everybody pretended that some human authority had made the decision."

"We knew you knew it was us," he said. "We knew that far from giving up, you'd chew on the problem for weeks. Months."

"We're up to years now," said Rho.

"I know. That's why I dropped everything the moment I recognized you and brought you in to converse with you."

"You don't still think we're dangerous, do you?" asked Rho.

"Of course we do. You haven't grown stupider, and you've gathered more data."

"So what if you interview me and decide we pose an immediate threat?" I asked.

"I'll report back to the council," said Strachey.

"You haven't really finished answering the question we asked a while ago. When we asked, 'What do you want us to call you,' you told us your name, your individual name. But what should we call all of you? As a group? We're humans... you're what?"

"In our own language, we call ourselves 'people.'"

"In English, that word is already taken," I said. "Any other suggestions?"

"'Aliens' has been doing well enough for a few years now."

"'Alien' means stranger or foreigner," I said. "But you're not strangers anymore."

"Oh, you think you know us?" asked Strachey.

"Better than we used to," said Rho. "But we also know you still have secrets."

"And we always will," said Strachey.

"'Always' is an extravagant claim," Rho said. "The reason we came here is to tell you what we've guessed about you, and ask you to tell us where we're right and where we're wrong."

"You're wrong when you imagine we *want* you to know more about us."

That's when I asked the question that had kept occurring to me ever since we came up with the postulate that the aliens were geneticists who fiddled with other animals' genomes. "Can your people interbreed with ours?"

Strachey shook his head. "Isn't reproduction a private matter?"

"I'm not asking anything personal," I said. "In general, can a conjoining of a human and an alien result in viable offspring?"

"We are content for our species to remain separate."

"You're still dodging," I said.

"Maybe they don't know," said Rho.

"They know," I said. "They've known for a long time."

Strachey bowed his head.

Another person came through his office door, closed it, and sat down with us. "Hello," he said. "I call myself Kublai Khan."

Rho said, "In Xanadu did Kubla Khan a stately pleasure dome decree."

"He did not," said Kublai Khan. "But we understand about poetic license."

"Why are you here?" asked Rho.

"Lytton invited me. He figured you would soon take him out of his depth. So, I'm here because I—"

"You're deeper?" I asked.

"Not really," said Kublai. "I simply am more familiar with the principles by which the council has tried to make decisions over the past forty years."

"So, you speak for them," said Rho.

"Nobody speaks for them. But I can advise you. I can even answer some questions, because I know when to stop."

"And Mr. Strachey here does not?" I asked.

"He knew that *he* needed to stop," said Kublai, "when he bowed his head and asked me to enter."

"Asked you?"

"We have a rudimentary mental communication that is undetectable to humans. It doesn't use a language; it's more like a series of signals."

"Like Morse Code?" I asked.

"Nothing like Morse Code," he said. "That's just a transcription of the spoken words in a transmissible form."

"More like monkey calls in the jungle?" asked Rho.

"I have never been in a jungle, nor seen a monkey," he answered.

"I don't care what your code is," I said. "May we ask our questions?"

"Yes."

I looked at Rho. She looked back at me with consternation, and shrugged.

Apparently, this would be *my* job, though it was her idea. "We believe you don't come from another planet. We don't believe your spaceships were real. In fact, we don't believe anything you've said about your origin and purpose."

"Very candid," said Kublai. "It saves so much time."

"Are we right?"

"Let me think," said Kublai.

"Don't answer," murmured Strachey, without raising his head.

"Dear Mr. Strachey," said Rho, "your statement *is* our answer. If we were wrong, you'd simply say so."

"Your species," I said, "evolved here on Earth."

"Well, not *here*," said Kublai. "But yes, on Earth. We are not of your species, which makes us aliens, after a fashion. But we have nothing like your ability to travel through space. It was hard enough for us to come up on land."

"That was the next question," said Rho. "We think your people evolved in the ocean."

"Do I look like a fish to you?" asked Kublai. "Does Lytton?"

"We think you're descended from the smartest of the nonchordates. Octopuses."

Again, Kublai pondered. "Octopuses are all geniuses compared to jellyfish and eels," he said. "They're about as smart as dolphins, though not nearly so playful, or helpful."

"You've been pretty damn helpful so far," I said.

"We wanted to come with gifts so you would like us and let us stay," said Kublai.

"Well, it worked," said Rho.

"Do you like us?" asked Kublai.

"Me personally? I don't know enough about you to say."

He looked at me quizzically, then turned to Rho.

"I don't trust you at all," she said. "You've lied to us constantly."

"Did our cancer cures not work? Do you still get pneumonia, meningitis, apoplexy?"

"You didn't lie about the benefits," said Rho. "And in a culture that has constant advertising, your claims were refreshingly accurate."

"So why don't you trust us?"

"Because you're concealing the truth from us. For instance, if you evolved from octopoda, how can you possibly look like us?" Rho asked.

"We look like you because anything else would have provoked fear. Besides, in our native form, we can't speak your language. Or any language. I mean, oral language. Our native language is gestural. Visual. And we also speak with color. The combinations and nuances have evolved, and they change over time. Language."

"Only one language?" I asked.

"There aren't as many of us as there are of you. We can't isolate a million of our people here and another million there, and then they never talk to each other so each group develops its own language. We are better served by all speaking the same language."

"There are fewer than a million of you," said Rho.

"I believe it is better for you and me if you drop the question of the size of our population."

"Less than a million," said Rho. "Markedly less."

Kublai looked at me. "You wanted answers. But if you're content with guesses, what do you need me for?"

"You're in disguise right now?" I asked.

"Not a disguise. This is really me. I'm present in this room with you. So is Lytton."

"I want to see," said Rho.

"See what?"

"How you transition from octopus to human," said Rho.

A couple of seconds. Then Strachey shouted, "No!"

But Kublai was already splitting in half up the middle. And then more bifurcations. His human appearance split with the parting of his arms. And then the colors changed, and he was a light-blue octopus. Well, no. He had ten arms. But not all equal.

I felt a little sick. It was like a slow explosion. My brain had accepted his human appearance. And I was familiar with the look and size of many species of octopus. But the changes of color in his skin were not blurred or washed out. They were sharp and clear, and they had looked like legs and arms.

"What about your face?" asked Rho.

"Can't a fellow keep a few secrets?" he asked.

I looked closely at the way his human neck and head grew up out of the crown of his arms. It dawned on me that maybe they had committed to coming on land to a greater degree than I had imagined. "That's not an illusion," I said. "Your head—you genetically altered yourselves to—"

"We needed to be able to breathe above the surface, yet still be able to go below."

"But your kind can breathe air through your skin."

"You're so well informed. But *my* kind needs lungs in the atmosphere. It's underwater now that we breathe through our skin. Like one massive gill."

"You altered your genes to have a human head?"

"Superficially," said Kublai. "We still use our own kind of brain, but we added parts to it to allow us to speak and listen."

"So, are you still an octopus?" asked Rho.

Abruptly Kublai's colors started changing. His legs stiffened and straightened, and where they joined they formed one continuous appearance of a normal human body. "This isn't an illusion, either," Kublai explained. "When I stand before you as a man, I'm standing. I'm really there. Unlike you, I don't hide behind clothing. I speak to you of my humanity through my colors and my shape."

"And you walk, like a biped."

"Very hard to learn," Kublai said. "But we don't let humans see any of us who haven't mastered bipedal locomotion."

"How could you evolve these abilities so quickly?" asked Rho.

"Quickly?" asked Kublai.

Strachey chuckled.

"Our kind, as you say," said Kublai, "first appeared about three hundred and thirty million years ago. And then we evolved by normal processes, shark-eat-ray, nature red in tooth and claw. When we went ashore, we began to extend our visits from a half hour to a whole day, sunup to sundown. It helped when we smart ones—forgive my vanity—began to be able to alter the genes of newly fertilized eggs. These changes normally bred true. The trick is to keep track of the changes so you can undo them when necessary."

"Now your children are born with human heads," said Rho.

"Oh, no, not *all.* It takes a special breeding and gestation regimen to get the skull to form, all the holes to connect correctly. The eustachian tubes are so difficult. And we don't bother with tonsils. Have to keep eyebrows and eyelashes or we'd look too weird. Swallowing and breathing through the same mouth? Hard. And chewing while breathing, and then holding our breath to swallow? There's a lot of choking when the head children are young."

"Why would you do that? Why not just contact us in your natural form?" asked Rho.

"We had to have vocal cords, an air column to sustain sound."

"But why do you need us at *all*?" I asked. "Our worlds don't overlap."

"Well, you do dribble a lot of effluent into the oceans and seas, and that feels like overlapping to us."

"You're dodging," I said. "You're trying very hard not to answer this question, and that's why I'm pretty sure that it's the most important one."

"Good manners would tell you that if your guest does not want to answer a question, you must stop asking it." Kublai looked quite self-satisfied with his answer.

Rho stepped to Strachey and ran her fingers through his hair. "Octopuses don't grow hair," she said.

"Most don't, some do," said Kublai. "Please don't torment the boy, Rho. He hasn't gotten used to physical contact with human females."

"What does male and female matter between species?" asked Rho.

But I knew. "They didn't evolve human heads or human lungs," I said.

She looked at me like I was insane. "They aren't transplanted on," she said.

"We asked about mating, and he dodged the question. He and all the others with heads, they come from human males mating with alien females."

Kublai lowered his head just as Strachey had.

"Are you going to have to kill us now?" I asked. I hoped I was joking.

"We didn't answer you," said Kublai, head still down.

"You most definitely did," said Rho. "In a gestural language that we also understand."

"Why did you assume it was human males with our women?"

"Even if you first met up with humans when the sailors thought you were mermaids," I said, "your females would have a much better chance of luring our males than the other way around."

"Sailors didn't mind mating with 'women' who didn't actually have faces," said Kublai. "Until we were able to regularly produce children with heads."

"That sounds so..." Rho began.

"Grotesque," I said. *Please don't ask what they did with the discards, Rho. Let's not insist on knowing everything.*

I don't think she heard my silent plea, but she did not ask that obvious follow-up question.

"Well," said Rho, "you've answered a lot of our questions. I suspect you answered more than you wanted to. I imagine you reached a point where you thought, well, we can never let them go back to the human race, so we might as well tell them the truth."

Kublai did not respond.

"So, we humans had three to five million years to evolve," I said, "depending on where you mark the starting line. You had a hundred times as many years. All without leaving a trace."

"We leave traces if you know where to look," said Strachey.

"Please don't tell us," said Rho.

"Let us go to Bolivia," said Kublai. "Not you, Lytton. You should go back to class."

Lytton got up and left the room.

"Bolivia?"

"You have no idea what most of our work has been. Getting to pass for human in order to help solve your problems, that was a massive undertaking. Learning an audible language. Learning to read."

"But you have other work?" asked Rho.

"Let's go to the airport," said Kublai.

"We can't afford tickets to Bolivia," I said.

Kublai laughed. "Airlines don't charge *us* for tickets, kids. And if you're with me, you won't have to pay, either."

"Do you reserve seats in advance?" I asked.

"We just show up, walk onto the plane," said Kublai, "choose the seats we want—and then the flight attendants break it to the people who just lost their seats, and give them nice airplane money to pay for later flights."

"Oh," said Rho. "You don't mind inconveniencing people?"

"No one has ever asked us to vacate our seats in favor of the actual ticket holders. If someone asked me, I'd get up and then stand in the aisle for the rest of the flight. Or go back into the bathroom, return to hyperpod form, and then slide out the door and get down into the luggage compartment. We try not to be seen, because the sight of a human head just scooting across the floor can be disconcerting."

It was an airplane flight. Nobody asked us to move out of our first-class seats. Discomfiting people who could afford first class didn't bother me as much as displacing someone in coach.

In Bolivia, I figured we were there for the mountains. But the guided tours were not for us. Kublai led the way on a strange combination of taxis and hiking until we came to a steep climb over a saddle between mountains.

"We really have to climb this?" I asked.

"Is that what is meant by whinging?" asked Kublai.

"Not in America. It's just whining there," said Rho.

We climbed. The air was thin. We were panting and had to keep stopping to rest. Kublai didn't bat an eye. We got to the top of the saddle.

Below us was a rather large valley, grassy but with more trees than I expected at that elevation.

I couldn't contemplate the flora for very long. Because two young women in very modern bikinis emerged from what I realized was a building. Or a tent. They were smiling. They were attractive.

"What are we here for?" asked Rho grimly.

"Not what Sigmund is wishing," said Kublai.

Rho glared at me.

"Kublai can't read my mind," I said mildly. "Admiring is not wishing."

Rho asked, "Who are these girls? They don't look like anybody local."

"In this place, nobody is local," said Kublai. "Nobody who looks like us or them. But yes, there are several families who live in this valley. You'll meet the... locals in a few minutes."

The ladies came over to the three of us, their eyes on me the whole way. I knew what game they were playing, from the way they made sure various body parts were shown to advantage as they wended their way along the meandering path.

"I hope you don't expect *me* to ever dress like that," said Rho under her breath.

"If I ever saw you wearing a swimsuit like that," I murmured back, "I'd have it off you in a second."

"So manly," she said softly. "So bold."

Since nobody had ever referred to me with those adjectives, I assumed she was mocking me. But I didn't mind much, because at least she wasn't distracting me from the approaching ladies.

"Why are they dressed like that?" asked Rho. "If they're like you, those bikinis are just a lighting effect on their skin."

"Indeed," said Kublai. "But they like to read women's magazines from many countries, and they try on different looks at different times. They were told that you know about our... shifting. Shape and color shifting. So they wanted to show you their best."

"A ball gown would have done just as well," said Rho.

"You said families live here?" I asked.

Kublai nodded. "They know you're here. They were up at the far end of the valley, around the bend, where the hunting is better. But listen—they're coming now. Oh. No, not *they*. Just one."

I tried to listen to what Kublai was hearing, and soon it became audible to me. Heavy footfalls, not too fast, but not too slow. *Very* heavy footfalls.

What could only be a Tyrannosaurus rex came around the bend. He was still a long way off but it was obvious he was huge. And his head was *way* too big for his body. But he was lumbering along with his head held high, looking forward, while his back was right in line with his tail, flat as a table.

I tried to figure out how a hyperpod's body could be reshaped to that extent—and at that size. I finally said, "Is that, like, four of your people moving together?"

"No, no, we shaped him, but we didn't make him out of ourselves," said Kublai.

"So, you did what, protected a few dinosaurs when all the rest of them were getting destroyed?" I asked, jumping to conclusions, because I'm human.

"We were busy surviving," said Kublai. "The meteor destroyed our world, too. But once we found a surviving breeding pair, yes, we did all we could to adapt them to the changing world."

"So, he's an authentic Tyrannosaurus rex," I said.

"When we helped dinosaurs adapt," said Kublai, with scorn in his imitation human voice, "we changed them genetically. Significant changes. They are no longer the same species."

I said, "They *look*—"

But Kublai interrupted me. "When humans caught glimpses of their ancestors during the phase when we had them flying, they gave them the name that in many human languages is translated as Wyrm, Wyvern—"

"Dragon," I said. Because I'm not stupid. And now that I had been alerted to the name change, I could see it. Definitely not a T. rex. "Where are the wings?"

Kublai chuckled. "Free-range dragons did not distinguish between humans and other mammalian prey animals. You were evolution's best experiment, and we didn't want them messing it up. Wings made them too difficult to control, so to protect your species, we restored the genes for T. rex forelimbs."

"Those nubby, useless—"

"Evolution doesn't preserve useless features," said Kublai. "And our dragons never breathed fire."

"Not that you didn't try," I suggested.

"Getting them to flare organically grown methane wasn't hard," said Kublai. "But we couldn't flameproof their own snouts. An evolutionary dead end."

"Only it wasn't evolutionary," I said. "It was genetic modification."

"Evolution is natural reinforcement of genetic variation. No matter what *causes* the genetic variation."

I rolled my eyes.

"Genetic modification, then," said Kublai. "But the current dragons don't know about many aspects of their genetic history. So, when he gets here, perhaps we can pursue a different topic of conversation."

"Why here?" I asked. "Dragons are legends all over the Old World, but isolated here in the Andes—"

"For a long time, we maintained our dragon populations in Persia, Nepal, Sumatra, places where there was plenty for them to feed on."

"Which does not describe *here*," I said.

"After the wings and the flames, we sequestered a bunch of eggs, brought them here, and hatched them out. Something as massive as that fellow," said Kublai, "and an obligate carnivore, there's no way they would find enough prey animals to stay alive. But *we're* from the sea. We made nets and brought an amazing amount of fish up the trails that our own treading made. We looked like ourselves, then. There were no primates when that meteor struck. We didn't plan to alter ourselves as radically as we have for you."

"And there are other dragons up here," said Rho.

"We've found that it's hard to maintain a breeding population of any vertebrate without at least one male and one female. We have twenty dragons, if you count the children, which we do."

"Still bringing fish?" I asked.

Kublai nodded. "We haul the nets using helicopters, now. Too many humans near the coast for us to make the hike on foot. This is about as high as helicopters can fly, with the air getting so thin."

The dragon was getting close, and it was slowing down. Now I saw that it did not have the head of a T. rex. "Where did the horns come from?"

"We did not start with a heavy predator like T. rex," said Kublai.

I thought a moment. Rho thought faster. "Ceratops," she said. "The horns bend backward now, but... Ceratops."

"I thought they were vegetarian," I said.

"They were strip-vegetarians," said Kublai. "They tended to kill the plants they fed on. And a carnivorous stomach and intestine are far easier to build. Shorter and faster."

The Ceratops was almost to us. Was it looking at me? The nubile young ladies? Rho? Kublai? Hard to tell where a reptile was looking, sometimes.

"So you're sure he's not hungry," said Rho.

"He's not what you think," said Kublai. "Once we had the world's entire population of dragons up here, we weren't content to let them live alone, without civilization, without language. We developed our own breathing so we weren't distressed by the elevation. And as long as there was a stream—which you see—we could wet down our skin several times a day. Oh, and by that time we had made ourselves warm-blooded. You have to eat more, but exposed to winter winds, we needed a means of not freezing."

"So, you bred them," said Rho.

"For intelligence," said Kublai. "Humans think a T. rex's brain was too small to hold much intelligence, but that's because they forget that dinosaurs are essentially birds. You English speakers used to have an epithet, 'birdbrain,' meaning somebody stupid, because birds had such tiny brains. All nonsense. The mammalian brain came down a separate line of evolution. Birds were fully operational. There are birds that are downright brilliant. Parrots, corvids. They have powerful brains because they're not organized like mammal brains. Much more intelligence is packed into each cubic centimeter. The same was true of the T. rex. And after we worked on them a little, our dragons were even smarter than the birds."

"Smarter than humans?" asked Rho.

"We don't make comparisons like that," said Kublai. Which I took to mean that he thought it would be rude to tell us the truth.

The dragon was now standing at the base of the cliff we were perched on. Its head was about even with the bikini girls, and it turned to them in greeting. It might even have been a dragonly version of a smile that he gave them. Nobody showed the slightest fear of this giant who was about four times my height, measuring from the ground it was standing on.

"Greetings," said the surprisingly high-pitched voice of the dragon.

"He talks," I said, like an idiot.

"I do talk," said the dragon. "The ten-armed fish-bringers insisted on teaching my ancestors, after *your* kind came along, and we continue the custom. With a few language changes along the way, because you humans can't stop changing the *way* you talk."

Kublai took over. "We used gestural codes before that. Poor Blossom, he hates learning a new language. But... you came from America and we discovered that neither of you speak Quechua or Spanish."

"I'm sorry to inconvenience you," I said to the improbably named Blossom.

"That was not a problem," he said. He? With a name like Blossom?

"He chose his own name," said Kublai. "Others try to sound fierce, but he named himself for the most beautiful things he saw. We've done similar health-preserving things for Blossom and his kin here in the valley, so they live considerably longer than their ancient predecessors. Blossom is over a hundred years old."

"And I feel every one of those years," said Blossom.

"You're in perfect shape," said Kublai. "You will outlive *me.* Though the People live far longer than *our* ancestors, too."

"I'm glad to be here," I said. "To know that these dinosaurs spawn—"

"Don't lump us all together," said Blossom. "Do we call you ape-spawn?"

"We call ourselves ape-spawn, when we feel like it," said Rho. "He's saying that he's glad to know that you're alive. And we're both surprised, I think, that you have acquired speech. Can you read?"

Blossom raised his short arms. "How can I reach down and pick up a book? How can I hold it where I can see it?"

"Short arms," said Rho. She looked at Kublai. "You couldn't give them useful arms?"

"They do fine with short arms."

"Do your arms have any function?" I asked.

"During mating," said Blossom, "the arms help males remain centered on the female's back."

I nodded, as Kublai said, "If not for that help in mating, the arms would probably have disappeared a hundred million years ago. But we want them to keep mating, so we're not going to mess with the arms."

"I've told them that if we have any babies with weird arms, we'll teach them to eat octopus," said Blossom. "These ten-armed manipulators think they have the right to alter anybody's genetics any way they please. This one thinks I don't know what they've done to us in the past."

"Blossom is very fierce," said Kublai. "Don't be deceived by his gentle name."

"We have reached a verdict," said Blossom.

"Verdict?" asked Rho.

"I spoke of our council," said Kublai. "Blossom is head of the council this season."

"You let dragons rule over you?" I asked Kublai. Blossom snorted.

"On land, it makes sense. They're very smart and powerful," said Kublai. "When we go in the water, we're not subject to the council at all."

"Verdict about what?" asked Rho.

"About you two," said Blossom. "We've been told how much Kublai here decided he had to tell you, because you had guessed so much."

"We don't let humans in on such information," said Kublai. "It would make us vulnerable to the wrong kind of human."

"So, the verdict is whether or not you let us go?" asked Rho. "Because we *do* know how to keep secrets."

"The verdict was whether to keep you here or end your lives," said

Blossom. "*Not* by chomping you, as I'm sure you both immediately imagined. We don't chomp sentient beings, which you both seem to be."

"Thank you," I said.

Rho just glared at him. "Why do you think you get to decide whether we live or die?"

"Told you," said Blossom.

"We tried to get you to stop looking for us," said Kublai.

"You didn't have to tell us we were right," said Rho. "You didn't have to bring us here to meet a dragon."

"Come on, Rho," I said. "This is the best thing that ever happened to you. Apart from meeting me."

"What good is it to live if I can't go home?" Rho asked.

"I'll be with you," I said encouragingly.

"So, for the rest of my life, you're the only genuine human company I'll have?"

"That sounds a little discouraging," I said, trying to pass my pain off as a joke. "Besides," I added, "we won't be staying here."

"We have cooperating condors who will hunt you down if you try to run," said Kublai. "Please don't make us do that."

"We won't run away. You'll be glad to send us back to civilization—no offense intended—because we can help protect your secret."

"What do you think our secret is?" asked Kublai.

"That you are from Earth, that your natural habitat is under the water, that only some of you have human heads, that you have a dragon retirement home in the Andes. That you can split your bodies into ten long, colorful arms and regain most of your hyperpod abilities. That all of you with human heads are the product of a mating between a human male and a female of your people."

"That's not complete, but as a list of reasons we can never let you go, it's pretty fair," said Kublai.

"Now it's time for you to listen," I said. "Really listen and consider."

"You tell 'em, Sigmund," said Rho, with fake encouragement.

"A few years ago," I said, "I was reading a book about ancient civilizations on Earth about three hundred million years ago. Rho thought it was complete nonsense, and I only half believed. But we were working on the Scan, so we took a week and a half to scan the oceans. We had decided you had to be in the ocean."

"So, this book writer," said Kublai. "How did he know all this?"

"He didn't *know* it. But that's the fun part."

"We're having fun?" asked Rho.

"A few people believed in the book—this man had quite a following—but serious people—people that other people pay attention to—*they* all despised the book, and anybody who said such outrageous things was branded as a kook, a loon. So, if anybody thought of the things in the book, they didn't dare tell anybody about their ideas because this book had poisoned the well. Nobody was going to take them seriously."

"Including us," said Rho. "Nobody paid the slightest attention to our ideas or our findings."

"So, if Rho and I go home and one of us writes a book or an article that tells about you, nobody will *dare* to admit they believe in it. So, by letting us go so we can write the book, we will make it even harder for any humans to put the slightest trust in the truth of your existence."

"By telling about us to humanity," said Kublai, "you'll make it impossible for anyone in power to admit they've even read it."

"Ridicule has stifled far more faith than evidence or argument ever did," I said.

Rho was nodding beside me.

"I'll tell you what," I said. "I'll write everything up in a fiction-style story. I'll try to get it published right away. If it gets illustrated, it will sell even better. And it will keep educated people from believing in your existence for at least a generation."

"It's a good plan," said one of the bikini girls.

"Humans are just that stubborn," said the other, as she climbed up on top of Blossom's head. The dragon didn't seem to mind when she settled down between his horns.

"You going to gamble with our lives, Kublai?" asked Blossom.

"Not gamble," said Kublai. "I know enough about the behavior of humans to be quite confident that they'll respond as Sigmund here predicts they will."

"It won't be perfect," I said. "There'll be some people who'll think there might be some truth in it. The way I was about that earlier book. I was even able to get Rho to cooperate with me in scanning for an ancient undersea or under-ice civilization in Antarctica."

"That's why we noticed you," said Kublai.

"But what you don't understand, *didn't* understand, is that we had given it up, only thinking about it now and then." I was really getting into this point, because if they didn't accept it, I didn't know how to avoid being prisoners or dead. "We posed *no* danger of actually discovering anything. Until your agents showed up, booted us out of our house, and completely trashed and erased our computers, wiping out our data. The fact that *you* thought our ideas were important was the first *real* evidence we had that we were on the right track. *That's* why we doubled down—because you had shown us that we were doing something you feared."

"But how could you go on, with everything erased?" asked Kublai.

"Your agents found two of my backups. Only Rho knew that I had at least two other backup sites. So, your foray into our house didn't actually wipe out the data. We still had it and could pick up exactly where we left off. But knowing, now, how important our work was to you, we knew that there was something for us to find. You showed us that."

Blossom laughed—a pretty powerful sound, with very bad breath. "You created your own nemesis, Kublai."

Kublai wasn't amused. "And what will stop somebody else from doing what *you* did?"

"First, the Scans are over. Nobody knows our data still exists. Second, *this* time if somebody seems to be researching your secrets, you won't be so obvious as to send alien agents to erase their computers, thereby confirming their suspicions."

"It sounds pretty convincing to me," said Blossom.

"We agree, as well," said a bikini girl. "You stepped on your own—"

The other bikini girl cut her off. "Sounds like you bit your own tongue, Kublai." A nicer metaphor.

"Tell you what," I said. "We'll stay right here while I write up my story. It'll be mostly true, including a lot of irrelevant details about

Rho's and my conversations, the amusing idea that I'm actually in love with her, all kinds of details that will make it sound like I'm not lying. My goal is for people to be kind of convinced that *I* think I'm telling the truth, that I'm just wrong. And I will tell about today. About Blossom and his family."

"And us?" asked a bikini girl.

"No," said Rho. Which made me decide that I would do exactly that. She shouldn't *always* get her way. It's not good for her.

So, Kublai, here is my story. I'll publish it in the most prominent place I can find and when people ask me if it's true, I'll say, "It's fiction. Do you know what fiction is?"

Take this to the council. Or let *me* take it and read it aloud.

No, I'm not saying you can't read it well, I'm saying it's *my* words and *my* story, so who can do a better job of sounding like me, than me?

So, you and the council decide. Oh, and I'll use a false name inside the story, so that nobody knows which member of the Scan wrote it. No reason to get the press on my trail—or any of the anti-alien groups that will want to silence me because they will believe every word is true, only I'm telling it disrespectfully.

I think I'm going to start calling your whole alien race "Skinny-Shins." And your Human–Skinny-Shins hybrids, I'll call "hardheads," because that's what you are, with your human head. With that skull inside.

I will also suggest one other strong possibility. You watched us come out of Africa on fragile boats and somehow navigate oceans well enough to spread all around the Indian Ocean. You had never built a boat. You wove things, but you couldn't make anything that required hard tools. Stone tools.

I know, you grip stone objects and tools quite readily now, because you've had a long time to create imitations of the human hand. You did that because you could see how, after only a million years or so, we humans had caught up with and surpassed your civilization. You were in awe of our monuments. Our chariots and carts. Our domesticated plants and animals—something you had never accomplished because you never thought of it.

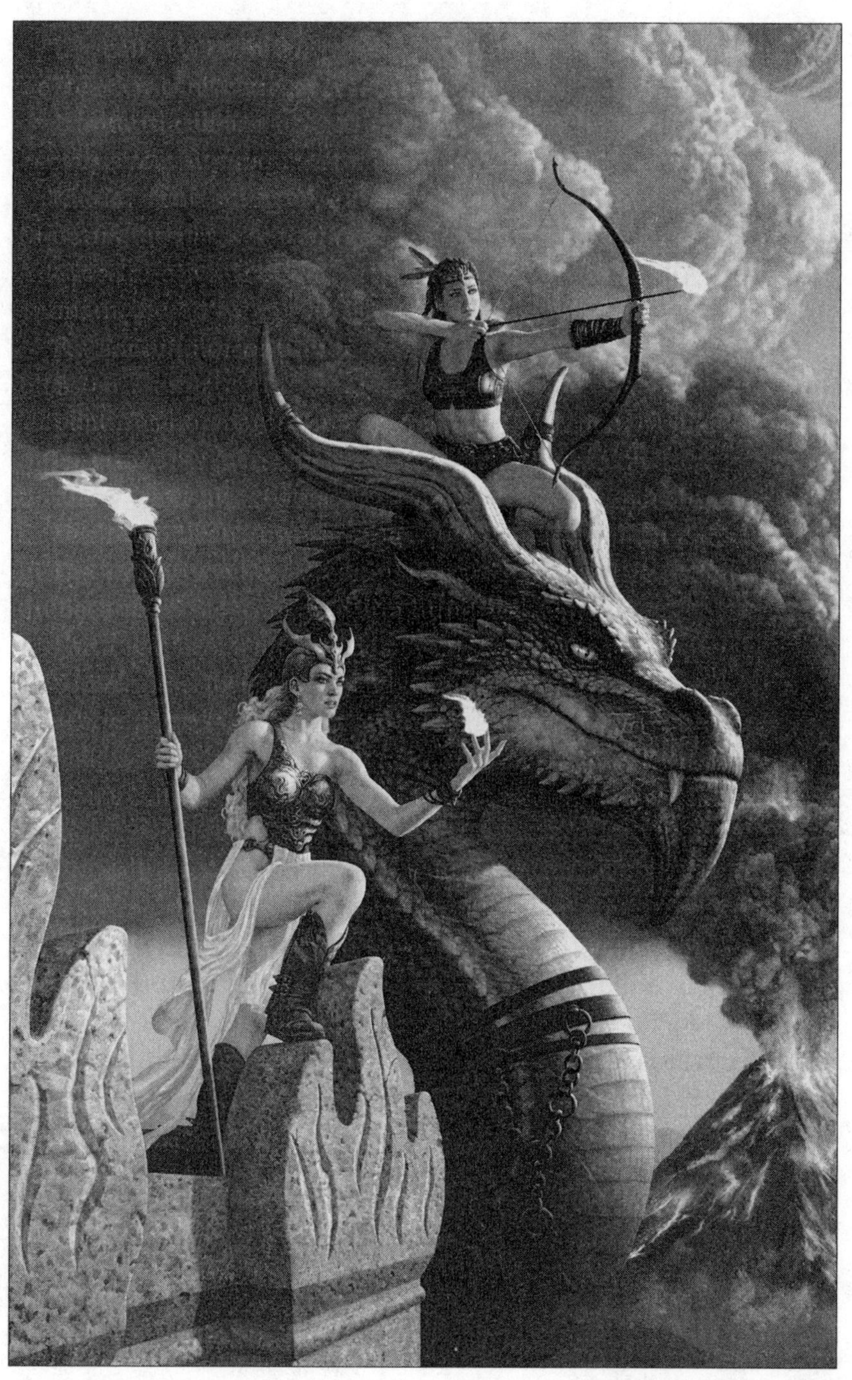

CIRUELO CABRAL

How do you build corrals underwater? How do you *irrigate* crops on the seafloor? On reefs?

So, while you were learning to reshape some of your arms in order to be able to produce working hands and later, feet, you also worked on learning to breathe in atmosphere, learning to *speak* in atmosphere, learning several human languages, learning to read and write. All the skills you already had when you first came to visit us openly.

You had ascertained the ills of our lives that had a biological solution, which in every possible case you supplied. It was generous. It changed and improved our lives. Our attitude toward you was respectful, then it changed to awe, and in very many cases it is now worship. Though of course there are quite a few humans who fear and resent you for the very same reasons.

What my fellow human beings do not understand is your deep plan. You were around more than three hundred million years before we popped up. You don't want to destroy us, defeat us, even limit us. You seem to be working to make the human body stronger, healthier, longer lived.

You have an ongoing program of inducing human males to mate with Skinny-Shins females. In those females, you've made or implanted or bred ova that are predisposed to join with human male DNA to create *you*, more hardheads like you. Able to breathe and live underwater, but on land as well. Able to pass for human when you want, and to disappear into the scenery when you want to escape. The children of Human–Skinny-Shins crossbreeding are all basically Skinny-Shins with add-ons. Because that's what you *are.* Crossbreeds between your original decapods and humans, and then more Skinny-Shins who mate with each other and breed true.

Simple demographics tell us your plan. Your people reproduce normally, increasing in number at a sensible rate. But by breeding with human males, you augment your numbers, while pure human-to-human breeding decreases each year, as there are fewer humans and more crossbreeds. Until someday it happens that there is only a small population of purebred humans and everybody else, *calling* themselves human now, are all *your* people.

Now, you're scientists and the extinction of species makes you sad. So, I think that just as you maintain a population of genetically modified tyrannosaurs, you still maintain a population of *your* people before the human-oriented changes were developed and bred into future generations. Sort of a lingering population of what you had become before we got in your face.

And you'll keep a population of purebred humans, who won't be lured in by your females, so they'll only breed with each other. And that way, you can tell yourself that you didn't wipe out the human species, because look, here they are. And anyway, the human genome lives on in you Skinny-Shins, because you have adopted our appearance *and* our culture.

This world will then be yours, while your ancestors and the remnant of our species are on closely guarded reservations, prisoners on our own planet, which is now utterly controlled by your hybrid descendants.

Hybrids. Hardhead Skinny-Shins. That's who will inherit the earth. Your genome, our genome, melded together with the best features of both. Will those future Skinny-Shins claim credit for all that their human ancestors achieved? And also for what *your* ancestors achieved?

And here is your answer—I'll write it for you. "*We* are something new. Two native species that evolved on Earth, now combined to achieve much greater things in the future. Not an alien invasion, but a new combined native species: the indigenes of Earth."

Since I can't stop you, I have to accept the near extinction of my species, and consider you to be our ultimate offspring. But in the meantime, I do believe that Rho and I will vote against you in the only way that's in our power. We'll produce a whole slew of babies who are purely human, to slow down the disappearance of humans on this Earth.

You treated this planet better than we did. Some might say you deserve this Earth more than we do. Certainly, you have every bit as much right as we do to prevail, and your demographic war of conquest is bound to succeed.

To the human readers of this story: Of course, everything that I've written here is purely made up. There is no ocean-born species imitating and then interbreeding with humans. There is no secret distribution of cures for all our fatal and annoying diseases. Antarctica is just rock and ice with a few penguins running around. Don't pay any attention to imaginary demographic emergencies. And if you hear rumors of sightings of living dragons, ignore them. Get rooted in reality, and stop wondering if imaginary stories might have a basis in truth.

A Ready-Made Bubble of Light

written by
Thomas R. Eggenberger

illustrated by
HAOTIAN ALLEN ZHANG

ABOUT THE AUTHOR

Thomas R. Eggenberger penned his first stories as a child in the small Ohio town where he was born. Continuing to write as he lived in Japan and France and traveled much of the world, he's now up to six manuscripts and a book's worth of short stories (and counting). When not hanging out with his lovely wife or spoiling his dogs, he works on unconventional fantasy trilogies and a wide range of science fiction stories.

He has previously been published in Fiction on the Web and Fairlight Shorts and received two Silver Honorable Mentions from the Writers of the Future Contest before placing first with this story, which is his first professional sale.

"A Ready-Made Bubble of Light" began with a simple curiosity: What would it look like if you actually stopped time (and not just people, as in most similar stories)? The answer, of course, is "dark," but things get interesting when you start mixing bubbles of stopped time and bubbles where time is running. And then it gets even more fascinating when you have a team of "timers" going into disasters frozen in stopped-time bubbles to save lives—and stumbling upon a threat to the foundations of our shared reality.

ABOUT THE ILLUSTRATOR

Haotian Allen Zhang was born in Baoding, Hebei, China. Encouraged by his family, he began studying traditional Chinese painting and calligraphy at a young age. In middle school, he started receiving formal training in sketching and painting which continued through high school. These years of practice built a solid artistic foundation for him.

After graduating from college, Allen worked in a related field, but it wasn't until he encountered concept art for games and films that something clicked. He was immediately drawn to the freedom it offered—the ability to build worlds, shape characters, and create atmospheres in ways his job at the time could never offer. The realization made it clear to him that concept design was the path he truly wanted to pursue. He resigned from his job, spent a year building a portfolio, and was eventually accepted into the College for Creative Studies to formally study concept design.

Allen is now a freelance concept designer and a portfolio coach for students. He continues to explore visual storytelling, design methods, and world-building, committed to turning imagination into vivid images and meaningful stories.

A Ready-Made Bubble of Light

The job had started fine.

We arrived at the scene fourteen minutes after the time-stop bubble had been triggered, knowing from vid capture and sensor data that it was a predicted car crash at seventy-six miles per hour, four riders, heading straight for an overpass exterior wall. Road was blocked, horizon secure. Clean bubble, just the target car inside. Bubble was two hundred and fifty milliseconds pre-impact, which gave us a runway of about twenty-five feet. More is always better, but it was enough to save them.

Xi sprayed gel to cool off an entry point while I prepped the heavy stuff with Madhuri. A cutter, a thumper, and four laser handsaws were far too much to carry, so I popped out a porter. Ours are custom-made, wired throwbacks; a wireless model could hiccup, and without precise control there's too much risk—clip any part of a body with a time-in bubble when there's momentum involved, and what happens isn't pretty. Even tethered, you have to keep the whole cord under time-in for it to work, and you have to know what you're up to. It looks simple, just a rack of shelves on wheels, but I'm the only one on the team licensed for it.

Katy was latching on her nanite tanks, struggling to get her shoulder-length blond bob clear of the straps.

I shook my head. "Not this time, Crunch." Katy got her nickname from the cereal she liked to pretend was a proper meal: it sounded like she was crunching gravel as she chewed.

"These nanites aren't for eating, Seb," said Katy. She rubbed her hawkish nose, sniffling. "I know that's too slow. They'll seed the steel with oxides, brittle it up."

"We need the hood in one piece. We're already at risk for shrapnel. Leave the nanites, we have to stick to mechanics on this one."

I couldn't blame her. It was her first crash. Even I'd only seen a handful in my career. Almost all had been on the racetrack, where the driving was still manual and the safeguards sometimes failed. A vehicle crash on the autoway was rare as a unicorn—and usually turned out to be a painted horse with a party hat. Malicious, I mean, not a real failure.

Madhuri unhooked the ceramic parasol from her suit and stowed it on the porter's top shelf. I grabbed it and handed it right back. "Come on, Mad Dog, you know better."

She played up a pout, short nose crinkling, long chin dimpling. "You called for cooling, right? So no plasma risk."

"Shrapnel."

Madhuri grunted unhappily but hooked it back on her suit. I knew she hated having it drag behind her; it wasn't made with short people in mind. But there was no way in hell I'd let her leave it. Complacency got timers killed.

We zipped up our navy-blue suits and ran final checks: oxygen tanks all in the green, no suit leaks detected, helmets snapped in, AR and com checks clean. Some timer teams leave an air supply optional—not me. Air doesn't move and mix in a time-stop; stop too long in one place or hit a pocket of exhaust, ozone, or chemicals, and you're in for a bad time.

Xi set down his cooling gel and led us through the cool patch he'd made on the bubble. We stayed close to keep our time-in bubbles linked, so we could see each other. I clicked on the AR wireframe and set the porter off to the side, then the four of us put our heads together for the huddle.

"Mad Dog and I have done a few of these, so monkey-see, monkey-do. Stop and ask if you have any doubt at all—one percent rule. One-tenth of one percent. This'll mess with your head even if you've done it before."

"We've all dealt with shrapnel, Seb," said Katy. "We're not kids."

"You're kids for your first five years, Crunch, no shortcuts—but everyone needs the reminder, even me. Shrapnel we can model and avoid. This vehicle we need to harness and control."

Katy kept her yap shut, thankfully.

"Occupants are at speed; there's no way to pull them. Only way out is through. We sever the hood section, push it out and spray mallow down the runway. We'll cut away as much else as we can, but don't take risks—absolutely zero tolerance for clipping here. Tight bubbles when you're near the vehicle, keep your arms in.

"First step is laser cutters, one on each side. We'll sever it just before the windshield, then light it up and let it hit. That will give us space for close work, cut away the interior components as much as we can. Then we spray the marshmallow and exit."

Xi raised his hand. He was the right kind of rookie—he listened more than he talked. Maybe his protruding ears and acne-scarred cheeks had made him shy. "We don't stay in?" he asked.

I shook my head. "We don't have shielding that can cover a scene like this, so I've asked our friendly housemates for a spare firetruck to get external cooling on the bubble. Should be right on our heels. No need for us to stay in since there'll be no plasma to protect from when the bubble blinks out. We go out where we came in, gel's on the porter. Other questions?"

None. I checked my watch. "Twenty-two minutes, plenty of time. No rushing. Cool heads. Crunch with me, Xi with Madhuri. Mad Dog and I'll use the cutters.

"One more thing—first time you light up anything, tell your partner and make sure they're clear. Vehicle is at seventy-six miles per hour, so anything that decides to peel off will mean a suit breach or worse. Madhuri and I will lead, we'll kiss the sides on the way up—if anything is going to fly, let's have it fly now."

"If we can bleed off momentum, why not just do that?" asked Katy. "Stop the car."

"What do you think, Crunch? Why not just do that?" I asked.

I could almost see her brain work as she remembered the people inside.

"Let's go," she mumbled.

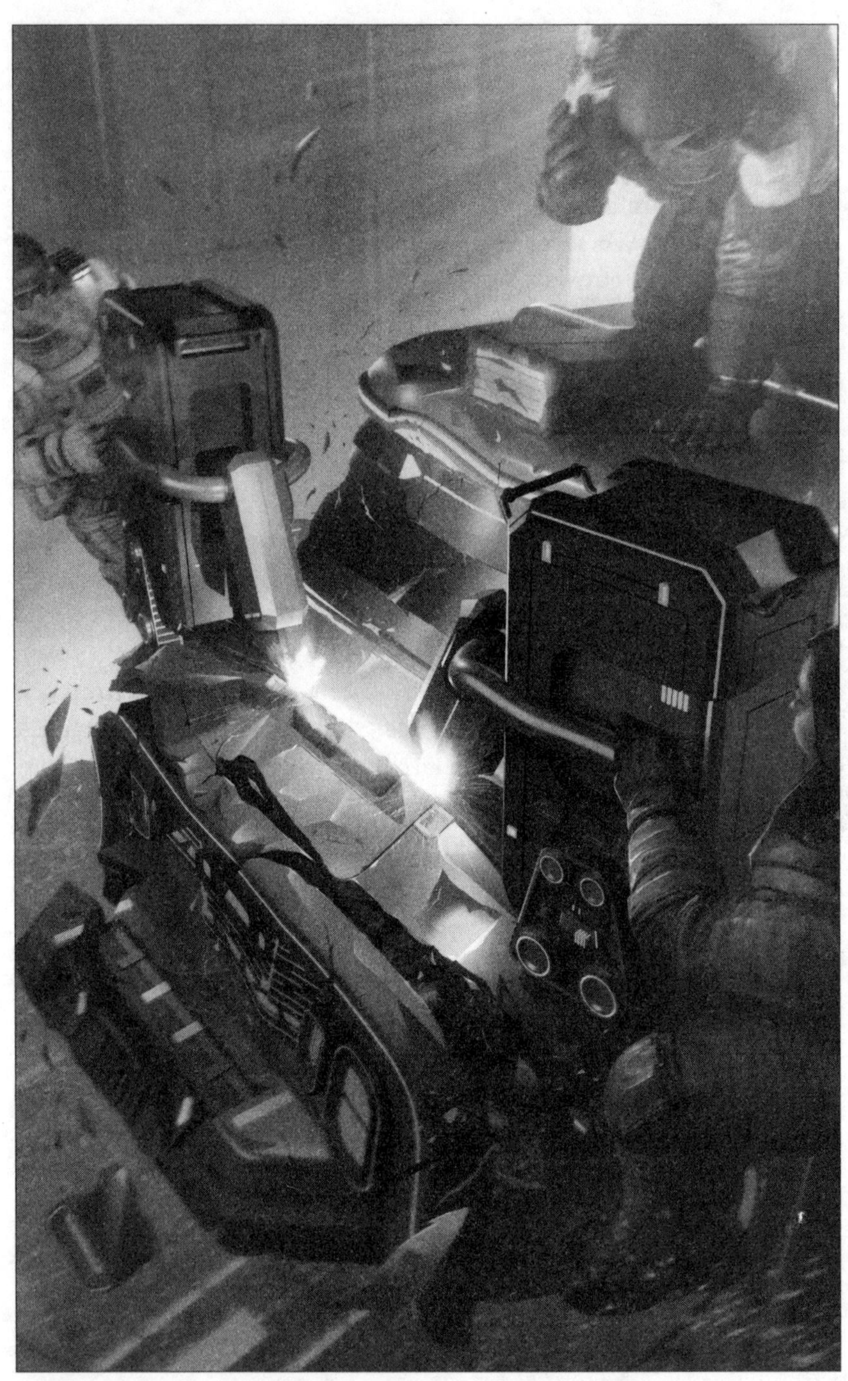

HAOTIAN ALLEN ZHANG

"That's right, Katy. There are four people at seventy-six miles an hour in that car, who might not enjoy glass and metal suddenly at rest."

"If we could even do it," chipped in Madhuri. "That's not how it works. We're talking about small stuff—pebbles churned up, devices in the door pockets."

Katy tried to break huddle, but I pushed her back in. "Listen to experience, Crunch. We've been over this."

"Yes, Chief."

"We've worked with victims at speed before, but this is the first time you'll be close enough to kill them if you screw up. Maybe you thought you were careful before. Double that. Triple it. Be hyper-aware of what you can see in your time-in bubble at all times. Any time you can see any part of the vehicle, stop and orient yourself, because a step in the wrong direction means—"

"Broken bones, internal bleeding, death," interrupted Xi.

Not a fan of being cut off, but Xi's heart was in the right place, so I let it slide. "Exactly. If photons are moving, everything you see is moving—at seventy-six miles an hour, while everything else is at a dead stop. You see a hand, that's now a broken wrist. You see an arm, that's now a fractured elbow. You see a head? Congrats, you've just broken their neck. Got it, Crunch?"

She nodded, daggers in her hazel eyes, but also looking appropriately chastened. A weird mix. It was always a weird mix with Katy.

We separated, stepping off in different directions, and the others immediately blinked out of existence as though they had never been.

Suddenly alone in a black expanse, I stepped over toward the AR wireframe of the porter. The AR overlay worked better for immobile objects like the vehicle and the porter than for the others on the team. It could only show last-known positions, which gave a ballpark but sometimes did more harm than good.

The wireframe snapped to reality as its time-in bubble intersected with mine and photons could bridge the gap again. Madhuri blinked into existence beside me as I tugged free a cutter and checked my watch. "Two minutes?" I suggested.

"To start the cut?" she asked.

"Yeah."

She nodded.

I handed off the cutter and rolled over the second for myself. They were bulky, mini fridges on wheels with an open jaw on one side. We set our watches, tapped them together and went off to our own worlds.

It was a good twenty seconds before Katy linked up at the rear left of the car—and she popped in ahead of me.

"Damn it, Crunch, are you trying to piss me off? Get behind, I need to—"

"Already skimmed it, Chief." She turned back toward the hood.

"Get behind, Katy."

She waved dismissively and blinked out.

Ninety seconds. No time for this nonsense. I stayed well clear of the car so I could hoof it without risk. The control wire for the porter tugged at my hip—the bubbles had decoupled. Had to walk back and collect it. Took it slow and steady after that.

Katy popped in at my feet while I was horizontal, pulling the cutter's jaw across the hood. "I can help, Seb."

"I've got thirty seconds to finish and clear." I eyed the wireframe as I wriggled sideways, tugging the top of the cutter's jaw along. The metal groaned beneath me, a trapped band of momentum. My bubble was so tight I'd had to tuck my legs up to keep them in.

"That's why I'm offering to help."

Too much distraction. "Get the hell clear."

She disappeared.

Twenty seconds. Almost there. Last link of the jaw, but it was caught on something. Underneath? Upper and lower parts of the jaw were wirelessly linked. I slammed it into the hood and scraped it along, biting into the paint, imagining the lower line of the jaw moving in tandem. It ground forward at an agonizing plod, then stopped. I wrapped the steel handle in both suited palms, sweat dripping onto my visor, and drove forward. No good.

Ten seconds. Was Madhuri on time? If she went and I didn't, the front could sheer when we lit the zone. An unpredictable impact put us all in danger. I abandoned the cutter, scrambled over the hood—and suddenly butted heads with Madhuri.

"Need something, Chief?"

I laughed and sighed with relief. "Hiya, Mad Dog."

"I had a feeling you were behind."

"Why's that?"

"Just a feeling. I listen to my gut."

"Thanks to your gut, then."

"And don't lecture me about safety, okay, Chief? I was watching for the end of your cutter."

I took a deep breath. "Deal."

"Is it stuck?"

I nodded. "Take a look."

Madhuri wormed past me. I heard her jiggle the jaw. I heard it scrape, then slide. She pulled it to meet the handle on her own cutter's upper jaw.

"All good, Chief." She winked. "Twenty seconds, to be sure?"

"Thirty. And I'll light the zone thirty seconds after."

Katy wasn't in the danger zone near the hood, so that was a plus. I walked to the rear and back without seeing her and had time to move the porter back before pushing the button.

Cutters are interesting things. They look rugged and mechanical, but they have some pretty advanced tech. The bottom jaw is lined with little foot-and-a-half time-in boxes, capable of the smallest bubbles currently possible—only twenty inches in diameter, though they can go bigger if needed. Each time you lengthen the top jaw by pulling out another segment, the matching box on the lower jaw goes live and the bubble widens until it finds the signal from the top jaw, then it uses the signal to match the length. If it doesn't match, it doesn't cut.

I squatted beside the power unit as I hit the switch, watching it work. There wasn't much to see—it hummed, and smoke wafted from the hood. I imagined the oscillating buzz saw of invisible laser lines, burning through metal.

Katy blinked in before me. "Are we behind schedule?"

"Maybe a minute. Ready to light the zone in a few more seconds."

"Good, I'll—"

"You'll stay right there, Crunch. Stop running around."

"It's my first crash, Seb. I wanted to have a look."

"You'll learn by watching me. It's not safe for you or these poor people to have you poking around in the middle of their car crash. Stick beside me so you can watch. I need to know where you are."

"Yes, Chief."

The cutter thumped off. "See," I said, "you missed the cutter." I tapped my watch to start a twenty-five-second countdown.

"I've seen the cutter."

"Did you see anything when you skimmed the car?"

"No."

"Nothing moved? Inside or out?"

"Nope."

"Good. No clipping?"

"I was careful, Seb. You can redo it yourself if you don't trust me."

She didn't get it. "Even once is dangerous."

"Then why do it?"

"Protocol. It helps keep us safe."

"So, we do something dangerous to keep us safe?"

"Not dangerous for us, Crunch, dangerous for them. It's a balance. If we can't finish, we can't help them."

My watch beeped. I readied the time-in box, diameter already calculated by computer based on the scans. There was actually an automated compressed-air thrower available too, to make sure the box landed just so. It was back at the station, where it always was. Humans are good enough computers to aim a toss.

"Watch and learn, Katy. I'm going to toss a time-in box onto the hood, now that we've cut it away from the rest of the vehicle. Keep me in sight."

I backed up a few steps and widened my bubble, eyeing the AR wireframe of the hood as I did a few practice hefts with the time-in box. My hand-eye had stayed sharp as I aged—plenty of practice kept it keen. I switched on the box and made the real toss.

It disappeared as soon as its bubble lost contact with my own, but

I could tell from the arc that it was good. It would clamp down on the center of the hood with its electromagnets, unleashing all the momentum in that hunk of steel, riding shotgun as the front of the car compacted with a smash against the concrete wall of the overpass.

I unplugged the porter. "Follow me in." I unhooked and opened my ceramic parasol.

This was where I felt every single one of my forty-eight years, crawling along the ground toward the impact site, pushing the parasol ahead. Projectiles were unlikely, but procedure was procedure. It seemed ages before we finally hit the bubble. I checked my watch: one minute, fifteen seconds. Aching knees will make anything seem like an eternity.

The overlay updated as wireless signal from the time-in box finally reached me. No danger from projectiles. The hood section had crumpled perfectly into a rectangular heap. It was a relief to sit up and tuck the parasol away. Katy helped pull me up.

"How are we on time?" she asked.

I checked my watch. "Still about fourteen. No need to rush. Stay here, I'll be back with the porter."

Madhuri and Xi were on the other side of the hood section's time-in bubble when I got back. I gave them a hearty wave, and they waved back.

The thumper was heavy to drag but quick to set up. Katy helped me position it just shy of the compacted hood. Pneumatic spikes dug into the asphalt to brace the machine. I gave Madhuri a thumbs-up.

She flashed two fingers as she released the electromagnets and grabbed the time-in box off the hood section—twenty seconds for Xi and her to get clear. I signaled her to round the car and rejoin us; probably no need to send them that far out of the hood's path, but better safe than sorry. Another quick thumbs-up, and then she cut the power on the time-in box.

I let the thumper go to work as I started the twenty-second countdown. Calculations showed that eleven hits should be enough to clear the compacted hood section from the path of the vehicle; I let it get in a few extra, since we had the time. Thirteen seconds later, I hefted another live time-in box onto the hood.

The accumulated force would send it sideways. The AR overlay showed that it would hit the corner of the overpass and stop in the middle of the road, clear of traffic now.

Madhuri and Xi popped in just after.

"You missed the toss," I said.

"It was amazing." Katy rolled her eyes. "Chief's really on fire."

"Sorry, Chief," said Madhuri. "Cutter slowed me down."

"Have Xi drag it next time. That's what kids are for."

"That, and spraying marshmallow."

"Yup." I pulled two tanks of the foam from the porter and held them out. "Leave us enough room to work."

Xi hefted his without complaint and headed over. Katy grimaced but took hers, too. They blinked out.

Madhuri helped me at the porter as we racked the thumper and the cutters and pulled out laser handsaws.

"Crunch giving you problems?" she asked.

I clipped a second handsaw to my belt. "Nothing I can't handle. She has a point, anyway—she's been here four years, but we hardly let her do any more than Xi."

"Give her my jobs, I don't mind."

"I'll think about it. But not on this one."

"No job is easy, Chief."

"Victims come first. And we're down to eleven minutes now, so shut up and start cutting."

Madhuri blinked out without a word.

We had a spare box, so I set the diameter wide enough to light the zone and carried it over to the middle. Spraying the pillowy marshmallow foam was easy enough—usually the first task we gave to kids—but it expanded, sometimes unpredictably, and an AR wireframe was no substitute for really eyeballing something.

The time-in box put the whole twenty-five-foot stretch in time-in, from where the car's hood had been all the way to the underpass wall, linking our bubbles so we could see each other.

First thing Katy did was throw a look. "I don't need to be micro-managed, Seb."

"I'm not." *I'm trying to help you.* Had the sense to bite my tongue on the latter.

"Good. Don't."

Xi was doing well, building up a cone of layers as he'd been taught. Katy was spraying too broadly, wasting time. I decided to tell her later, since it wouldn't hurt anything; better for her to stay out of my hair.

"Ten minutes, folks." I tightened up my bubble and joined Madhuri near the windshield. It would have to come out, but first things first. She was already opening her ceramic parasol as I reached for mine. We kneeled and braced.

Madhuri counted up to synchronize as we expanded our bubbles. The AR overlay showed the time-in boundary creep forward until it hit the windshield and crossed behind. A keening wobble, some cracking, nothing more. The cut of the hood had been clean, and the laminate held the glass together with only a few fractures and faint spider-webbing. We packed the parasols, dialed back the time-in boundary, then cut a couple of holes in the windshield with the laser handsaws and yanked it out manually.

I was tempted to heave it out of the way, but I knew better: it would get stuck at the time-out boundary and might endanger us on the way out. Madhuri and I walked it over to the side, out of the vehicle's path, and lowered it onto the grass of the autoway median.

Xi waved and came over when we got back. His marshmallowing was damned near perfect, shaped so the car would be enfolded and embraced. Each microscopic chamber in the aerated foam would resist but collapse, stealing momentum in a cascade of infinitesimal collisions. He'd stopped roughly a yard away, which was about right. Katy would be done in a few seconds.

I checked my watch. Eight minutes. No rush, but no time to dawdle. At least two minutes to exit, count it as three to leave a buffer. Three more to finish the padding and give the victims their best shot at surviving the crash.

"Xi, give me your foam. Start cooling the exit. Gel is on the porter. Expect us in three minutes." He glanced at the saw at my hip, originally meant for him, but handed off his sprayer without a word and went for the porter.

"Katy, give your foam to Mad Dog when you're done—"

"Done!" she sang and jogged her sprayer to Madhuri with a grin.

"And stay here to observe."

"Can't I help? There's more foam on the porter."

"Observation only. You haven't done this before."

"I've sprayed plenty, Chief."

"Not like this."

Katy made a sour face but moved back.

Madhuri stood waiting, saw hanging at her hip, thumb on the sprayer.

I stepped closer. "We're done with the saws. Taking out more won't help." I gestured around at the AR overlay. "These two are facing backward, we can't do anything about the seats. The two in the back are facing forward, but there's not much more we can do for them. All four have their belts on.

"We'll max out the mallow, get some inside. It won't fill to fit, but it's the best we can do. The main help is what we've already done."

Madhuri eyed the canister skeptically. "Can they breathe in foam? Can they exit?"

"They'll be fine."

"You sure?"

"You've just volunteered for the foam dunk tank at our next 3F."

"Crap."

"Experience is the best teacher." I used my watch to turn off the extra time-in box that had been lighting the whole scene. Madhuri and I checked the AR and agreed on angles, then slowly retreated as we sprayed mallow, leaving the foam in time-out in midair—otherwise it would just build up at the time-in border, expand and fall. Katy stayed just close enough behind me for our bubbles to intersect, watching with bored eyes and crossed arms.

When we were almost done, I handed off my foam to Katy.

"Thanks." Sarcasm thicker than mallow.

I left Madhuri and her to it and walked the porter to the exit.

Xi was waiting. Cooling gel oozed down the side of the bubble. I expanded my time-in boundary so I could peek outside.

Two firetrucks in sight, fountaining water up and out of sight; curtains of water washed through my time-in. I was sure Xi knew, but he gamely sprayed away with his gel. At least my mantra had taken with him, then: procedure first. Maybe today we didn't need to cool the exit, but tomorrow we would, and there had to be zero room for lapses. Do it whether you need to or not, do it until your body remembers and you can follow procedure half-asleep.

I herded the team out, then punctured the wall of gel myself. We made a quick visit to the nearest firetruck, and they sprayed us down with glee—too much and too forcefully, of course. It's a friendly rivalry.

We were back at our truck with two minutes left on the bubble. I pinged the captain on the scene and started pulling off my suit as the firetrucks cut their cooling water and moved away.

It always makes my head spin when a bubble blinks out—the bigger it is, the more surreal. In an instant the looming black was gone, the landscape whole. The car slapped into the waiting foam and tilled through it, coming to a stop a full yard before the concrete wall. Marshmallow wept from the roof and tided away from the sides.

There was no movement in the car. Suit hanging at my waist, I peeled my eyes and took a step on instinct. Paramedics hustled in.

More mallow washed out as they pulled the doors. Snowy bodies hefted out. Limp. Still on the pavement. Scans, bags, injects. Nothing.

Madhuri was muttering. I blinked wet. Had we screwed up?

"You said they could breathe."

"They can. They could."

"Doesn't look—"

"It was seconds, Mad Dog. That's not it."

What was it, then? I ran back through the job, entry to skimming to cutters to thumper to saws to mallow foam, and everything should have been fine. I trusted Madhuri. It could have been the kids not taking care with clipping—Katy had disappeared for a stretch—but that didn't make sense either: no signs of trauma that I could see.

Xi shuffled over, arms crossed. He looked stricken. "Did we lose them?"

The EMTs were giving up. Baffled shrugs. Unrolling body bags.

"The hell?" said Katy.

She spoke for us all.

I can accept failure. I can accept deaths happening on my watch. No way to do the job otherwise. What I can't accept is not knowing why.

Every failure needs to make us better; every loss has to teach us how to avoid the next. If we don't know what happened, I make it my mission to find out. So, I paid a visit to the coroner the next day.

Joon and I get along fine, though we'll never be friends—she likes to talk more than I like to listen. She jabbed a finger the moment I stepped into the room.

"I know exactly what you want."

Not impressed. Anyone could see my badge stalking around the building map, call up my recent cases. But I smiled genially and let her have her moment.

"I've already sent it," she continued.

"That'll be great when I'm back at the house."

"You don't have your AR?"

"Nope."

"What kind of timer are you?" she asked.

I shrugged.

"An o——"

"Old-timer, got it."

"It's funny, Seb. You should laugh. It's not illegal to laugh."

"It is, if it's the tenth time you've heard it."

"And you still forgot the punch line. You're proving my point."

"Are you going to abuse me, or show me what you found?"

"A little of column A, a little of column B." She beckoned me over to her screen.

"What'd you find?"

Joon brought up images onscreen. Brain scans.

"Were they dead before the stop?" I asked.

"Listen for once, Seb. You don't have epilepsy, right?"

She asked me every time. I shook my head. Joon clicked on the screen's hologram effect. I didn't have epilepsy, but holograms gave me a headache. I squinted at it.

"This is one of the more obvious ones. Female, thirties, sitting in the front, facing the back. Do you see the problem?"

I pointed to a wide swathe near the base of the brain that glowed white. "Is that a problem with the hologram?"

"It's not. It's a problem with this poor woman's brain." Joon rotated the brain, pointing to several black squares near the surface. "These are augments. Not unusual, though she had money if she had this many."

She rotated the brain to show the bottom and pointed to another black rectangle—this one at an awkward angle, a sharp corner jutting. "And this one seems to be the butler of the bunch."

"The butler?"

"It's always the butler. Read a book sometime."

"That augment killed her?"

"It's embedded in her meninges, indicating that it came loose inside her brain at a very high velocity—high enough to shred her amygdala, her hippocampus, and her brain stem, then cut all the way into the dura mater."

"Seventy-six mph?"

"What?"

"Seventy-six miles per hour. Would that do it?"

Joon shrugged and pushed her long black hair back over her shoulder.

"You don't know how it happened?"

"I do the *what*, Seb, not the *how*. That's your department, right? Or the police? Oh, but there's more." She skipped the next one, went to the third. "The second one is pretty much the same: female, forties, identical injury from an identical augment.

"The third and the fourth are also alike—both male, both in their thirties, sitting in the back facing forward. And theyyyy..." She rotated the brain hologram to a front view. "Do you see it? They have the same injury, but in reverse."

I peered at the brain. The bottom of the prefrontal cortex was a smooth, luminous white, a striking contrast to the wrinkled gray above. "So, these two have brain damage from the hippocampus, through the middle here—"

"The amygdala," Joon jumped in.

"And then up into the prefrontal cortex?"

"It seems to have bounced off the meninges instead of cutting into it, maybe because the angle was different. It went up into the prefrontal cortex and caused even more damage. The brain stem seems to have been minimally affected in the males, but I suppose their brains were scrambled enough to kill them."

"Very scientific."

"That's as good as you get, old-timer. The brain is the last great frontier."

She dragged the brain around with a finger so I could see where the augment had lodged, low in the center of the prefrontal cortex. "There's the culprit. And that's it." Joon turned off the hologram effect and quickly tapped through the scan of the fourth victim on her way back to the main screen.

My breath caught. "Back." I leaned in over her shoulder. "Show me the fourth victim."

She did it without question. Maybe it was the intensity in my voice. I couldn't stop blinking.

And there it was: the distinct C-coil of a time-in box.

Joon clicked on the hologram view. I rotated the scan to view the augment on both sides, but it was just a flat black square. I reached past her and clicked off the hologram. There it was again, on the flat screen—the C-coil.

I went back through the others. It was there each time—visible on the other side for the first and third victims, but always there. When I clicked on the hologram, it disappeared. Too fine a detail? Score one for old tech.

"What is it?"

"Those look like time-in boxes."

"Ah. Interesting."

"You don't get it." My mouth felt thick. "Those don't exist. That tech doesn't exist. It's impossible."

"It's a new model?"

"Our best tech is a foot-and-a-half time-in box. This is what, an inch?"

"A bit less."

"It's impossible. No one has this. It doesn't exist."

"We're looking at it."

I shook my head. I couldn't believe my own eyes.

Joon looked thoughtful. "It would explain the sudden velocity."

She was right. A live time-in box of that size, in the brain, at seventy-six miles per hour, with everything around it suddenly stopped out? It would ping around the brain like a bullet.

Our time-stop hadn't saved them. It had killed them.

There was no next step with Joon. She couldn't cut out an augment for me, she couldn't tell me who the victims were. Procedure. "Check the news," she'd said. "You'll know soon enough."

I didn't doubt it—car crashes are rare enough for easy headline fodder. But days passed, and nada. News about the crash, yes, but no info on the vics. Gave Joon a ping and she shrugged at me in her VR cottage, English setter nuzzling her knee. "Can't release details until we notify the next of kin." So, the next of kin was hard to find? She wouldn't give me more.

I kept what I'd learned from the team. We were horse wranglers, and this was a centaur: it would confuse, not illuminate. Maybe I'd tell them later, if I could find out enough to be useful.

In the meantime, we had other calls. Disaster has no holiday.

Some were simple. Guy took a header off a high balcony, box pinged the velocity and stopped him out; we gave him a soft landing. Knew he had to be poor, a rich guy would just get an aug to make him happy. Tickle the right part of the brain in the right way at the right times, you feel fine. But you have to pay for it—humans didn't evolve that way. Used to help the tribe to have anxious insomniacs, up to sound the alarm. Now they just suffer, and nobody gives a damn. No reparations for unlucky genes.

Some were tough. Poor woman got sprayed with some med-grade nanobots, designed to break down bio waste. Didn't get stopped out till half her face was gone. We juiced her with the standard 100-volt,

100-millisecond charge to kill the bots, but had to light her up to do it. She screamed like hell until we put her back in time-out. I asked the paramedics after if she would be all right, and they weren't optimistic—problems from the dead nanobots in her system would linger, and the trauma would take a heavy toll; she would struggle to accept a new-grown face. They gave her fifty-fifty odds that she'd off herself by end of year. At least the prick who'd done it got a fitting end: her ex was found at home a few hours later, nothing but a pile of goo. Nanobots are hard to contain.

Through it all, no matter if it was a hard case or some feel-good fluff, Katy pushed to take on more. She wanted equal weight with Madhuri and me and pressed hard to show she could do it. It ground my gears, but privately I admired her pluck. No matter how much I tried to rein her in, no matter how I punished her back at the house, she'd jump the gun on the next call so she could get her taste.

She was sulking in the kitchen, morning after I'd yelled at her for running ahead. Hunched at our end of the long dining table we shared with the firefighters, nose in a book. Munching on that cereal she loved, the one that had given rise to her nickname.

I sat across. Clearly, she didn't want to talk. Too bad. "Crunch."

She tossed out a grunt, stayed glued to her book.

"Gotta talk it out, Katy. Come on."

She took an extra-big bite. Sounded like she was chewing glass. But she closed her book and leaned back. "Talk what out, Chief? I'm a screw-up, right? You don't trust me."

"That's not it, Katy. It's really not."

"Then what, Seb? I've been here more than four years, and you treat me like a kid."

"You do more than Xi."

"Hardly. And he's barely a year in. He *is* a kid."

"Crunch, you know how long I've been doing this?"

"Since the dinosaurs roamed the earth."

"Near enough. And Mad Dog's got half that, give or take. Between us, we still don't know a quarter of what we should. We see something new each week."

"So, you want me to ride along for another decade?"

"Nah." I rubbed an eyebrow. Felt like one of them wanted to fall. "Ten years later, I'd just tell you the same thing."

Katy crossed her arms.

"We always have to put the victims first. That means the ones with the most experience have to lead. The ones with the most practice. Say you're stopped out in a car crash, an explosion, a building coming down on top of you—who would you put in charge of saving you? Me or you?"

"I've heard this before, Seb."

"And you'll hear it again. You know I'm right."

"Times change. Tech changes. You're not as steady as you used to be, even four years ago. I see it—your knees, your elbows. Hell, you're slow with the porter, and you hitch your throws sometimes."

I shrugged. "You see your chance. I'm halfway to being old."

She snorted. "Halfway."

"Stick here till I'm hobbling around in diapers, then. I don't want you to leave, Crunch. You've got talent, anybody can see that. You think you haven't learned anything the last four years? Look at Xi and look at you. He's a kid. You're a—teenager, or something. You know just enough to be dangerous. But it needs to be second nature to you, and you're not there yet."

Katy leaned over the table. Her forgotten cereal slumped away from crisp. "Just let me help Madhuri. She's up for it, you know that. She'll still take lead, but I'll get more hands-on. You get the same result, and the team's not dead when you retire."

I reclined and thumbed my chin, pretending to think about it. Can't give in too easy or they get cocky. "What's wrong with me? You can keep helping me out. We make a good team."

"Mad Dog's not a control freak like you, Chief. Come on."

"All right. Done. Just one condition: you make sure she volunteers for the foam dunk tank at the 3F fundraiser."

We shook on it.

Finally, almost two weeks later, the news hit: MERCER HEIR DEAD IN AUTOWAY WRECK.

I'm hardly one for celebrity worship, but even my heart skipped a beat. That was Madeleine Mercer's daughter in the car? From the article, it was the one in her thirties. The woman in her forties was a corporate lawyer; the two men were bodyguards.

You'd expect a quadrillionaire's daughter to skyhop, not squeak along on the autoway like some jill headed to a construct gig. The security and the lawyer seemed pretty standard, but the autoway ride rubbed me wrong. And that aug. That impossible aug.

Knowing it was Mercer's daughter, though, the impossible looked a lot less so. If anyone could have built something like that, it would have been Madeleine Mercer. She'd have the genius, the tech, the technicians to do it—and the legal force to keep it quiet, the finances to hoard it instead of selling.

But I couldn't comprehend why she wouldn't sell a breakthrough like that. It would make her another quadrillion for sure, and help millions of people along the way. And why the hell would she implant it in people's heads—including her own daughter's?

Time-in boxes that small would be like going from pickaxes to scalpels. We could solve our clipping problems, save lives and limbs with perfect control over how we put vics in time-in. Nano injects would be a breeze. And that was just for us; there'd be a whole slew of other uses that would never even occur to me.

I'd be lying if I said that wasn't part of my motivation. I wanted the tech. Not just for the good it would do, either—my inner cowboy was still in there, giddy over bragging rights.

But, more than that, I wanted to know why Madeleine Mercer was holding it back. And I wanted to know why she'd murdered her own daughter.

The augs were gone, of course. Joon and I had both flagged the hell out of those corpses, for autopsies and evidence collection and every cause-of-death addendum with a half-cocked link, but it didn't mean a jot once Madeleine Mercer made her wishes known. The official cause of death was brain trauma from the crash, and that was that. Bodies released, bodies torched, augs gone.

Silently, I had contemplated, in the long nights of those two short weeks, asking Joon to slice out an aug and seal the body back up. The words never made it out of my mouth. Process was process. I couldn't have her risk her job. When the ID came, I was doubly glad I'd squashed the ask. Mercer would scan the bodies, dead cert.

I pushed the cops to take it up. For Jane Doe they had zero interest; for Larkin Mercer, they had far less. I teased them with scans showing the impossibly small time-in box, assured them we'd done a proud job at the crash. They very generously did not laugh.

Still knew some old cowboys from my freelancing days, so I went fishing. Most had gone straight, most as timers on asteroid mines, but there are always side hustles and old friends. None had seen tech like that, and all thought I was joking around—or that I'd gone mallow-brained, believing every cooked pic in my feed.

No idea what came next, I broached it with Madhuri. Trusted her to keep her mouth shut, even if she didn't believe.

Madhuri took it with a calm little shrug. "Explains a lot, Chief."

I nodded. "Think I should give up?"

"Nope."

"Yeah?"

"My hero is Lilly Miura."

"I know."

"She didn't give up."

"Deep."

"Piss off." She smiled, a twinkle in her eye. "Let's do it together."

"Do what?"

"Go to the source."

I snorted. Mercer herself? Gutsy.

"You have a better idea?" Madhuri asked.

"Nope. Hell of a risk, though."

"Every day's a risk, Seb."

I rubbed my cheek. "You sure?"

"Lilly Miura."

"Fair point." It was, too. Belt mining wouldn't exist if she'd stopped short.

"Besides," Madhuri said, playfully elbowing me in the ribs, "we're a team, doofus."

It wasn't easy, of course, to nail down an appointment with a quadrillionaire. We had to play hardball.

Madhuri took lead. She likes books over talk and time-stops over teatime, but she can be a charmer when she sets her mind to it. I dry up with people I don't know.

Madhuri's infectious grin and mischievous winks didn't do it, though. Even pointing out we'd worked Larkin's crash was waved off. It took a diamond-edged hint that we had scans of the augs to even move the needle, and then it had to work up the chain. When we got the green light, it wasn't from Madeleine directly, or her PA, just some schlub in security handed a script and a phone.

But we had our meet.

It wasn't worth a damn; we both knew that. Madeleine wouldn't tell us anything, and she wouldn't be the sort to let things slip. She'd take a few minutes to feel us out, then have us tossed. The scans wouldn't be believed, not in an age where damn near anything could be ginned up. I found myself wondering why she was wasting her time with us.

We mic'ed up and lensed, knowing it would be pointless—and knowing we had to try.

Madhuri and I took time off. We both had plenty of days saved up—we were married to the job, and it was a good marriage. Latifa came over from Clovis to fill in, and Carlos from West Park. Both solid, mid-level experience, ready for more responsibility but not starved for it. They would piss off Katy, but hell, anyone would piss her off right now.

My mind was at ease as we took a screamer to Bismarck, North Dakota—not just because we had two solid timers to fill in, but because I expected nothing from the visit. I'd given up on the whole thing before we even walked in the door. I wasn't going through with it for any particular result; the deck was stacked and I'm no fool. I just wanted to be able to tell myself I'd done all I could. I wanted an end to my restless nights.

Madhuri convinced me to spring for the skyhop. It was barely thirty minutes on the autoway, and it cost ten times more by air, but I gave in and she was right: better to give the impression of means, and see Mercer from above on the way in.

The town on the company's address was called Wing, and it was nothing but a patch of hunkered-down bungalows, eyeing each other warily over open grass and a modest asphalt grid. The Mercer complex was miles away, its fingers of shadow grasping the flatland. Halfway to October, the air was already cool.

It wasn't what I had expected. The complex was huddled together around a massive central tower. There were no green spaces, no athletic fields, no techs ruminating on rooftop gardens. It was a company in a defensive crouch, warding away outsiders. Built for control, not collaboration.

I saw from the air a glittering corner, a glass madness of pyramids at war, and I thought I recognized it from the vids. It was where she took her interviews. I'd always assumed it was how the whole place looked, but that was obviously untrue. At the least, I thought it must be the lobby—but as we swirled around the other side and lurched down to a brutalist facade, I realized it was at the back. Something private, then. Or a cultivated lie.

The personnel mirrored the landscape. It was minutes before anyone emerged from the panel of red industrial folding doors that marked the entrance, and they seemed put out, unused to visitors. We had a long wait, standing out in the bitter dregs of Dakota summer, as they confirmed we actually did have an appointment—and then came the tunnel.

I thought at the time that employees must have a back door, but it's not true: the only way in or out is through that mad nautilus shell of a hall, eternally curving just so slightly to the right, low ceiling pressing, warm light taunting from LEDs at the floor joins. Sloping downward just enough to make you stumble now and then. When it's not buzzing it's snapping, when it's not snapping it's whining, and at any given moment your hair might stand on end or your fingers might zing what they touch or a strobe will turn the world to stop-motion. There are sections with blue lights, sections with red lights, dim parts that make your eyeballs ache with pressure—and confounding stretches of antiseptic, dazzling white, like mental palate cleansers before the next taste of the insane.

The worst part are the checkpoints, sterile white metal doors keeping you trapped as your knuckles ache and blood pounds hot in your ears, there's grit in your eyes you can't blink away, your heart is skittering beats, ozone invades your lungs—and only then, when you start to believe the real world was a dream, does some sullen sensor rouse itself to jog the doors and let you shuffle, blinking, into the next hell.

They take everything off you before you go in—even transfer the BitC from your wrist wallet and give it back when you exit the building, I guess sometimes the wallet gets zapped—so I have no idea how much time it takes to endure the tunnel. They tell you four minutes thirty. Madhuri says that seems right; I think they're off by a couple hours.

We emerged to face a single, enormous elevator, which spread wide its maw with a joyful ding. I hobbled in on legs of rubber, trying not to use Madhuri for support. It closed and shot up on its own.

There were no windows, not in the elevator and not as we followed a slouching, unsmiling intern through claustrophobic corridors somewhere above—floor forty-four, judging by the numbers on the rooms we passed. The whole building was probably one big Faraday cage, even though it seemed a pointless gesture after the tunnel. I was sure our lenses and mics were cooked.

We were deposited with a desultory wave in a room at the end of a hall, amusingly marked Visitor Lobby. It was nothing but a doublewide coffin, lined with two slim couches—faux leather over concrete—a handful of drooping plants, and a Mr. Chef machine after they'd taken all our BitC. At least there was a window. Madhuri and I leaned by it and took in the endless waves of distant grain, rippling in the wind.

We talked over Katy's attitude yet again, chewed gossip, and finally slumped in silence on the sofas. I must have dozed off.

Sunset was stabbing high along the wall when I woke with a start.

An attractive young brunette was staring down at me, all doe eyes and pursed lips—I nearly laughed.

"Hello?" I said, mouth pasty, voice thick.

"You were snoring!" called Madhuri from the other side.

I righted myself, surreptitiously wiped my mouth with my sleeve.

"This way, please," said the young woman, gesturing toward the door.

This time the journey was a quick one, out the door and right into the adjoining conference room. It was a palace after the Visitor Lobby. Should have felt pinched without windows, but I didn't mind. There was even a little plate with thin, twisted breadsticks. I eyed them hungrily but held back. We stood around next to the enormous oval conference table, ignoring the chairs.

"Those are Dot's," said the brunette, as though it should mean something.

"Yeah, Chief, those are Dot's," said Madhuri. She nudged me with her shoulder. "Not for you."

The young woman spread a rich, thick laughter over the room. "Dot's Pretzels," she said. "A local treat. Most of us aren't from around here, but those have wormed their way into our hearts. And our thighs, and our hips, and our tummies." She reached out and drew them closer.

"Are you from around here?" I asked.

"Me? Oh, no. I'm from the South. Dakota. South Dakota. The other one." She flashed a nervous grin, grabbed a pretzel stick and took down half in one bite.

"I'm sure it's very different," said Madhuri. I tried not to laugh and ended up yawning.

"Where are— Oh! My name's Xanthia." She hastily wiped her fingers on her pencil skirt and held out her hand for a shake. I'll admit I was surprised at the firm grip.

"I'm Seb."

"I'm Madhuri."

"I know. I mean, nice to meet you. And I was going to ask where you're from, but I already know you flew in from California."

"Land of dreams," I offered up after a few seconds.

"'Cause reality sucks," Madhuri threw in.

Xanthia pinched another pretzel. "I have family there," she said, crunching. "I've been a few times. Going for another visit in a few weeks. I like the palm trees."

I yawned again. She reeled as though I'd struck her. "Coffee. Coffee! I'm so sorry. You must be exhausted. You want coffee?"

"That would be great," I said.

"Both of you?"

Madhuri nodded with a wincing smile, like she'd just asked a toddler to make her pancakes.

As the brunette scurried off, I reached for a pretzel-stick thing.

Madhuri slapped my hand away. "If they're going to kill us, at least don't make it easy."

"She was eating them," I protested. My belly rumbled. Last thing I'd had was a sandwich before the screamer, hours and hours ago now.

"Of course! She has the antidote. That's why she sped off like that."

"If they wanted us dead, they would just shoot us," I said. It seemed likely, here in Fort Mercer. She could make us disappear, I was sure. Make it look like we'd never come. By comparison, my insomnia didn't seem so bad.

"You're right," announced a crisp voice that made us both jump. In strode Madeleine Mercer, looking just how she looked on TV—except she was a tad more wrinkled, more than a tad shorter, and her shock of white hair was unkempt. The piercing, luminous brown eyes that made her seem nine feet tall, though, those were the same.

She commanded the room with those eyes, holding us in place as she stood between us. "The older I get, the less I can stomach beating around the bush. If I'd wanted you dead, you'd be dead. It would be trivial to dispose of you. Less trivial to make it look like you'd never come, but we could manage."

Madeleine peered up at me and clucked. "I saw those yawns, Mr. Bouchard. Wake up. You'll have to be quick on the uptake, or this will take entirely too long."

I nodded, still tired from my nap, from the lack of food and caffeine.

Madeleine rounded the table and sank into a chair with a relieved sigh. Her shoulders barely cleared the table. She herself looked suddenly exhausted, deeply and truly spent in a way that eats away your core. I remembered feeling how she looked. That's when I'd left the Wild West of freelancing and got on the government payroll.

She waved dismissively. "Your recording gear-ups are dead. The tunnel eats that for breakfast." She glanced at Madhuri. "Get some prophylactic nanites for your mammaries, Ms. Anand," she said. "And you"—pointing at me—"really need some for your colon—and your prostate."

I opened my mouth and she looked me to silence. "I know you have questions. I don't care. They'll be answered, but they'll be answered on my schedule." She melted farther into the chair with a deflating exhale, flicked a hand at the door. "Close that."

Madhuri had it halfway shut when Xanthia bumped it open again, carrying a tray with two coffees and a box of fudge-striped cookies. Seeming blissfully unaware of Mercer's glare, she set the tray atop the table and presented one of the mugs to me in an odd two-handed sort of curtsy, waving a palm over and then under the cup as though performing a magic trick.

The coffee smelled heavenly. One of the new varietals, probably, cherries and almonds on the nose. But then, as I raised it toward my lips—the scent of burnt plastic, just a whiff.

I set it down on the table and sent a warning look to Madhuri—already luxuriating in her first sip. All that air spent on denials of poisoning, just to make it easier to actually do it?

"Xan! Out!" said Madeleine.

Some sort of salute from the brunette. She flounced out with a flash of teeth, no hint of guilt. The door clicked shut behind her. Madhuri made questioning eyes at my mug.

Hell with it. Either we lived in a civilized society, or we didn't. Lack of creamer pointed to the latter, but it was more than drinkable black. I had a seat and wolfed two cookies before Madeleine cleared her throat. Madhuri settled into the chair next to me, barely taller than Mercer.

"I do hope you're comfortable, and I mean that sincerely. I can be short when I'm tired." She gave me a sidelong look, as though daring me to make the obvious joke.

"If those aren't enough"—she gestured to the cookies with disdain—"I can order up real food. I might indulge as well. They have terrible taste here, but it's hard to mess up a good steak."

I shared a shrug with Madhuri. "Sure," she answered for us. "That would be nice. Thank you."

"You're here to—" Madeleine cut off as I reached for another cookie. "Don't ruin your appetite. It won't be long. I find AR distracting, but someone's always listening." She wriggled her fingers at the ceiling.

I showed my palms and leaned back, though my belly still ached for sustenance.

Mercer pulled her lips to tight lines. A hint of moisture at her eyes. "You are here," she blurted out fiercely, "to judge me."

Because you killed your daughter? I wanted to ask, but the rule was clear—no questions.

"Humans are problematic," said Madeleine. "We're not idiots, exactly—intelligence is certainly in the mix—but our brains are a jumble of legacy devices competing for bandwidth, and there's no one to iron out the bugs."

She did something like jazz hands, and I coughed to cover a smile.

"Sure, you can rearrange all sorts of genes—or aug yourself—to improve memory and whatnot, you can zen yourself to hell and back to improve executive function, but none of that changes the fundamentals. Whether I start by telling you good things or bad things about me will shape the lens you see me through. That would warp what I'm trying to do here.

"So, I'm going to start as neutrally as I can. I'm just going to go from the beginning and let time guide the tale." A sour smirk stained her lips, just for an instant. "You want to know if I killed Larkin. You want to know about the time-in boxes you think you saw. You'll have those answers, but you'll have them in context or not at all.

"Aw, screw it." She nearly climbed on the table pulling over the cookies. "Any time now, with the steaks," she announced to the room, mouth full. "Can I?" She gestured at my coffee. I handed it over so she could have a slurp. No poison, then. A plus. "Leave some water in here next time, Xan. Make sure all the conference rooms have water."

Madeleine settled back into her chair. "Hypoglycemia, just a touch.

Long day. Plus, I'm queen of the castle here and I've hardly left for years, which makes my socialization a bit of a wild card. If I offend you, you'll just have suck it up and try not to cry."

I found myself liking her. I was sure Madhuri did, too.

"This whole thing was Lark's idea. We're insular here—highly insular—and she wanted to make sure we weren't crawling so far up our own asses that we didn't know a flashlight from the sun. We do a sanity check once a year. This year, I've invited you to sit in judgment, because you're timers and because you've been a pain in the ass. One stone, two birds.

"So. The beginning. Do you remember climate change? They still teach that?"

I nodded. Madhuri nodded. We locked eyes and raised eyebrows.

"Yes, yes. Humor the old bat. So, you know things got out of hand. They had plenty of warning and plenty of opportunity to rope it in, and instead they messed around and had a big old banquet of denial and magical thinking. Sure, we clawed things back later, but we couldn't put the genie back in the bottle. Where we live, the food we grow and how and where we grow it, the constant invasion and mutation of new plagues, an unstable sea and unsteady air making travel harder, the unbearable summers and brutal winters—we're used to it now, but the world looks a hell of a lot different than it did two hundred years ago.

"Now we could try to push it back to that baseline, and people scream about it. Too risky! Can't trust science, they did this to us in the first place! Trust nature!" She rolled her eyes. "Intelligence is in the mix, but it can be pretty hard to find. My point is that we don't like change. We're wired that way. And we don't plan well—which is why we don't even have a damn bottle of water in here," she said, raising her voice and speaking at the ceiling. "As you drive over a cliff, most of the idiots in the car will be arguing over who to blame instead of figuring out how to deploy the parachute.

"Anything in there you disagree with?" she asked, raising an eyebrow at us each in turn.

I shrugged.

"I didn't realize this was interactive," said Madhuri.

"It is when I say it is. Answer the question."

"Nope," said Madhuri.

I shook my head.

"Good. That was an easy part. Keep up, now." She leaned back and steepled her hands, index fingers curling and straightening. Her lips bulged as she ran her tongue over her teeth. "About fifteen years ago, we were testing the behavior of our entangled com systems—you've heard of those, right? They use quantum entanglement to let people communicate in real time anywhere in the universe. Well, as much of the universe as we've seen, but there doesn't seem to be a limit."

My breath caught when she mentioned entangled coms, and I let her explanation wash over me as my mind buzzed with possibilities. I'd never considered using them, but it seemed like they could work in a time-out bubble. It would solve so many of the problems we had in time-out if we could just talk to each other. Had it worked?

Madeleine wasn't stopping for questions. "So, we were testing the behavior of our entangled com systems in time-out and time-in bubbles when we noticed something odd. There was slippage when starting and stopping any bubble, time-in or time-out. Microseconds, but it was there. We'd put ten people in time-in bubbles inside a time-out bubble and they'd come back with different times of day—microseconds, like I said, but there. And they didn't normalize when they exited the bubble. The time-out bubbles too, they'd lose a few microseconds between opening and closing.

"I'm sure you can imagine what this means. Friction in a system that needs to be frictionless. Space-time rubbing against its own legs like a cat in heat—until one day it trips and breaks its neck. Our calculations showed the possibility of a full second of slippage within a few hundred years, then more and more over the centuries.

"I went to people. High up. Government, science, industry, entertainment. I have access, I'm sure you can imagine. Every single one of them laughed me out of the room. It wasn't proof of anything, it's too far in the future to care, we'll figure out how to fix it later. People would be happy to lose a few seconds now and then, they laughed—anything to pass the time.

"But the real issue isn't lost time, it's causality. I'm sure a couple of timers can appreciate the importance of causality."

I nodded. Madhuri nodded.

Madeleine nodded back. "Good. Without it, the world stops making sense. Observation fails, so science fails." She wagged a finger. "Not like in the movies. Nobody's going to wake up with a thousand BitC and *then* win the lottery. I mean unpredictable time slippage. Clocks would be erratic, each giving a slightly different time. Anything that relies on precise coordination at the microsecond or nanosecond level would experience continuous, unpredictable faults—yes, I mean computers. Everything would turn into a glitchy bugbox, if it worked at all. We would be back in the stone ages long before we hit slippage of a full second.

"That's the level of danger we're in. If we don't fix this before it happens, we'll lose our ability to fix it." She locked eyes with Madhuri and then with me, a warm intensity flooding through her gaze. "Are you following? You can ask questions, if you want. It's important you understand."

Madhuri coughed. "Even if time is stuttering, it won't break anything if everyone experiences it in the same way. Lights flicker but we don't notice. Electrons in a CPU wouldn't magically go haywire—if they flickered at the same rate as everything else, we wouldn't even observe any variance in speed, and calculations would be fine. Or do I have something wrong?"

Madeleine reached out and rapped a brief rhythm on the table with her fingers. "Good question!" she announced. "Common question. And, yes, you have it wrong. Observers reported different times, remember. Slippage was observer-based. If I showed a flashing light to a room of people, they would see flashes at different timings, or it would stutter at different times for different people. We would lose our shared version of observed reality."

She gave us a pregnant pause before opening her mouth to start again. I raised my hand.

Her eyes nearly rolled out of their sockets. "Ask, Mr. Bouchard."

I cleared my throat. "You're acting like this is generalized. I don't—I don't know if that's the right word, my background is application, not theory. But you're acting like this slippage would happen constantly, when you've only seen it around time-in and time-out."

There was a knock at the door. "Yes!" called Madeleine, and it swung inward to admit an invading bustle of men and women in tight-buttoned uniforms, who swiftly deposited trays, poured wine, lifted covers and snowed down white pepper before exiting with exact choreography and pulling shut the latch with a gentle click.

I was first into my steak, stomach growling pinched like a hyena, but my ears were perked as Mercer chewed and talked. "Fair point, I forgot to cover that. We've noticed slippage happening in normal time. Picoseconds, difficult to even measure."

She slurped at her pinot. "We compared values on both sides of the planet to values near Europa, by the Belt, and out past Pluto—using entangled coms to coordinate, of course. We see differences in femtoseconds between hemispheres on Earth, growing to picoseconds by the Belt and getting larger with distance. Still in the picosecond range at Pluto, but worrying, wouldn't you say?"

I shrugged, scooping up buttery mashed potatoes. It felt incredibly satisfying to fill my belly after so long without food, a warm serotonin blanket.

Madeleine burped. "You should be absolutely terrified, because that tells us two things. One, time can be a local phenomenon, which validates the observed differences we see here for time-in and time-out. Two, the way in which time seems to be universal allows for those differences to be exaggerated by non-local impacts. There's no reason for slippage to be any different at Europa and Pluto, really, since we're only using time boxes around Earth and the Belt, unless it's something like a fabric that we're rippling or a pool we're tossing rocks into.

"I like gelatin molds better as a model—they're 3-D, and that's a lot more intuitive than talking about something 2-D like fabric. The world's not 2-D. Plus, they still make them here for dessert from time to time. Have you ever had one?"

I shook my head.

"Have you ever seen one?"

Again, I shook my head, finally taking a breather from my meal to have a drink.

"Well, I'm sure you know what they are. It's just a big bowl of gelatin with fruit in it, turned upside down. It holds its shape because of the amino acid bonds in the protein—but those are easy to break with heat, which is why you can make a mold in the first place and why you have to keep the damned thing in the fridge.

"We're heating it up. Every time we use a time-in box, a time-out box, we're adding a minuscule—but meaningful—amount of heat into the system, attacking those bonds. Do it long enough and you end up with nothing but a puddle of fruit water. That's what we're doing to the reality we live in. And nobody would listen."

"So, you decided to make better time boxes?" asked Madhuri, barely discernible, mouth thick with potato.

Madeleine pushed back her half-eaten steak, poured herself more wine and leaned back with a satisfied air. "Not exactly. This is where you shut up again." She sniffed the wine, swirled it, sipped it.

"What I *decided* was that I needed to fix things. I'm ambivalent about humanity, but I don't want to screw up time for some other poor intelligent species. And humanity's not all bad, we have wagyu and espresso and those videos where random strangers help trapped animals. Whatever the reason, I decided to play the hero. I decided to save us from ourselves.

"Nobody would listen, right? And these were the smart ones. Your average Madhur or Madhuri—"

Madhuri grinned and gave her a thumbs-up for the shout-out.

"Couldn't give two squeaky pebbles about this, even if they understood it. Time boxes were suddenly everywhere. We were running tests because they started using big, clunky, expensive industrial time boxes out in the Belt, and that's one of our big markets. Those were some of the first. It took what, ten years? Fifteen? Before they had them deployed for public use. We've seen the curves for miniaturization and cost efficiency and adoption rates on similar tech, and each degree on those arcs means a whole new slew of use cases.

"I'm sure you've seen how they use them for trips to the Belt now, firing up in overlaps. You fall asleep in orbit and wake up seven minutes later at Ceres. Step that down and people start bubbling when they take a screamer across hemispheres, then on the commute to work, then while they're waiting for the dentist. Some crazy rockheads are copying the rigs from Belt transports for home use, just because they want to see the world in a thousand years. Keep shrinking it down and you'll have uses in sports, games, drugs, sex, looking younger, feeling better, quack quack quack." She said the last part using her hand as a puppet mouth. I assumed it was a regional quirk—or she was referring to bad docs. "Humans are annoyingly inventive, hopelessly epicurean and easily duped. The perfect combination to spread time-box tech far and wide, and damn the consequences.

"So, I set myself against them. I set the whole company against them. Well, not many know the whole scope of what we're doing—can't trust 'em, really—but the core group. We threw this up," she said, gesturing at the walls around them, "in a few months and transferred all our equipment over a few weeks. Lark was a teenager, and I'm sure you can imagine how happy she was to leave all her friends behind in Oregon and move out here to Nowhereland.

"We needed space to breathe. Even just building something like the tunnel would have raised eyebrows on the coast. Here they know how to take the money and keep their mouths shut. That doesn't mean no rumors, that doesn't mean asshole reporters don't come sniffing around now and then—but it does make things a bit easier. For the rest, well, we keep to ourselves. We live here, work here, eat here, sleep here, schools are here, gyms are here, shops and entertainment, everything you'd want. Built up after, most of it, a lot of it underground, but it's here now and we don't have leaks. And we don't let outsiders in, usually. You're lucky."

I shared a quick look with Madhuri. I don't think either of us were feeling particularly lucky. This seemed like a villain's soliloquy—right before they put you in the ground for knowing too much.

Madeleine smiled ruefully and shook her head. "Like I keep telling you, you'd be dead already if I wanted to kill you. You're safe. The whole point of this—well, there are two points: I get to unburden myself, and you get to judge me."

Madhuri was getting into her wine. She leaned forward, eyes glassy, lips curled. "There's no point in judging you, though, is there? You won't stop just if we say to. And if we go and talk to a reporter or something, we'll have no proof—even if you have done something wrong, which you haven't even admitted yet."

Mercer shot me a look, like she was expecting me to pipe up. I shrugged, and she nodded with respect in her eyes. "You'll have to let me finish," she said, turning back to Madhuri, "if you want answers to those questions. And stop interrupting."

Madhuri made a floral, mocking curtsy from her seat but kept her mouth shut.

"So, we've focused intensely on time-box tech since the move. We still produce entangled coms, we still work on improving them. That's our bread-and-butter, and our time-box program needs to eat. And I need my steaks. We've gotten very good at secrecy over the years. Even in Oregon we had it down to a science. Something you may not know is that we don't have a patent for our entangled coms systems. I bet early on that our method was too outside the scientific mainstream to be discovered by anyone else within at least thirty years, and so far I've been right. Patents only protect for a fraction of that, and practically you're giving away your tech to bootleggers the world over. We kept it proprietary, held it as our secret, and reaped the rewards. Funding in that research space has almost entirely dried up, now that our systems are everywhere and our prices are reasonable. Well, they're used to paying our prices, and they don't know our costs, so it seems reasonable to them. It's highway robbery, of course. That's just capitalism.

"The point is that we know how to keep a secret. The tunnel helps, but the tunnel is just our cornerback. We've got a whole field filled with other defenders you'd never see. One of those is what was in Lark's skull.

"Yes, it's a time-in box. Yes, it's really that small—and it works. Like with our coms, we made a breakthrough, and we've kept it to ourselves. I'm sure it's clear to you by now why we haven't been selling them, though it would make me another few quadrillion, easy.

"It might not have occurred to you yet why we'd be researching time-box tech if we're *against* time-box tech." She looked us in the eye in turn, gaze searching, smirk tugging at her lips.

I raised my hand. "To counter it? To figure out how to counter it, I mean."

Madhuri nodded approvingly.

Madeleine snorted. "That's always the guess. It's always wrong."

That stung. I was used to being the one handing out wisdom.

If Madeleine saw my scowl, she didn't care. She threw back the last of her wine, poured out the dregs of the bottle into her glass.

Immediately, the door opened, and the same group of buttoned-up waitstaff bustled away our trays, delicately putting aside our wine glasses and soundlessly setting another bottle atop the table before shuffling out.

Mercer topped up her glass. "Like I said, breaking causality would leave us without options. Unlike with climate change, there's absolutely nothing that could put that genie back in its bottle. We have to stop it before it's too late.

"So, we're gearing up for war—of a sort. We're going to use time-box tech to destroy time-box tech. And, more importantly, the means of production."

Madhuri was grinning. I didn't see the joke. Madeleine beckoned for her to speak.

"I think you'll need bigger time boxes."

Madeleine brushed away the joke. "Ah, yes. I'm getting a bit scattershot. I'm tired. The wine isn't helping." She had another sip. "We needed to understand the entirety of phenomena around time-in and time-out. It's easier to study at small scales, so we went small before we went big. But we have some big boys in orbit already, and placed strategically earthside, with more to come. The tiny ones give us options for infiltration—and for when we have disloyalty in the fold."

"Is that why you killed her?" I blurted out, cringing inside as I broke the rules. "Was she disloyal?"

"Ah, Mr. Bouchard, happy to see you've realized you're not in middle school. You want to know if I killed Larkin?" She took a lengthy draught of wine. Her eyes glimmered; she was hoarse when she spoke. "I did. I didn't push the button, I didn't give the order, but I made the system. I made the rules. She left without authorization, with one of our lawyers and two bodyguards, and they all left their phones behind. They met the fate prescribed. I miss her every damn day, but she knew what would happen. They all did.

"Maybe she thought we couldn't touch her on the autoway. Maybe she thought she'd found a way to cheat the failsafe. She gambled, and she lost. I can't weigh my daughter against the future of all humanity, Mr. Bouchard. She'll always come up short. Anyone would."

She wiped her eyes with her sleeve. "That's all. I can tell you more about the plan, if you want. The science. Whatever you want. Ask your questions, if you have any left."

There was a long silence. Madeleine stared sightlessly at the door as she imbibed her wine.

I found myself blinking, swallowing, mind encased in ice. I had found my answers and did not feel satisfied. My vindication was a gnat.

Madhuri seemed no more fit to ask anything. She sat with pursed lips, face scrunched, eyeing Mercer as though she were a perplexing specimen.

I felt a low panic rise, and a sense of indignation. The miracles we performed daily—she would see them gone. Katy and Xi pouring their hearts into the job, just to be cut off. Mercer saw only evil as we did good—and she had murdered her own daughter for her warped cause. She had no right to judge.

"You have no proof," I blurted, clenching trembling hands.

"Of what? The problem with time-box tech? I do, Mr. Bouchard. I have copious amounts of proof."

"Trends, just trends. Random measurements. Nothing confirmed. You said yourself you won't know for hundreds of years if you're even right."

Madeleine leaned back, took me in with a deep breath. She held my eyes with her luminous umber orbs, and it felt like she was tonguing the folds of my brain. "I freely offer myself up for judgment. I ask in return only to be judged fairly. If you're worried about your job, maybe we can find something for you here."

"I didn't say anything about a job. Don't try to bribe me. I don't want to destroy, anyway, I want to create. I want to help."

"Where's the proof?" asked Madhuri.

"Here, of course. Reams of it. Triple-checked by a room of scientists and mathematicians eminently more qualified than you—myself not least among them. I'm not sitting here all night while you pretend to understand the numbers, and I'm not letting you walk out with any of it, so you'll just have to take my word for it."

"That doesn't work for me," I said, crossing my arms and shaking my head stubbornly. "That's not evidence, that's assertion."

"Mr. Bouchard, do you really think I would turn my life around, dedicate all my resources—kill my daughter—for some sort of misguided whim?"

"I think you could be wrong. Anybody can be wrong. They used to bleed people as a cure."

"So, I should do nothing? Let the problem compound? Wait until it's too late?"

"Wait until it's *time*," Madhuri jumped in. "Two hundred years from now, they'll know things we've only dreamed of. They'll find the problem. They'll fix it."

Madeleine frowned into her wine, then set down the glass. "And if they don't?"

"Then it wasn't meant to be," said Madhuri, spreading her open palms.

"What? Humanity?"

"Yes."

"Hm. How poetic." Mercer thumbed at the engraved swoops on her glass, then grabbed up the wine and poured it down her throat with a dramatic swallow. She launched the wine glass over my head; I flinched. It pinged off the wall and bounded off the carpet. "I'm thinking this was a mistake," she said.

"Because we're not telling you what you want to hear?" I realized with a lurch in my belly that I'd forgotten where we were, who she was. She'd said we were safe but I didn't trust it, didn't trust her, didn't trust any of it. My indignation was no less loud, but I resolved to keep my mouth shut.

"Because you're timers," said Madeleine sourly. "And you think you know more than you do. You save lives and harness forces using tools that people like me create, and you mistake yourselves for gods. You're technicians, nothing more."

"That's not an insult," said Madhuri. "It just means we operate within our limits."

"It means you don't—can't—see what's possible."

"It means we see our constraints. I would never kill my own child. Nothing is worth that."

I nodded. I'd never had a kid, never wanted a kid. Same for Madhuri. But we knew better.

Madeleine looked my way. She swallowed. She rubbed her face with her hands. "This isn't—fourteen times, I've done this. All kinds of people. All kinds. It was a sanity check, that's all. It was a no-brainer: you save the human race, or you let it melt away in a puddle of lost causality. You're timers. You were supposed to understand."

Madhuri opened her mouth, but I shot her a warning look.

Madeleine caught it. "Let her speak."

Madhuri cleared her throat. "I won't justify Larkin's murder for you."

Mercer leaned back in her chair. She slowly nodded. "That's fair." She sighed. "That's fair." For a long moment, she stared somewhere between us, eyes lost.

Madhuri and I shared a look.

"That's all," said Madeleine, almost a whisper.

Doors punctured the room with sharp claps. It wasn't security storming in, though, it was the waitstaff from before. They rounded on us from the far door, sweeping their arms and pointing toward the now-open portal behind us, an unmistakable invitation to clear out. I complied, one last glance at the unmoving Madeleine, and found myself hustled toward the hallway, Madhuri at my side, surrounded by this horde of severe catering staff.

"Wait!" Madeleine called, and the whole procession immediately froze. The waitstaff bustled out of our way so we could see her.

"All this way, and I almost forgot why I'd picked you in the first place." Her eyes lingered on my face. "How did she look, at the end?" she asked, voice growing hoarse. "My Larkin?"

I blinked, mouth dry, mind going to those foam-slathered lumps dragged from the car. Already limp, already dead.

Madhuri was quicker. "She was beautiful. Peaceful."

Madeleine smiled gratefully, eyes wet. "Thank you." She cleared her throat. "It's bull, but thank you."

She flicked her fingers, and the staff hurried us out of the room. I heard it latch gently behind us.

Xanthia stepped before us. She grinned, and it was warm. "Would you like some Dot's for the journey home?"

Madhuri shook her head. I took two.

It turned out the tunnel was a both-ways thing. No exceptions. They gave me blackout shades when I balked, but that was it. Madhuri led me through it by the hand. The trip seemed two hours shorter with the shades.

We sat the skyhop and the screamer in silence.

It was the easy quiet of old friends, deep with learned instinct. We were both mulling over what had happened, what to do. We both knew it. There would be plenty of time to talk when we had come to the ends of our separate roads.

That turned out to be three days later in the kitchen, just after a marathon session helping control a fire. The firefighters were still out there mopping up, Xi and Katy were in the showers. We had the place to ourselves post-chow, stinking of sweat, half-shed suits hanging off our waists. Bodies pining for sleep, minds wired.

"You really believe that?" I asked. Out of nowhere, but she knew what I meant.

Madhuri nodded. "I wasn't just backing you up, Chief. You know I wouldn't do that."

"And here I thought you liked me."

"Not that much." She shook her head. "I don't want that crazy old witch trying to save me with her messed-up ideas."

"Leave the future to the future."

"Exactly. And don't kill your family."

I snorted. "Unless they're like Crunch." She had been doubly difficult last time-out.

"Unless they're like Crunch," Madhuri agreed.

My knees creaked as I rose, gripping the table for support.

"Wait, Chief."

I settled back with a happy groan. "Yeah?"

"Are we sure there's nothing we can do?"

I shook my head. I'd already played through it all—zero evidence, hearsay, us versus a quadrillionaire. We might prove we were in the building through geo logs, but they'd never confirm we met with her. And there was no way in hell they had records of what was said sitting around to be subpoenaed. She'd unburdened herself, we'd gotten our answers, that would have to be the end. Even though, knowing what I knew, I slept worse now.

Madhuri chewed her lip. "What about the satellites? That would be proof."

"Not really. And her lawyers wouldn't let the feds near them, anyway."

"We could just post it. Get it out in the open."

I shrugged. "Congrats, now you've got a few hundred conspiraquacks dogging you. What do you want to do with them?"

"I mean the confirmed feeds."

"Funny thing about those, they need confirmation."

"You're no fun, Chief."

I grinned. "I am, Mad Dog, but only in my world—where my knees ache and everything goes to hell."

That was the last word on it till we found the pattern.

It was on my jacket, the one I'd worn to Mercer, and Xi tried to brush it off as I came in from lunch a few days later. Then he rubbed his fingers over it again, and pushed into my chest, lips pursed.

I put up my hands. "I'm flattered, Xi, but it wouldn't be appropriate."

"Is that a logo?" He eyed the spot critically.

"What?"

He stepped back, pointing. "Feel that."

I followed his line, tilling the fabric with my fingertips. There was a bumpy, squarish thing, about the size of a pinkie nail.

At that moment, in that context, I had no idea what it was. I told Xi I must have snagged it on something, shrugged it off, snapped a close-up at my bunk. It looked like some sort of pattern, lumpy and irregular, but I couldn't fathom why it had suddenly appeared.

It was the next morning when it finally clicked. Madhuri came into the kitchen as I was having coffee with the kids, and it smacked me in the face. My scalp tingled. The intern at Mercer, when she'd handed me the coffee—the showmanship had been sleight of hand, the burning had been my jacket. A shot in the gut with a burst of laser, to mark me with whatever this was.

"It's an R2 code, Seb," said Madhuri, later that morning. She arched an eyebrow. "You really didn't know that?"

"Should I?"

"You've never backed up your wallet?"

"I've never had enough to care. I get my meals and my bunk from the great state of California."

"Exactly, you have no expenses. What are you spending on?"

It was an excellent question. I had no answer—at least none I wanted to share. "That's beside the point," I deflected.

Madhuri squinted at the photo. "You're sure it was Xanthia?"

Ah, yes, that was her name. I nodded.

"Let's see if we can read it, then."

We could—after a bit of topographical manipulation, turning the peaks black and the valleys white. It rang a bell when it looked how it should.

The code opened a BitC transaction hash, one anonymized address to another. There was a comment on the hash: "AxeCrypt," followed by a string of random characters.

Madhuri was quick to find the app and plug in the string. It demanded an encryption key.

Madhuri grabbed my wrist, pulled it close. The app took my public wallet address and spat out text: "3020 TULARE. 20:00. OCT 9."

There had been no question in the last three weeks. We both knew we would go. Curiosity had bested us. The risk was obvious, but there was nothing extra to it; Mercer could easily find us, any day of the week.

The main roads are unsettling at dusk. The long swing toward night has a short and sudden peak, morphing in a blink from a line of uniform readycars in a ballet of synchronous motion—so smooth and regular you'd swear the buildings were rolling by on conveyor belts—to an endless string of floating cabins, occupants exposed to the world. There used to be an option to turn the lights off, but there was too much funny business. I felt naked; more than that, I felt forced to be a voyeur.

The car nicked over at 3200 Tulare. We stumbled out to find a brightly lit family restaurant, somehow warped into our reality from a century before. It was an Ollie's Pancake House, proclaimed in bold white over royal blue. Below, a less proud sign in a different font marked it a pillbar.

Xanthia spotted us from inside and waved us in with undeserved enthusiasm. She'd probably already popped a pill. I wanted to do the same, take the edge off my anxiety, but it was hard to focus on the menu for all the chatter.

"You know what was here before?" Xanthia asked and then kept on talking without an answer. "It was an Ollie's! And before that was an Ollie's, but they tore it down to build a new Ollie's. And before that was another Ollie's but it was different, and before that was a different Ollie's, and another and another and another. They just kept making Ollie's Pancake Houses here, like a temple or something. A temple of pancakes. They should call it Temp-Ollie's! Or maybe Etern-Ollie's, since it just keeps coming back. I only get to come here twice a year—not here, the one in Bismarck—when we have breaks. I want to come here every day.

"Oh, and this one has pancakes, too! It's a pillbar, but they have pancakes too, isn't that floogy? Goolfy? I'm trying to be cool but there aren't any cool kids in Wing, so it's easy to forget. It's cool, is what I mean. Oh! You should look at the menu! You should take a Blue Sunrise, that's what I took. It feels suuuuuper floogy? Goolfy?" She looked confused, but in a fun way.

I decided against the Blue Sunrise, with prejudice. A simple Pillsner for me; Madhuri got something called a Bean Wiggler. Then we both ordered plate-cracking pancake stacks with obscene heaps of sugared fruit. When in Rome.

I downed my pill with filtered water and felt myself mellow. The screeching laughs and blaring conversations faded; gentle, diffused consciousness melted in. I was aware of everything happening under those harsh white lights, in those tall, red pleather booths, swaying and tromping around on the blue checkerboard floor, and pleasantly distant from all those things. Madhuri giggled beside me. She had the dirt-eating grin and glow of good humor that marked all her best nights out. I wished she'd gone a bit lighter—we still didn't know what kind of meetup this was—but I didn't feel too harsh about it.

"I already ordered my pancakes, guys, don't worry," said Xanthia.

"I was sooooo worried," said Madhuri, reaching out to grasp her hand.

"The pancakes here are sooooo good," said Xanthia.

"Is that why we're here?" I asked.

"Sure," she said. "I come here every time I visit."

"I'm gonna come here every week," said Madhuri, swaying. "Every week till the end of time." She burst out laughing, and Xanthia joined in. I leaned back with a shrugging grin, waited it out.

"Oh!" said Xanthia brightly. She swooped into her bag and smacked plastic on the table. I blinked down at a memchip, memories welling. Hadn't seen one since my freelancing days. "And to give you that."

Our orders arrived. I stowed the chip in a pocket in haste—too much haste. The waiter eyeballed me as he rained down plates and then lingered, lips pursing and unpursing. "No outside pills allowed," he said, unprompted. "It's policy, and they're pretty strict about it. Just FYI."

Xanthia wobbled her hand in a mocking puppet as he melted away. They had a good laugh; I was feeling more serious. We all dug into our pancakes, overrun with fake fruit syrup and oat-whip.

"What's on it?" I asked.

"Stuff," said Xanthia, waving the question away. "Everything you need. Oh!" She dug into her bag again. "I brought Dot's for you! You took two so I think you must have really liked them, and really they should be something you can buy anywhere but somehow they only seem to sell them in the Dakotas. Or maybe Minnesota, too? Have you ever been to Minnesota? Or Wisconsin? Is Wisconsin a thing, or am I making it up? It sounds too funny to be real. Wisssss-con-sin. Wiss——"

"What do you mean, everything I need?"

"Plans and stuff. That's why I said there's stuff on it, because there is. Lots of stuff. Did you know 'stuff' is a German word? I didn't know that." She tossed a pack of Dot's at me. I caught it and found myself feeling genuinely pleased. I had enjoyed them.

"So we can catch her, Seb. So we can prove it." Madhuri grinned at me, then loudly whispered at Xanthia, "Sorry, he's a bit slow sometimes."

Maybe I was slow. I was struggling to process it. The Pillsner was weak, which didn't help; the raucous, overwhelming interior was stabbing back in. "Why?" I asked.

"Why not?" said Xanthia, mouth full.

"You'll lose your job," I pointed out.

"Who cares? I'm just a PA, and she treats me like crap. Crap is a German word, too!"

"Plus, she killed Lark," Madhuri cut in. "Lark was cool, man. She was super floogsy."

"Goolfsy?" Xanthia asked herself, brows knitted.

"You don't believe in it anymore, then?" I asked.

"In what?"

"Destroying the time boxes."

"I believe in a lot of stuff—I mean, a lot of crap—I mean, damn, I'm trying to say it in English. You probably don't even speak German, do you?"

"I speak German," I lied, internally cursing Blue Sunrise.

"Oh, cool. So, I believe in a lot of crap, but I've also seen some really weird things—sorry, weird crap. I think maybe she's locking up science people, and doing weird stuff to stop time. And now, she killed Lark, so I'm thinking maybe she's bad news—sorry, *neue-sprachen*. And that's it." She happily returned to her pancakes.

Madhuri put down her utensils and craned her neck, peering suspiciously at Xanthia. "Wait, did you just say she can *stop time*?"

"Lots of people can stop time, Mad Dog," I reminded her. "We do it every day."

She looked over at me, wide-eyed, then all at once relaxed with a sly grin. She shot me with a finger. "Gotcha, Chief. It's all cool."

I never got a better answer out of Xanthia.

They talked a lot of nonsense for the better part of two hours, slowly coming down with the help of tankards of mineral water and a shared bacon platter. I exited my Pillsner with a grump, feeling the room too raw, and sat in contemplation with a mild ache to my skull and the memchip burning in my pocket. It was hard to make sense of it—any of it. Why she'd have it, why she'd do it, why Mercer would let any of it happen. My gut said it was a trap, but I was so wired and skittish that my gut couldn't be trusted. Doom was everywhere.

Xanthia went for another Blue Sunrise and Madhuri seemed set to join. I hastily butted my way out of the booth, pushing Madhuri along.

"Can't," I said. "Sorry. We're on call from eleven."

Madhuri looked up at me, but only shrugged.

Xanthia shrugged, too. She downed the pill, exhaustion glimmering in her depths as she looked up at me. "Have fun, then, guys! It was great to see you again!" She waved. Madhuri waved back.

I didn't. I fled the garish lights, the endless chatter and pill-fueled, nightmarish nonsense, only remembering at a look from the bouncer to tap my wallet on the way out.

The readycar cut-out was empty. I watched people float by, trapped in their individual, pleather-wrapped worlds, engines quiet as they were carried along. They didn't seem to see me.

Madhuri joined me in silence, crossing her arms against the chill.

"It'll be a few for the car," I said.

She nodded, then abruptly turned to face me. "We're going to the police now, right?"

"What?"

"We should go now, we can explain it all."

"Why doesn't she go herself?"

Madhuri shrugged. "She doesn't want the risk. Or she thinks they'll believe us more."

"Why?"

"We're old? We're timers? I don't know."

"You really think Mercer didn't see that code on my way out? You really think she couldn't decrypt it? Do you really think *Xanthia* would have access to those files, and find a clever way to sneak them out?"

Madhuri grabbed my arm and pulled me around to face her. "Do you think she didn't know where to find us this whole time? If we die on the autoway, they'll find that chip."

"Like we found the augs? We don't even know what's on the chip yet."

"She could kill us whenever she wanted, Seb." She gestured back at the pillbar. "Why do all that?"

"Because we live in a world without accidents—at least not the easy kind." I pulled out the chip. "This is probably full of fake files, fake reasons for someone *else* to want me dead." I wound back, ready to hurl it into the night.

Madhuri snatched my wrist. "Then I'll take it. I'll go alone."

I relented. She pocketed the chip.

"I believe her, Seb. People are what they are."

Our readycar pulled up, flashed a friendly greeting. Slid open the door as Madhuri approached.

"Aw, hell," I muttered, heading in after her. "Aw, hell," I said louder, so she could hear.

We sped away, in our ready-made bubble of light.

Thickly

written by
Dorothy de Kok

illustrated by
TRACY EIRE

ABOUT THE AUTHOR

Dorothy de Kok skims author bios with mild suspicion—aware they matter, but quietly convinced they are proof that even the greatest writers have writer's block when they have to write about themselves... and here we are.

Her own storytelling journey began at twelve, when she attempted her first novel: an earnest and spectacularly terrible fan fiction of Enid Blyton's Magic Faraway Tree series. She finished it, reread it proudly, then lost it, which is just as well, as it was a threat to great literature.

Since then, Dorothy has collected an unusually broad résumé: high school English teacher, academic editor, safe-house director, real estate agent, and hopeful but horrendous gardener. She has spent years listening to people's stories—students, clients, and survivors—and those voices sometimes find their way into her fiction.

She now lives in the small Karoo village of Bedford, South Africa, where the power supply is erratic and the potholes are legislated, and where inspiration tends to wander in before the first morning screech of the hadeda. She is also, by her own admission, unofficially blacklisted from owning a library card in several provinces due to her unfortunate habit of becoming emotionally attached to borrowed books and "forgetting" to return them.

"Thickly," her Writers of the Future entry, explores what happens when the desire to be seen becomes literal—a body horror tale about enhancement, erasure, and the price of visibility in a world that demands women transform themselves to matter. She crafted it in the quiet hours before sunrise, in a South African township setting she knows intimately, hoping to catch something magical before daybreak. This time, the magic stayed.

ABOUT THE ILLUSTRATOR

Tracy Eire also illustrated "Shell Game." For more information about her, please see page 137.

Thickly

Nomsa bought the blister pack at the spaza on a Wednesday, when the fridges were off again and the air smelled faintly of paraffin from the lamps. Abush, the Ethiopian from Tigray, sat behind the counter with a little battery fan whining beside him, its head jerking left and right as if telling her not to do it.

Above the shelf of Lennon products and beneath the fading Tigray flag hung a glossy pamphlet stamped with the logo of a foreign biotech firm: *Eterna-Life™ Trials—Now Available in South Africa.* The purple foil packet below it looked far too proud for something that cost only twenty rand. THICKLY, printed with little flowers around the name. And the warning: ONLY ONE TABLET A DAY. DO NOT REPEAT WITHIN SEVEN DAYS.

"Safe," Abush said when Nomsa's fingers hesitated. His accent stretched the word until it meant more than it should. "Very safe if you follow instructions. Tested in America. Works fast."

Mama said someone was bound to burst. No matter how beautiful, there had to be a limit to how fat a woman should get. But Nomsa slid the coins across the counter anyway. She wasn't thin, but a little more flesh would make her feel more *seen*.

Outside, the air was bright with frying vetkoek and battered snoek. A child chased a yellow plastic bag that rose out of reach each time his hands nearly closed around it. Nomsa clutched the packet all the way home, in case it was snatched by the wind.

At the communal tap that afternoon, women queued with rinsed-out cooking-oil cans, gossip spilling faster than the water showering the sunflowers on the plastic containers. They tugged at their blouses, laughing, flaunting their expanding hips as if the whole street was a glamour gala.

Zanele was loudest, as always. She lived two doors down, a neighbour more than a friend, the type who never missed a chance to show off. Her handbag was always too bright for her outfit, her nails too long and too pink.

"Yoh, look at me," Zanele said, twisting her waist so everyone could see. It was fuller now, more generous. "Thickly is making me a real wife now-now. Men are calling like I'm giving away beer."

The queue laughed so loudly the tap sputtered and ran dry without anyone complaining. Nomsa smiled politely, but a flush of hot envy warmed her cheeks. Zanele's laugh was so loud, so sure it deserved to be heard. The extra space Zanele now took up in the world pushed everyone else slightly aside, more than before, and for a moment Nomsa wanted nothing more than to be *that* kind of heavy. That kind of round.

That evening, when she mentioned it, Mama sniffed. "That noisy girl is bold but has no respect. Always showing herself. If anyone bursts, it will be her."

What if Mama was right? What if Zanele, or anyone, burst? Nomsa hid the blister pack in her church shoes at the bottom of her wardrobe.

By the weekend, there were purple foils in handbags and bedside drawers all over the township. The rush felt like Christmas but better, with no one bothering to hide what they had bought. Rumours spread that Abush was making so much money he would soon move out of the room at the back of the shop and into a real house.

The salon was festive as customers boasted about their new shapes. Hair extensions were attached, relaxer fumes made the air sharp, hot combs hissed against braids, and women laughed like plovers. Even the hairdresser had tried it: "Eish, my hands are finished from braiding, but maybe Thickly will fatten them too." Everyone howled.

That month, the Child Grant payout—the government's small monthly allowance for mothers—became a pageant. Old women in headscarves grumbled that the queue had grown fatter and slower.

Young mothers showed off their shrinking dresses before they reached the ATM, as though the people behind them were judges with scorecards.

The urge became a pressure in Nomsa's chest. She thought of the blister pack hidden in her church shoes. She thought of being seen, of being bold like Zanele.

On Monday, Nomsa took the taxi to her casual job at Pep. The taxi was packed the way taxis always are: sixteen people in seats made for fourteen, knees pressed into knees, shopping bags cutting into ankles, a baby crying in its mother's lap. Music leaked from the cracked speakers, a hip-hop track booming between bursts of static. The driver shouted at another car in front, then at a passenger in the back who still hadn't passed up the fare.

Zanele was in the middle row, taking lipstick from the pearl-white Louis Vuitton handbag gleaming on her lap like it wasn't a knockoff, talking too loudly about her cousin in Durban and how the men there stared like wolves. She applied the pink lipstick, dropped it into her bag and clicked her nails against the seatback as if she were the one driving.

Then her voice cut off mid-sentence.

At first, it didn't look like much—just a ripple under the fabric of her dress, the kind you get when your stomach cramps. She pressed her palm to it, laughed nervously, tried to keep talking. But the ripple came again, longer, running across her belly like a catfish turning in shallow water.

The woman next to her pulled away. "Eish, what's wrong with you?"

Zanele shook her head, eyes darting, lips moving without sound. Sweat rolled down her temples. The ripple grew into a bulge. Her dress stretched tight across her belly, the fabric straining. The taxi jolted over a pothole, and everyone screamed as the bulge shifted again, rolling upward, pressing against her ribs.

"Yoh, *sisi*, are you pregnant?" someone joked, but the laugh caught in their throat.

The air changed. It became hot, sweaty, heavy with fear. People leaned away from Zanele, and the two old sisters prayed the Lord's

Prayer as though they were conducting an exorcism. The baby at the back began to wail.

The driver's eyes flashed in the rearview mirror. He looked back over his shoulder and shouted, "Not in my taxi! Not here!" He slammed the steering wheel like it was her fault, then barely missing an oncoming truck. Passengers were still screaming when the taxi curved back into its own lane.

Zanele clawed at her stomach, nails tearing the cheap cotton. A seam opened in the fabric, but it wasn't just cloth splitting. Her skin stretched so thin it shone, then it tore with a sound like a ripe boil bursting under pressure.

Something pushed through. A hand first, slick and perfectly manicured. Then a shoulder, a face. Another Zanele, glossy and luminous, forcing herself out with terrifying ease.

The first Zanele slumped against the seat, colour draining from her cheeks, her eyes going grey and flat. Her mouth opened like a pet shop goldfish—*help, help*—but her voice was no louder than breath.

The second Zanele slid free, landing wetly on the vinyl seat. She lifted her chin, smiled with white teeth in a mouth already painted, hair stylishly braided, dress brighter than the first's, the fabric uncreased, as though she'd stepped out of a mirror instead of a body. But beneath the first Zanele, a dark stain was already spreading on the seat, seeping from the tear in her side. She wasn't just emptying of personality but of something more essential.

"She's beautiful," someone whispered, reverent. All eyes were on the new Zanele. The fear melted into relief, almost applause.

The driver saw the mess on his seat. He slammed the brakes, jumped out and yanked the sliding door open. "Out, *wena*. Out!"

Nomsa watched as the old Zanele stumbled onto the gravel. The taxi wove away between the potholes, taking everyone home—except the old Zanele.

Nomsa sat frozen, thinking of Mama's words: If anyone bursts, it will be her.

She had burst, but now people loved her new version even more.

By the time the taxi looped back to the rank, everyone was talking. News moves faster than a Toyota Quantum.

"She burst like a watermelon," said one man, waving his hands wide, making the sound of tearing fruit.

"No, no," corrected a woman in a red headscarf. "It was neat-neat. Like a caesarean. The better one just stepped out, smiling."

A group of schoolgirls clustered by the cooldrink stand acted it out, one clutching her stomach, the others laughing and clapping when she straightened her dress and posed. People doubled over, choking on their vetkoek.

The taxi drivers got in on it too, shouting destinations with new lines tucked in: *"Fort Beaufort! Adelaide! Town! Last one! Going now-now! Make sure you don't burst, sisi!"*

Someone swore they saw the old Zanele drifting past the scrapyard, a shadow against the corrugated iron. Someone else said that was nonsense, she had never existed—it was only the new one all along.

By late afternoon, the story had bent itself into a lesson. "Don't be greedy with Thickly," warned an auntie, wagging her finger. "Some bodies keep it as fat, some as fire. If you force too much, it will split you open. Just like Zanele."

But others leaned close and whispered the opposite: "You see? The bursting is how it works. That's when you know you've arrived."

When Nomsa got home the news had already reached the stoep. Mama was slicing beans, a neighbour perched beside her, their voices low but urgent.

"They say it was Zanele," the neighbour whispered, eyes wide.

Mama clicked her tongue. "Of course it was. That girl was always too loud. Always pushing herself forward. That's what happens when a girl is greedy. Of course, it's better than being a mouse, I suppose. At least people saw her. Nobody sees a daughter who's a mouse."

"They say she bought Thickly twice, took double. Didn't follow the instructions. But they say the new one looks... better," the neighbour insisted, voice dropping to almost reverence. "Like *she* was meant for this world."

Mama's hands didn't stop moving, snapping the beans with sharp little cracks. "Better doesn't mean real. In our time, a fat young woman meant her father had cows. A fat wife meant her husband had a job. Thin women were for funerals, not weddings. But a woman who *splits*?" She clicked her tongue again. "You cannot make something from nothing. That pill doesn't create new flesh from air; it steals it from your future. It burns your life fast, and when it's done, the original is just ash. Poof, gone." She dropped the last handful into the dish and wiped her hands against her apron.

The neighbour shivered, muttered something about praying harder, then excused herself into the dusk.

Mama was right about Zanele being the one to burst. But people were saying she took two pills a day instead of one—that she overdosed. They were right. Nomsa had seen the empty blister packs in Zanele's bag.

Nomsa felt in the dark wardrobe for her black shoes. The blister pack greeted her fingers with a sharp edge. She read the insert, but the words were too long and too difficult—*hyperplastic cellular effusion*?

She swallowed the pill with a cup of lukewarm tap water. She held the empty cup, wondering if she should cough the tablet back up. It was already gone, though—dissolving, deciding for her. She slipped the remaining pack under her pillow, the fine print folded inside.

The house was quiet. Mama's little RDP—one of the government's matchbox houses—was neat but cramped, two rooms pressed together as if built in a hurry before the contractor left with money still in his pocket. Outside, the walls were a stubborn pink, though the colour had long since dulled under dust. Inside, the walls were still raw brick, collecting dust that only budged if you took a broom to it.

Thin vinyl covered most of the cement floor in the bedroom, its patterns worn pale where her feet always landed when she woke. In the kitchen, the units she had lay-byed from Lewis still smelled faintly of chipboard glue, the drawers sliding stiffly, handles bright against the house's tired fittings. A single hotplate balanced on top where a stove hob had been in the brochure.

She slept with the blister pack under her pillow, the way you do with money you don't trust to a cupboard, or your church shoes.

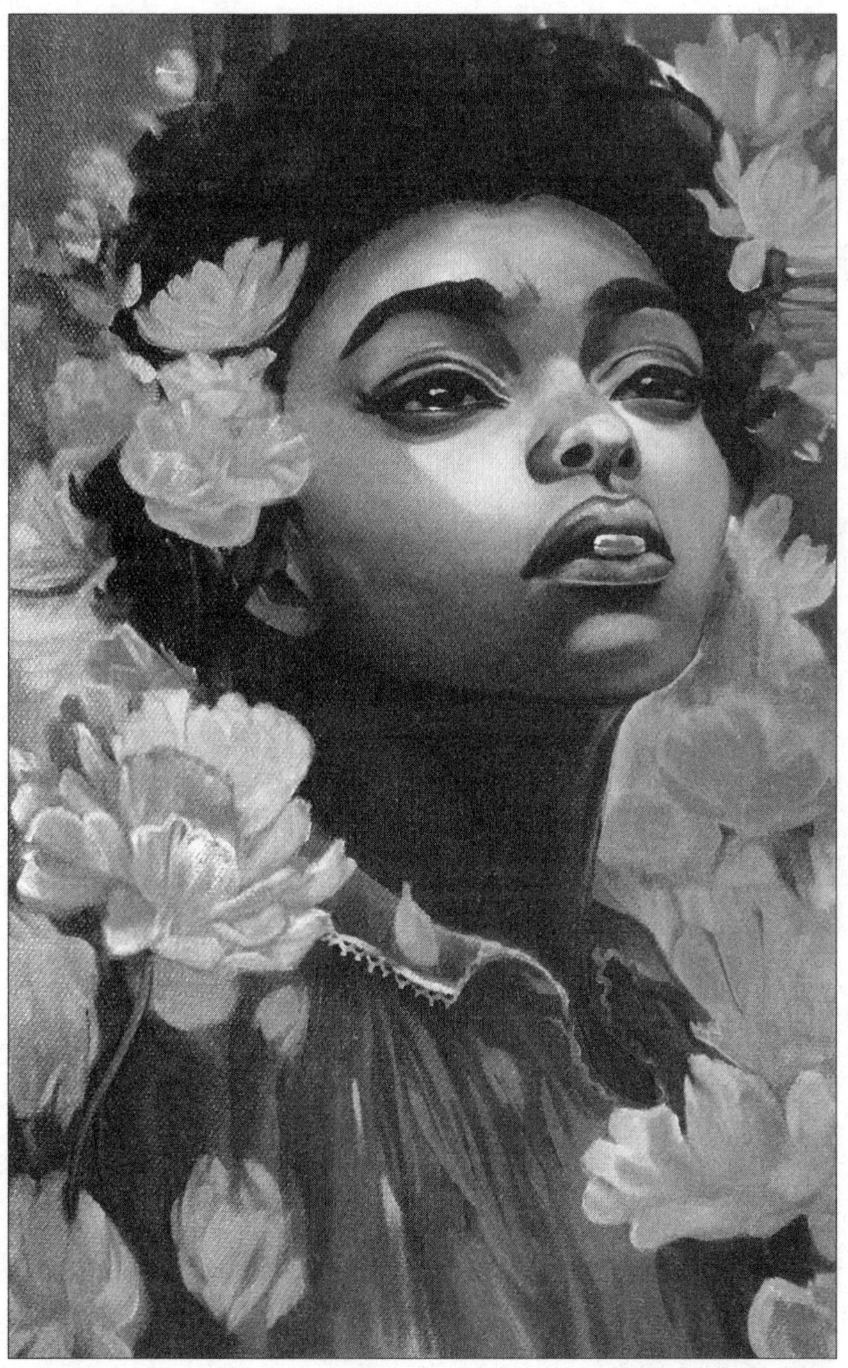

TRACY EIRE

Her dreams were of hands pressing from the inside, like a baby's kicks, but faster and everywhere at once.

She woke before dawn, the air heavy with sewage breathing beneath the concrete manhole cover on the street corner. She stretched in her bed, and for a moment she thought she was taller. The cracked mirror inside the wardrobe door showed her hips curved wider. And her cheeks were fuller, as though she were chewing meat that needed longer cooking. Her face glowed like she'd just come in from a long run, though she hadn't run since she left school. Across the mirror, her words were still there, written months ago in red lipstick, the letters neat and square and red:

I AM LOVED.

I AM BEAUTIFUL.

I AM BRAVE.

She had gone through a stage of reading them every night before she went to bed, hoping they would be true by morning. For the first time, she believed they might be.

The kitchen window rattled in the morning breeze while the shiny kettle sputtered on the glowing spiral. Her face reflected on the side of the kettle, and she almost didn't recognise the woman there, the way the light favoured her now. Her mouth turned up at the corners, just slightly, before she remembered herself.

On her walk to the taxi rank, the fruit seller, who had never greeted her before, called her *sisi* with warmth and pressed a peach into her hand.

By the time she reached the taxis, heads were already turning, even while the usual chaos surrounded her.

Hooters blasted sporadically, drivers shouted destinations in voices that cut through everything else: *"Adelaide! Adelaide! Town! Last one! Going now-now!"* The air smelled of oil, sweat, and roasted mielies sold from a drum that hissed each time juices dripped onto the coals.

Women shoved forward with shopping bags in their hands and babies tied to their backs. Taxi doors slammed, tinny music blasted from the speakers, and everyone's shoulder pressed into everyone else's space.

Usually Nomsa hated it—the jostling, the elbows, the feeling of being carted like luggage. But this morning was different. When she squeezed through the crush, men shifted aside for her. The driver's eyes slid down her figure before he clicked his tongue and waved her forward with something like respect.

Inside the taxi, a girl she half-knew leaned over and whispered, "Eish, *sisi*, Thickly is working on you fast-fast." Her tone was part envy, part awe. Behind her, another woman smirked, tugging her own blouse to show how her curves had grown.

Nomsa smiled but said nothing. She kept the peach in her lap, its skin cool against her palm—proof she was being seen.

Then she noticed the important changes. At Pep, the discount clothing store where she sometimes unpacked stock for a day's wage, the manager suddenly gave her the front till in the shop. Customers came in for one item at a time—a handkerchief, a pair of socks—just for the excuse to stand at her till. One man leaned on the counter far too long, grinning, until the queue behind him hissed for him to move. Another, in oil-stained overalls, asked if she was interested in applying for a wind-farm bursary programme, as if higher education had always been hers for the taking.

On the way home, the driver who used to bark now nodded respectfully, calling her to the front seat. The usual shoving still happened behind her—elbows, groceries, a headless chicken—but somehow space opened around her body. Men insisted on carrying her shopping bag, slipping compliments between offers of airtime, already asking for her phone number so they could load credit onto her phone. At the tavern, where the walls sweated with beer fumes and kwaito pounded from bass-heavy speakers, someone delivered a cider to her table and left without waiting for thanks.

Even the aunties who had always scolded her for not settling down began to soften. They shook their heads and smiled knowingly. It was like they were saying, *"Nomsa, you are looking ready now. Maybe God is finally preparing you."*

The township hummed like a power line with a steady connection.

That night, a new restlessness grew beneath her skin. The house felt smaller, as if the walls had crept in during supper, when Mama had smiled more than usual, had seemed *pleased* with her, for no reason Nomsa could think of.

The vinyl flooring was cold and silent beneath her feet, but the wardrobe door creaked as she moved past. In the kitchen, the chipboard cupboards stood in the silence like paid witnesses in a courtroom.

Nomsa lay on her bed, the blister pack still under her pillow. The restlessness came like a whisper when she turned the light off. At first it was only a breath, a suggestion in her own rhythm of breathing. Then it shaped itself into words.

Let me out, it said.

She sat up sharply. The window rattled once.

I will do it better.

Her hands went to her stomach without thinking, pressing against the new softness there. It was warm, almost hot, and as though a second pulse beat beneath her own.

"Not yet," she whispered, as if refusing a second helping of Mama's vinegar pudding.

Waiting is what you do, the voice replied, firmer now. *Winning is mine.*

The house was too quiet. Even the dogs outside were not barking. She lay down again but kept her hands clamped across her belly, as though she could hold herself together by force of will. In her mind she saw the lipstick words on the mirror—*I am brave*—but the whisper beneath her ribs said: *No, I am.*

Sleep came late, thick and uneasy. When it did, she dreamed she was in the taxi again, only this time it was *her* skin rippling, *her* dress tearing while strangers clapped.

The next night was worse.

Nomsa lay stiff on her back, the vinyl flooring glowing faintly in the moonlight where the curtain didn't quite close. She told herself she would not sleep; she would listen for the whisper and stop it before it began.

But sleep came anyway, sudden as unscheduled load shedding.

In the dream, she was in the Pep storeroom, boxes stacked around her like walls. Each carton pulsed as if something inside breathed. The air smelled of plastic and new school shoes. She tore a carton open, and her own face looked back at her from inside, smiling. Another box split, and another Nomsa crawled out. This one was fuller, brighter, more certain.

They surrounded her, not unkind, but insistent. *Step aside,* they hissed in a single voice. *You've done enough waiting. Let us work now. Step aside.*

She tried to shout, but her throat just hissed like a kettle running dry. Her other selves laughed with the ease of women certain of their beauty.

She woke gasping, the taste of cardboard and school shoes in her mouth. The whisper was there, waiting, curled against her ribs.

Why are you afraid of yourself? It asked, almost gently. *You will thank me when I am free.*

Nomsa pressed her face into the pillow and cried soundlessly, so Mama wouldn't hear through the thin wall.

While some women followed the tablets' insert directions closely and just got fatter, others doubled up, and they began appearing in the taverns and in the Spar. Eureka split open behind the till, and her new, improved version took over like she'd always been there, smiling at the customers and counting out their change, while the shy Eureka quietly disappeared out the door and into the wind.

They even began to appear in places of power.

Ward Councillor Ndileka made her move in broad daylight. Everyone remembered her cousin's assassination, the way people in the taverns had whispered about tender fraud and kickbacks. People said she was no different, except that she wore the ANC headscarf while he'd worn the T-shirt.

But then she ruptured during a meeting at the community hall. Right there—between a complaint about rain flooding their homes because of blocked stormwater drains and a fight over the soccer field—her body rippled, then split, and the new Ndileka stepped out. The old one staggered away, barely noticed. The new one stood tall in a beautifully styled ANC dress, skin glowing, voice steady.

And strangely, she listened.

When the aunties spoke about dry community taps, she took notes herself instead of waving them at a secretary. When the youth demanded more lights on the soccer pitch, she said, *"We'll find a budget, even if I must cut my own."*

By the end of the meeting, people were clapping, even those who had sworn never again to vote ANC. A neighbour leaned close to Nomsa and whispered, "Maybe Thickly is fixing our leaders, hey. Maybe the new ones are better."

Nomsa didn't answer. Her stomach was alive with a pulse that wasn't hers, and in her ears the whisper was louder than the clapping.

Let me out.

On the following Sunday, the church was full, pews heavy with two kinds of women: those who had started Thickly and those still thinking about it. The air was close, bodies pressed shoulder to shoulder, hip to hip, perfume mixing with body lotion and dust.

Pastor Kasper leaned into the pulpit, eyes bright, voice rolling like God's thunder.

"Jesus said you must be born again," he cried. "But this"—he held up a depleted blister pack of Thickly—"my brothers and sisters, is the devil's born again. This is Western witchcraft wrapped in purple foil. And I will not have women in my church tearing themselves open like goats at the slaughter. This house is of the Lord, not of Thickly."

A murmur rippled through the congregation. Some women lowered their eyes. Others shifted proudly in their seats, new curves filling their Sunday best.

Two rows from the front sat Councillor Ndileka, her new self radiant. She stood, lifting her voice so all could hear.

"Pastor, with respect," she said, "this is not witchcraft. This is liberation. For too long women have been told to shrink, to be less, to wait while men decide. With Thickly, we are finding ourselves. We are standing taller. We are visible. Is that not also God's will?"

Applause broke out, sharp and eager. The pastor's face tightened, his Bible thumped once on the wooden pulpit.

In the middle pew, Nomsa's hands were clenched on her lap. The whisper inside her was no longer just a voice—it was pressure, a roiling wave under her skin. Her stomach shifted when she breathed too deeply. Her back prickled as though another spine pressed against it.

She bent her head and prayed silently, *No, no, no.* She had taken only one a day. But her dress felt tighter than it had that morning. Her veins throbbed with heat, and her pulse doubled, beating in two places at once.

On Monday, the new Zanele died on her way home. The ambulance driver, who had to ask for help to lift her body into the ambulance, said it might be a heart attack. An autopsy was needed.

On Wednesday, the church was packed for the funeral, women squeezed into pews, men crowding the doors, children running in the aisle until they were hushed. The coffin sat at the front, draped in purple cloth, and it was so hot that even the plastic arum lilies sagged. The air buzzed with whispers—some said she died cursed, others that her transformation had only been cut short, that the real Zanele lived on somewhere brighter. Some even said that the old Zanele had followed the new one home and stabbed her in the night in a fit of jealousy.

The autopsy said natural causes.

But the whisper that coiled like a snake in Nomsa's belly was simpler: greed bursts more than the body. Zanele had taken a double dose, and you cannot pack two souls into one skin and expect the seams to hold. The body was not a suitcase. Sooner or later, it would tear for good.

Nomsa was glad she had been following the instructions carefully, and Abush had said it was safe. Very safe.

Pastor Kasper stepped up to the pulpit, Bible held high. His voice cracked against the ceiling.

"Sisters, you saw it with your own eyes. Zanele, one day loud at the tap, the next day split in a taxi like a woman giving birth to herself. And now—dead. This is not a blessing but a curse! The Gospel prepares us for the *Rapture*, not this devil's *Rupture*! Now this young woman is torn from our midst."

Gasps, mutters, a ripple of shame.

Kasper raised his hands. "You cannot divide your soul! The first shall be last, and the last shall be cast out! You must repent, daughters of Zion. Leave this witchcraft. Leave this Thickly. Leave it now!"

Ndileka stood up. "Pastor, with all due respect, this is not a crusade. And I want to say something about our Zanele."

Pastor Kasper was tired, and he stepped aside from the pulpit to make room for the ward councillor.

"Zanele carried not a curse but liberation," Ndileka declared, her voice carrying more forcefully than Kasper's ever had. "Zanele showed us the way. She became more than the world allowed. Do not let fear bury her twice. Find yourselves. Be free. That's what she would have wanted, my sisters."

Applause rose inside the little hall, sharp and unashamed. The women rose to their feet, still clapping loudly. Kasper's hands had balled into tight fists.

Nomsa sat trembling beside Mama. Her stomach rolled, her veins throbbed with double beats. The whisper was no longer hidden—it pulsed in her throat, in her skin, in every breath.

Now. Stand. Become what you are meant to be.

She leaned forward, legs ready to rise. Mama's elbow jabbed her hard in the ribs. "Sit," she hissed.

The pressure broke anyway. Her dress strained, seams crying out. She clamped her arms across herself, but the skin split, slow then sudden, opening with a sound that silenced the applause. All heads turned, and all eyes were on Nomsa.

Her new self stepped out, shining, hair immaculate, smile assured. A gasp went through the crowd, then ululations and clapping—as though Zanele had just risen from the dead. Mama's hand flew to her mouth, pride and grief warring in her eyes like light through broken glass. The old Nomsa, thin and pale as a ghost, slid from the pew unnoticed. Her voice, when she tried to call her Mama, had no weight at all.

Ndileka cried, "See! It is a sign!"

Kasper thundered, "Satan is among us!" He was neither Catholic nor Anglican, but he made the sign of the cross many times.

Outside, the funeral spilled into feasting. Pots of pap steamed on open fires, men passed bottles, children ran with chicken bones in greasy hands. Laughter rose, as it always did, because funerals were community gatherings, and community gatherings always ended in eating and drinking. And this time they had a reason to celebrate, too—the new Nomsa, no longer too timid or too reserved.

At the edge of it all, when the sun sank lower and the music turned up, the old Nomsa stood in the shadows, unseen. She watched Owen, the bottle collector with warts thick on his body, folding cardboard boxes behind the hardware store into a bed. She saw the cows threading their way home through the dusty dusk, their bodies heavy with calves or milk.

Her eyes were drawn to the cemetery on the hill, its crosses dark against the orange sky. She'd overheard the grey-haired men at the funeral say the original Zanele had gone there to wait, to sleep in the empty grave until her bright, new half joined her in the earth. Nomsa wanted to cry, but now her tears belonged to someone else.

Once she had wanted to be noticed. Now she was forgotten, less than a shadow. The whisper was gone, and silence filled her like a hollow bowl.

The new Nomsa didn't just take her place; she improved it. The house was cleaner. Conversations with Mama were brighter, filled with plans. She wore the same clothes, but they looked better on her. The old Nomsa, thinning at the edges like an abandoned rumour, sat in the shadows of the alley across the street. A dog sniffed her hand, searching for the memory of a meal.

Not quite a ghost, she was haunting her own life.

A week later, Bedford was buzzing again. Banners flapped above the town hall, BEDFORD BETTER, in bold green letters. Councillor Ndileka stood on the steps, microphone in hand, the new Nomsa at her side in a bright blouse, hair shining, floral skirt flapping in the summer breeze. Cameras flashed, and a journalist from the *Daily Maverick* scribbled notes.

"Today," Ndileka declared, "we turn a page. No more bucket toilets—flush toilets for every home. And you can already see the painters at work. Bedford is becoming what it was always meant to be!"

Behind them, workers in overalls spread white paint across the greyed walls of the town hall. A truck rattled past carrying porcelain bowls, and municipal officials waved from the trailing vehicles escorting the gleaming cargo.

Applause thundered. Ululations pierced the air. The new Nomsa smiled for the cameras, her fist raised in the Amandla salute, the darling of progress.

Mama didn't see Nomsa wave, even though she was standing right in front of her, and no one in the crowd heard her call out, "Mama, Mama." No one in the crowd saw the old Nomsa slip around the building, a piece of chalk in her hand.

After the speech, when the painters returned to their work, they swore it was a ghost that wrote on the wall, because they never saw anyone. The cameras and journalists quickly found the half-painted wall. The large letters were neat and square and red:

MAMA, I AM STILL HERE.

Ghost Dog

written by
Mark McWaters

illustrated by
ANNA MALONE

ABOUT THE AUTHOR

Mark McWaters has long been a fan of all things that go bump in the night, scratch at the door, or blow cold air on the back of your neck. From a very young age, he carried a pad and pencil around with him, composing poems to give to girls. He devoured all the Hardy Boys and Doc Savage books he could get his hands on and expanded his reading horizons from there. Robert Heinlein, Asimov, and Bradbury whetted his appetite for sci-fi. Anne Rice's Interview with the Vampire *blew his mind. A scene from Stephen King's* Salem's Lot *haunts him to this day. And* Watchers *by Dean Koontz made him a lifelong fan.*

He earned an MFA in creative writing from the University of North Carolina at Greensboro, became an award-winning advertising copywriter and creative director, and earned enough Clios, ADDYs, and Communication Arts awards for bragging rights.

As a writer, Mark discovered the Florida Writers Association, critique groups, and writing competitions. He's won a steady stream of Florida Writers Association Royal Palm Literary Awards for his unpublished short stories and novels. The inspiration behind "Ghost Dog" came while reading entry rules for a magazine looking for unusual spins on traditional horror. Houses, people, dolls, even cars get haunted. So, he thought, why not dogs? Bentley, a West Highland White Terrier who sleeps under Mark's desk while he writes, agreed.

ABOUT THE ILLUSTRATOR

Anna Malone is an aspiring illustrator and concept artist whose work centers on world-building, storytelling, and characters. Growing up in a rural town in southeast Missouri, she developed a deep appreciation for creativity as a form of escape and self-expression, fueling her desire to build new worlds through her art. Skilled in both digital media and traditional painting, she combines expressive figurative techniques with atmospheric environments to create artwork that invites viewers into narrative spaces. Her visual language is shaped by the dynamic elegance of J. C. Leyendecker, the stylized fluidity of Alphonse Mucha, and the painterly nuance of John Singer Sargent.

Malone traveled to Columbus, Ohio, to pursue her passion professionally and is currently studying for her bachelor of fine arts at the Columbus College of Art and Design. With a growing portfolio of concept art and fantasy illustration, she continues to expand her creative practice through thoughtful experimentation, narrative research, and ongoing professional growth. Her long-term aspiration is to write and illustrate her own graphic novel, transforming her characters and imaginative universes into fully realized stories that others can explore and connect with.

Ghost Dog

My bedroom doorway, six feet from the foot of my bed, is a black rectangle at night. It faces onto a hallway even blacker. On most nights, Bentley, my stocky little Westie, sleeps on the bed beside me. On this night, something woke me from a sound sleep. I raised my head to see Bentley standing at attention, focused on the doorway, ears pricked and carrot tail shivering erect. His silhouette showed light gray in the alarm clock glow.

"Come on, dude. I took you out at midnight. It's three in the morn——"

A scraping noise came rasping through the walls, like something trying to get out. Or in?

Bentley woofed low, tentative. If I had four feet and a tail, I would've joined him.

"What is it, boy? Hmm?"

The scraping noise came again, insistent. My arm hairs tingled.

He woofed louder this time. Then whined and shivered. Nervous front feet shifted for purchase on the bed's edge. All his attention focused on that black doorway.

"What do you see, little man?"

I ran my hand along his back to comfort him. He turned and licked it, and I felt him shaking, the fur along the ridge of his back turned to stiff bristle.

"It's okay, Bentley. Doors are locked. Alarm's still set. No boogeyman here."

It didn't calm him. If anything, he grew more frantic—his growls, louder.

"You're freaking me out, buddy. Chill."

The sound came again. Bentley launched off the bed and disappeared through the doorway. Nails scraped and clicked on the hardwood floors as he vanished, barking. I listened to his progress as he careened through the house, slipping and sliding. His barks rose in pitch and now, turned to yelps, they came from one spot.

The back door?

I got up and followed my dog at a much slower pace. I stoked my courage by flipping on lights and roaring at Bentley to *Shut up!* in my deepest voice. But pajamas are scant armor at 3:00 a.m. The only real weapon I owned was a baseball bat. Which did me zero good leaning against shelves out in my garage.

Bentley's barking ceased the second I turned the corner into my family room.

He looked back at me once before refocusing his attention on the back patio French doors. The scraping sounds came again, this time accompanied by full-throated growls and wet, snuffling noises. The back doors rattled and shook in their frame, and my little terrier went insane.

Tail erect, his entire body quivering, Bentley hurled himself at those patio doors. Scratching, snarling, his scrabbling paws ripped huge gaps in the blinds of both doors before my shouts got through to him. He stopped and backed off long enough for me to bend down and grab his collar. I pulled him back behind me and crouched down to his level to peer through the openings in my newly redesigned blinds. And I froze.

I saw a large black shape pace back and forth outside. More transparent than solid, its form shifted as it moved. Where its head should have been, if it had a head, twin red orbs floated three feet off the ground. They left brilliant red streaks in the night's blackness as the thing passed up and back outside my doors.

What the heck? I tried to swallow and failed. *Could be a bear, right? We get a lot of black bears here in central Florida. Though, I never had one act like this.*

I still heard scratching and scraping, but this time it came from

Bentley's churning little legs, trying to pull himself free of my grasp. He barked and strained at his collar. No way was I going to let go.

"Ssh, Bentley. Don't piss that thing off. Whatever it is, we're staying here."

As frantic as Bentley got, the thing outside matched him. The scrape of its claws started up again—louder this time. They sounded more frenzied and grew in spurts, and then relaxed. Frenzied. Relaxed. Frenzied. Relaxed.

When they finally stopped, I breathed in a deep, grateful breath and felt my thumping heart slow. My dry throat craved cold water, and I eyed the fridge. Its chilled bottles of water called to me. *Five yards away. Easy peasy. I can make that.* I licked dry lips and prepared to let go of Bentley.

Then the thing outside switched its attention to the door handles.

My "handles" were levers I'd changed to when I moved in. Levers were more convenient and easier to open than knobs. And now my easy-open door hardware rattled in place along with the doors. Both levers flicked up and down at random while the thing tried to figure them out. Surely, it wasn't that smart. Was it?

What *did* I know? The only thing between me and *it* was a pair of double-paned glass doors held shut by a deadbolt I'd chosen for its looks.

I let go of Bentley and whirled for the garage, sweaty bare feet squeaking on the wood floor. When I returned seconds later, baseball bat in hand, the noise had stopped.

My little man stood stock-still, panting. His brave little Westie tail pointed straight at the ceiling.

I slapped my Louisville Slugger in my palm and crept forward, a middle-aged ex-surfer, ready to take on whatever might chew its way through those French doors.

I got as far as Bentley and stopped. He looked up at me and whined, once, before resuming guard duty.

Through gaps in the ruined blinds, I saw that the thing outside had stopped pacing. No more red streaks passed by in space. Instead, the floating red orbs held fast and resolved into a pair of bright, red-orange eyes. Brilliant as hot coals, they stared at us, blinking slowly after long, still intervals.

Bentley stuck his nose in the air and howled. A thin, quavery howl that prickled hairs on the back of my neck. His howl trailed off until he took a breath and howled again. And again.

The dark thing blinked a few more times before it turned its head and disappeared.

Bentley stopped howling, and I put down my bat.

"Who's a good boy? Who's my little man!" I rubbed and scratched him along his back and under his belly. "You scared him, little dude! Good boy! Good little warrior. We weren't scared, were we? Huh? No sir!"

Bentley took his praise in stride and walked over to lap at his water bowl.

Howling and scaring off night monsters is thirsty work.

"Good idea, buddy." I joined him with an ice-cold water from the fridge.

I waited until morning to open that patio door and walk outside, puffed up with more confidence than I felt. I remembered something I saw on a nature show about surviving the wilds in Alaska: Make yourself look as big as you can when facing a large predator. I had no idea what my *predator* looked like, nor how large it might be. But deep gouges scratched into the outside of my French doors reached up higher than my head. They told me more than I really wanted to know.

Nothing else happened over the next few days. I took that as a good sign and began an active investigation. I scoured my yard for animal tracks and any other evidence. Nothing. I asked my neighbors if they'd had any bear visitors lately. "Nope. Just keep Bentley in. You'll be fine." No one heard anything or saw anything. Okay, then. Our creature of the night had either been a bear or a bad dream. But my ruined patio doors mocked me every time I took Bentley outside to do his business. They wanted an explanation that made sense, and I had none.

I tried to stay busy; spend less time dreaming up boogeymen

and more time getting on with life. I took on another shift at the roofing company. Joined extra Bible classes at church. Even smoked a brisket for the weekend. It almost worked. Life had almost returned to normal when our night creature came back again. This time, after the late news.

I'd just turned Jimmy Kimmel off on the TV in the den and got up from my recliner. Bentley rose, stretching, from his doggie bed beside me. Bentley watches TV with me, waiting for dog food commercials. We made it all the way to the bedroom when I heard the crash.

Bentley took off right away, barking.

I grabbed my Hank Aaron Signature Edition from its new location beside the bed.

This time, Bentley's frantic yelps brought me to the very heart of my home—the kitchen.

Bentley stood at the edge of a circle of trash, focused on the sliding door leading outside from the breakfast nook. Silent now, he stood panting, as if waiting for backup.

Behind him, my kitchen trash can lay overturned, surrounded by last night's fried chicken leftovers, ripped up wrappers, and chewed paper towels. The mess covered every square foot of the kitchen floor. Like it had been arranged for maximum impact. Garbage art.

Either something had invaded my home—*It got inside my house!*— or there was a simpler answer. For sanity's sake, I chose simpler.

"What did you do? Huh, Bentley? Bad boy!"

His tail drooped, but his focus never wavered. I strode past him, checked the lock on the sliding door, and looked outside. Nothing out there but night. Black as pitch. Thank God.

"Dammit man, you had me spooked! All this to cover up your crime? Since when are you a trash diver, huh?"

I leaned my bat against the wall beside the sliding door, waded into the kitchen, and righted the trash can. I scooped up the mess and let the mindless task still my heartbeat as papers and cans and food scraps assumed their rightful place. This obvious explanation for the loud crash calmed me, but my stubborn mind refused to let it go. *But you heard the crash, right?... Bentley was beside you when you heard it... If* he *didn't do it...?*

I ignored the persistent voice in my head and pulled a broom and dustpan from the closet. I took my time scooping up the confetti-sized fragments of wrappers and foam meat trays spread to every corner.

I kicked a tuna can as I swept and picked it up, as well as a soup can, canned ham, and a can of green beans to add to my haul.

All four cans were dented and crushed. Round puncture holes all along their lengths hinted that something very large and very hungry had done the mischief here. Bentley's little mouth had neither the canines nor the strength for it.

Still not ready to face reality, I held out a flattened spiced ham can for Bentley to smell. "Did you do this, buddy? Hmm?" He walked over and extended his nose, then growled and backed off. He lay down and looked up at me like the answer was obvious. He knew. Didn't *I* know?

No. I did not. I didn't want to know. Rather than indulge an imagination raised on the *Walking Dead* and Hollywood's fascination with all things demonic and ghostly, I refused to guess.

Trash cleaned up and floor swept, I returned the broom and pan to their closet and weighed my choices: Coffee? Or bed? I chose bed. Then, after thrashing for over an hour and tying my bedcovers in a knot, I changed my mind and chose door number one. The smell of brewing coffee centered me and woke me from what I told myself was a nightmare not worth pursuing.

I took my time, sipping blistering cups of breakfast blend, until Saturday morning brightened my windows. I shook off my lack of sleep with a cold shower, bent all that caffeine to my will, and pulled out a months-old, dog-eared to-do list. Lucky me, the two-acre yard surrounding my Florida ranch home gave me plenty of room to work off my desperate energy. I spent most of the day tinkering with a temperamental lawnmower, mowing, raking, and bagging clippings. All handy metaphors for whistling in the dark.

Days passed and fogged my memories of our night visitor. Suitably brave again, I worked on several versions of plausible explanations: *Probably some neighbor's big ass German shepherd... if it's not a bear. But a bear is certainly more likely... What else could it be...? Moonless nights make it hard to see clearly... Besides, all dogs are trash bandits, aren't they?... Bentley's a tough little dog... He could easily bite through a can of spiced ham if he wanted to... Couldn't he? Sure, he could. Yeah. Good talk.*

Bentley calmed down the very next day, except for occasional side glances at the kitchen trash can. The baseball bat leaning against my headboard soon faded into the wallpaper. And the slavering jaws of bloodthirsty night demons faded from my nightmares.

Until it happened once again. This time, with feeling.

Bentley's Mr. Snake squeaky toy got the ball rolling. I've stepped on that ratty thing plenty of times, and Bentley always comes running like he's shot from a dog catapult. Then, I pretend like I'm stealing his favorite toy, which requires several minutes of pitch and fetch. It's his game. We play it because I know he likes it way more than I do.

This night, he and I had almost finished dinner. I was trying to convince him broccoli was perfect for dogs. He stubbornly disagreed.

Then Mr. Snake squeaked.

Bentley took off running, and I jumped up, expecting the barks and yelps to start up any second. But they didn't. No barks. No yelps. Not this time. Silence met me as I tiptoed after my dog. My mind, however, hit fifth gear before I ever left the dining room.

"Bentley? Hey, boy, where are you? Say something. Speak! SPEAK!"

It's the one trick we do that guarantees him a treat.

Not this time.

"Bentley!"

I ran through the house, ducked into rooms, threw open doors. My heart climbed into my throat, and I could barely catch my breath.

"Bentley. BENTLEY!"

Then, I saw the guest bedroom door. I keep that door closed at all times. No mysterious reason. It prevents the messes Bentley and I make from ending up in there and guarantees one room I'll never have to clean.

Now the door stood wide open, and I sprinted for it.

I stopped in the doorway and saw Bentley sitting on the floor beside the guest bed. Panting, tongue hanging out like he'd chased a yard full of rabbits, all his attention focused on the bed.

"Bentley? Hey, boy. Didn't you hear me?"

He whined when he saw me but turned back to stare up at the bed.

In the center of the spread, piled in a pyramid of bright colors, floppy shapes, and balls of every size and material, all of Bentley's dog toys were stacked in a dog toy pyramid. Mr. Snake drooped across the orderly array like ribbon on a present. I felt the cold breath of something chill the back of my neck, and I jumped and whirled around. Nothing there.

"What is going on, here? What do you want? Stop it. STOP IT!"

My heart pounded in my chest, and my breaths came fast and shallow. All my senses dialed to eleven. I couldn't swallow if I had to. Bentley looked back at me, and I realized my panting matched his. We were the ones making noise, breathing like we'd run a marathon.

But there was nothing else. No noise. No movement.

Then, why did I feel something watching me? And waiting?

Waiting for what? A reaction?

"Fine. You win. You got my attention!" I strode forward and plucked Mr. Snake off the pile of dog toys and hurled it back out the door. "Go get him, Bentley. Go get your snake!"

Bentley looked at me and walked, *walked*, through the door.

Not good. Mr. Snake always got him running. Bentley approached Mr. Snake, stopped, and sniffed at it. This new behavior scared me more than the artful toy display.

I walked over to the queen-sized guest bed and examined its spread for prints, indentations, dirt streaks, any signs to show how that dog toy mountain might have been put there. More than anything, I wanted to see Bentley's little paw prints. I'd rather believe my little guy had suddenly become a dog toy sculptural savant than entertain thoughts of an intruder that could pass through walls.

But, no. The spread looked as pulled tight and wrinkle-free as the day I last made the bed, smoothed it flat, and shut the door behind me.

Screw it.

I leaned over and scooped up the dog toy pyramid. Then carried the armful out to the family room and looked for Bentley's toy basket. I keep it on the fireplace hearth for him to root around in when he gets bored and wants to switch toys. The basket is rattan, and its woven wood is a survivor of several doggie generations before Bentley. Lots of memories are bound up in that bundle of sticks.

Or were.

I stood there looking at the spot where the basket always sat and felt violated. Which quickly became pissed off. Then, morphed into a feverish, pounding headache behind my eyes. To have memories of my beloved dogs reduced to a few gnawed sticks? Their basket, and now Bentley's, strewn about the room in pieces? The chewed-up, shredded bits of wood not much bigger than pencil nubs taunted me.

Uh-uh. No way. I dropped my armload of dog toys on the hearth and screamed.

"That's enough! Stop it! NO MORE!"

Bentley stood off to the side with his head cocked.

"Just quit it! Leave us alone!"

I knew yelling at the universe probably wouldn't do anything. But it helped me feel better… Until the next time.

I needed to figure out the who, the what, and the why of it fast, before these intrusions became dangerous. Before the intruder chewed on us?

I'm a rational, reasonable guy, most days. I could solve this. Right? I had internet access, didn't I? And a semi-open mind.

The Hotel Cassadaga, built in the 1920s, looked its age with convincing art deco architecture on the outside. Inside, the theme repeated, with uneven wood floors and plenty of creaks and squeaks per square foot. Its early association with a spiritualist camp that settled in the area in the late nineteenth century established the hotel's historical roots as the center for all things supernatural in this tiny, central Florida hamlet.

"Welcome to the Hotel Cassadaga! My name is Mary Alice, and I'll be your spiritual guide for the day."

The woman behind the hotel gift shop desk looked cheery and earnest, with smears of rouge on each cheek. A retiree dressed in boho chic, with bracelets on her arms and trinkets dangling from a creamy lace blouse, she fit the image of the hotel like a sepia print come to life, an ideal ambassador.

"I'm Mark, and this is Bentley." Bentley pressed tight against my leg. A new behavior. "Sign outside said you're pet friendly, so..."

"Absolutely. Dogs are people too. Am I right? May I interest you in a reading? Our psychics and mediums provide body, mind, and spiritual balance for whatever may be bothering you."

She tapped a finger to her lips. "Hmm. You look troubled, sir. Perhaps you'd like to speak with a dear one who's passed on? Maybe you're bothered by certain problems of the spirit?"

"Trouble is," I hesitated, because I'd never tried to put my problem into words, "I don't know *what's* bothering me. The internet was no help. And churches? No, thank you. I've seen *The Exorcist*. I'm here because I couldn't come up with any other ideas."

"My goodness, yes." She nodded, feigning sincere interest. I could tell she hadn't heard a word I said. "That does sound mysterious. Perhaps if you could describe—"

"We're being haunted."

"I'm sorry?"

"HAUNTED."

Blurting the word out loud snapped Mary Alice to attention in her chair and stopped the few tourists in the gift shop like hitting pause on a movie. They froze, shopping for crystals, pithy wall plaques, and potions, all forgotten.

"But I have no idea by who. Or by what. I can't figure out why it's happening. My house isn't built on top of a graveyard. No murders occurred there. It's Florida, so I don't even *have* a basement. Yet, things keep happening I can't explain."

I stopped to breathe and realized I'd twisted my "take one" Hotel Cassadaga brochure into a sweaty, mangled mess.

The shopping tourists leaned closer.

Mary Alice pressed a hand to her chest and cleared her throat.

"Mr. Mark, I want to assure you, you have come to the right place. We have a celebrated roster of spiritualists, psychics, and mediums to guide you through your current situation. But, just between you and I—"

The floor creaked as everyone in the shop took a step forward.

"The one person I'd recommend for you is Francie Athena."

"Francie is—a what? Is he a palm reader? Or a psychic?"

She raised her eyebrows, surprised at my ignorance. "*She*, Mr. Mark, is a medium. You may have seen her on TV or read about her. You are in luck, as a matter of fact. She recently came off a national parapsychology tour of college campuses."

"Awesome. Sign me up. Is she in?"

"Let me see, let me see." Her bracelets jangled as she leafed through a calendar. "Ah! We're in luck!" She stabbed a finger at a page in her book. "Looks like Francie has an opening two weeks from today!"

"No."

"I'm sorry?"

"I can't wait that long. This thing is wrecking my home. It comes inside at will. I can't sleep at night. I can't nap in the afternoon because every little noise wakes me up. I've lost ten pounds in two weeks. Bentley here is so freaked out he shakes and whines for no reason. Look at him. He's like this when I'm home because he won't leave my side. When I'm not home, he hides under the bed."

By now, I'd gathered a crowd. Several began shaking their heads when I mentioned Bentley. Scaring *me*? No big deal. But terrorizing a poor, defenseless animal? That was absolutely going too far.

"I'm afraid I—*we*—can't last another two weeks," I said. "This thing is escalating its attacks, and I came here because I have nowhere else to turn." My throat tightened, and my voice rose an octave. "I'm afraid of what might happen next."

"Come on, Mary. Help him out!" a voice shouted from the back.

"Yeah. Can't you see he's desperate? And poor Bentley," another said.

"God, yes. Save the *dog*!"

"All right, all right." Poor Mary Alice threw up her hands, resigned to dealing with a moron. "I suppose I can consider this a special case. Let me see what I can do." She pulled a cell phone from her purse and began dialing. "Folks," she looked up, "if you'd like to continue shopping. I'm sure you'll find many things more interesting than this poor man's problems. Go on, now." She shooed them away. "A little privacy, please."

Most of my crowd dispersed. A few hung back, pretending to shop.

Mary Alice looked up at me and shook her head. "I am sorry about—oh, hey! You're there. Well, I have a sort of situation…" She bent over her phone and whispered, so I only heard snippets…"I know, and I wouldn't ask, but— If you could that would be so— You are a lifesaver, Francie. Okay. All right then. Thank you."

She hung up her phone and raised her eyes heavenward. "Mr. Mark, you have an angel looking out for you." She stashed the phone in her purse. "Francie will see you."

"Great! When will I be able to—"

"Right away." She consulted an antique watch on a chain around her neck. "In, I'd say, thirty seconds or so. Francie had a cancellation that freed up some time for us."

"Thank God! I guess I do have an angel looking out for me."

"Oh, I don't know, Mr. Thomas. I'm good. But I'm no angel."

I must have jumped. One second Mary Alice and I were talking. A second later, this young woman appears beside me. "So sorry to startle." She laughed. "I wore my sneaking-up-on-people shoes today."

I laughed back. But then, like an idiot, all I could do was stare. "Are you sure about that *no angel* thing?" I stumbled all over my oh-so-witty comeback. Suddenly, I felt awkward as a teenager. To my eyes, *angel* did not do this heavenly creature justice. A blonde Jennifer Love Hewitt, dressed in a white, satin tuxedo she wore like skin. She wore her hair pulled back in a ponytail, with two curls framing sapphire eyes. This vision could have sat for any one of the old masters.

"Quite sure." She smiled and stuck out her hand. "How do you do? I'm Francie Athena."

I shook her cool, firm grip and tried to untie my tongue.

"Mark, um, Thomas."

"Pleased to meet you Mark, *um, Thomas*. And this handsome young man?"

"Bentley."

She bent over and scratched under Bentley's chin. Then froze for a second and looked up at me. "Let's go to my office, shall we? We'll have more privacy there."

She inclined her head, and we followed her to a staircase beside the check-in counter. "Please be careful on this. It's original, and by that, I mean old."

Bentley took the steps like a pro, and we walked out onto a dark hallway. Sconces on the walls, flowered wallpaper, shades of *The Shining*. The floor creaked as we followed her to an office at the end.

"Come in, come in." She sat behind a plain, wooden desk and directed me to the chair facing her. Bentley lay down beside me. The office was small with a blue sky and clouds painted on the ceiling. A rainbow of scented candles perfumed the air, while the obligatory crystal ball gathered dust atop a filing cabinet. A mobile of moons in various phases dangled from a Tiffany glass ceiling light. I supposed it was all very medium-like.

I glanced up at the blue-sky ceiling. "So, this is where the angels come from?"

"Good one." She chuckled. "No, as a matter of fact. That was here when I moved in. I don't even see it anymore. Now, I don't mean to rush you, but I have a pretty jammed schedule so—"

"We're being haunted."

She closed her eyes and nodded. "I thought as much."

"Already? How did you—?"

"Let's be sure." Francie walked around her desk and knelt beside Bentley. She rubbed his ears, ran her hands along his sides, scratched his belly. Then rocked back on her heels.

"You poor dear." Francie stood up and paced—three steps up, three steps back—in the tiny office.

"What's wrong?" I asked. "Is it Bentley?"

"It's both of you. How long have you had your, let's call it, your *Visitor*?"

"My Visitor?"

She rubbed her forehead and stopped. "Have you seen it yet?"

"Sort of."

"Tell me."

So, I did.

Two anxious days later, I answered the doorbell Sunday morning to find Francie Athena standing on my front porch. Her white tux outfit from the Hotel Cassadaga replaced by blue jeans, a Kansas City Chiefs jersey, and a ball cap with *Gaia* stitched in pink on the front. I couldn't take my eyes off her.

She grinned and hefted a canvas satchel with Crystal Power printed in lavender on the outside. "My tools. I'm not moving in, I promise." Her other hand held a potted plant ablaze with bright yellow flowers.

Bentley took that moment to worm his way out the door. It gave me a graceful excuse to quit staring at Francie like a total douche. "*Bentley!* Please. Come in. *Bentley, manners!* I can't believe you make house calls."

"This is my first. Hey there, Bentley." She dropped her bag and knelt to scratch Bentley under the chin. "I'd recognize this guy anywhere." Bentley promptly flopped over on his side.

"I must warn you. He's a connoisseur of belly rubs."

"Challenge accepted. Here. Hold this. St. John's wort." She gave me the plant and went double-time on Bentley's belly. "For protection. We'll pick a good spot for it later. Somewhere this little guy won't eat the flowers."

The three of us sat in my breakfast nook a few minutes later. Bentley lay at our feet half-asleep while I sat and watched Francie pull item after item from her bag.

She placed an unopened container of salt on the table. "Salt is a purifier. Some use it as a barrier to keep out negative energies."

A bundle of sage. "A little smudging never hurts."

A handful of crystals. "Selenite crystals. More protection. I hope I brought enough for every room."

A long strand of garlic bulbs. "Before you say anything. No, it's not just for vampires."

A hefty, ornate bronze cross. "This was my grandmother's. We're going to need all the help we can get. If you don't mind?" She took it over to the fireplace and blew a kiss heavenward. "Thank you, Mimi. Watch over this house." She set the cross on the mantle and stepped back to admire her work.

"Sorry, but I can't help feeling like I'm in a B movie." I gestured from the mantle to my overburdened breakfast table. "Seriously? I don't mean to sound ungrateful, but—"

"Hold on, I'm not done." Francie came back to the table and took one last item from her bag: a big, blue jug of glass cleaner.

"Seriously?"

She raised her chin. "When's the last time you cleaned your house?"

"Uh—"

"That's what I thought." She plopped it down on the table. "I get it. You're a guy. I bet the tags are still on your vacuum cleaner." She made a face and looked around. "Do you even *have* a mop?"

"No fair. You read minds and stuff."

She laughed. "I do not read minds. Not exactly anyway. All these are for cleansing. Spiritually and literally. A clean, tidy environment encourages positive energy flow and discourages negative energies or spirits. Window cleaner isn't sexy, I'll grant you. That's why you'll never see it in a B movie." She air-quoted B movie. "But you have a serious problem. Believe me, I don't *usually* make house calls carrying cleaning supplies."

"I guess it's not what I expected, is all."

"That makes two of us. Can you pull up Bentley's photos on your phone for me?"

"I can do that. All of them?"

"Yes, please. I want to show you something."

I tapped a few times, brought up my Bentley library, and scrolled, pausing here and there on a particularly great photo.

Francie reached out and snapped her fingers. "Give. Come on, now. I know he's your baby and he's super cute. I want to show you something."

She plucked the phone from my hand and smiled as she scrolled, stopping every so often to tap the screen. Like I did.

I cleared my throat. "I thought you were going to show me something?"

"He is a cutie, that's for sure. Ah, here we go." She tapped and scrolled, up and back, several times before stopping. Those blue eyes looked at me, satisfied. "Ready?"

"Uh, yes? I guess?"

"Good. Mind if we sit on your sofa?" She headed for it without waiting for an answer.

"That's what it's made for— Okay."

She sat and patted the cushion beside her. "Sit."

We sat thigh to thigh, and I confess, looking at Bentley pictures was the second thing on my mind.

She sat on the edge of the cushion and brushed a lock of hair off her face, the consummate professional. "I am what's called a physical medium. All mediums receive their information in different ways. Physical mediums, like me, get theirs from *things*. You've probably seen the movies and TV shows where a medium uses a watch, or wallet, some personal possession, to find a missing person."

"With mixed results."

"True. There are no guarantees with this stuff. Now, let's talk about your little buddy. In my case, *Bentley* was your personal possession. I read him immediately, back at the Cassadaga Hotel."

She held my phone so we could both watch as she scrolled through his photos. "The good news? Your phone's pictures are chronological." She scrolled up. "These are earlier." She scrolled down. "These are later. Not too long ago, by these dates, something happened to Bentley. Watch."

She slowed her scrolling, tapping each photo she passed, and wincing every time. Until she got to pictures of Bentley asleep in his bed. She tapped on it with no reaction. "Here, he's a happy boy in dreamland." She moved one photo down. "Here, he and *you* have a problem. What happened?"

"Nothing. He looks like my same little guy."

"He's not. Please. Check your calendar. Something happened between these two photos. Did you go anywhere? Have visitors? Did he get in a scrap with a neighbor's dog?"

I thumbed through the dates on my calendar, and there it was. "Oh. Crap."

"What *crap*?"

"Golf weekend with some buddies. I boarded Bentley at a local kennel. No dogs allowed at the golf resort. I didn't like it. But—"

"Well, *something* liked it very much."

"What do you mean?"

"He, and by *he*, I mean Bentley, picked up a hitchhiker at the kennel."

"Like fleas?"

"There's no flea bath for what Bentley came home with, unfortunately."

"Let me see if I get this. Bentley stays at a haunted kennel. Then afterward, brings home a ghost that took a liking to him. Seriously? First of all, I didn't know ghosts even haunted kennels. Second of all—"

"Second of all—what came home with Bentley isn't a *ghost*." Her voice turned snippy. "Not in the way you're thinking. And just so you know, ghosts are all around us. All the time."

"That's not creepy at all."

"As are *spirits*. What grabbed ahold of your boy is a spirit—and a malevolent one. It's not a demon, thank the Goddess." She rubbed her face. "We can relax as far as that's concerned."

"I haven't relaxed in two weeks. But, ignoring that for a minute. Why did this thing pick on Bentley?"

"That one's easy—because of you."

"Oh, come on."

"Quit scowling. You love your dog?"

"More than anything. More than most people, no offense."

"None taken. I'm the same way with my pets." She bounced up and began pacing, concentrating on her words. "Listen." She stopped and held out a palm. "Your love and your spirit have become part of Bentley." She held out the other palm. "As *he* has become a part of *you*. You've heard how some owners start to resemble their pets?"

"Yes."

"That's truer than you know. Our bonds with our pets go spirit deep. Your two spirits are linked." She clapped her hands together and squeezed them tight. "By the way, before you ask, almost all religions acknowledge that animals have spirits. That malevolence that came home with Bentley senses the love, the spirit, between you two and—" She shrugged.

"What?"

"It wants what Bentley has."

"Which is—?"

"You, Mark. It wants *you*."

Francie traced her finger down the gouges in my back doors. "You two must have been scared out of your minds."

"Bentley's a terrier. Attack is his *default* position. Pretty sure I was more freaked out than he was."

She wiped her hands on her knees.

"Weird thing is," I said, "I have cameras set up. Front and back. They didn't pick up a thing. My neighbors didn't hear anything. Of course, my nearest neighbor isn't really all that close. Homesites in this neighborhood are well over an acre."

"Not surprising. Your cameras, like your neighbors, wouldn't see or hear anything. This Visitor of yours is only interested in you, not your neighbors. On occasion, manifestations like yours might move furniture or rustle bushes. But physical contact with anything in our material world is exceedingly rare."

I pointed to my scratched-up door. "And yet—"

"And yet," she pointed at the damaged doors. "There are always exceptions. I'm glad I saw this firsthand. It provides context. I'd say, without a doubt, yours is the most determined spirit I've ever run across."

"Oh, goody." I'm sure I looked anything but pleased.

"We have work to do."

We started by burning sage in every room to purify the house. "I'm so used to this now. I actually love the smell." Francie blew on the glowing clutch of straw in her fist and waved smoke around her as she walked.

I sneezed and opened a window.

Next, St. John's wort joined Francie's cross on the mantel. "We brought you company, Mimi. I knew you wouldn't mind."

We poured salt crystals along every threshold and windowsill. "Believe it or not, plain old table salt is powerful magic in the supernatural world. Witches and shamans swear by it."

"You sound like a National Geographic travelogue."

She stared at me, unimpressed.

"I'm afraid your ride on the spirit train has only begun. Once your malevolent friend discovers we've blocked its entry, I imagine it won't be too happy about it."

I stopped jouncing the selenite crystals in my hand. "What happens if it's not happy?"

"Easy with those," Francie said. "Here, let me." She cupped her hands, and I transferred the sweaty crystals from my nervous hands to hers. "Hard to say. I mean, your Visitor is quite capable of entering your home at will. It's the gouges on the outside of that patio door that have me worried. You know how much force it took for a *spirit entity* to do that?"

I chewed a thumbnail. "Gee, thanks a lot, Francie. I'm worried enough."

"I want us to be prepared, that's all."

"Us?" I looked at her. Was she blushing?

"You don't think I'm going to let you and Bentley face this thing alone, do you?"

"I guess my thinking didn't get that far."

She shrugged and smiled. "Mine did. I have a change of clothes in my car. I'm only telling you that now because I didn't want to scare you before."

"I wouldn't have been scared." I chuckled and hoped like hell my chuckle sounded brave.

"Really? What would you have thought if I showed up on your porch with a suitcase?"

Dammit. I felt my face getting redder by the second. Now we're both blushing.

"Uh, huh," she said. "That's what I mean. Come on now. We still have a ton of work to do. The crystals go in every room. We'll place them as close as we can to every window. After that, it's window-cleaner time. I brought paper towels if we need them."

Afterward, we sat on the couch, Bentley curled up between us, and sipped iced chamomile tea. Francie brought the tea with her. "I thought we could use a little bit of calm." I had to say, watching her rummage around in my kitchen to brew put a happy glow in my stomach.

I breathed in the cold, friendly smell of ammonia. "My house smells so clean now."

Francie laughed. "I should hope so. You might want to look into buying cleaning supplies in bulk." A single blonde lock hung down on her beautiful forehead. It'd been a tough day, and this woman absolutely glowed in her Chiefs jersey and jeans. Me? I felt like I'd run two marathons.

I still held the near-empty jug of cleaner. I shook it and grinned. "How much does a maid service cost?"

"*Mark*. One, good old-fashioned scrubbing and you're ready to throw in the towel?"

"Every blessed one of them. I had no idea Bentley could dirty up our house so much."

"Bentley, huh?" She stifled a grin. "You know, Westies don't shed, and he doesn't use the kitchen or the bathrooms. So…"

"Okay, I see where this is going. I'm a slob, I admit it."

"Oh, you have your good points." She smiled at me, then immediately busied herself by swirling her glass of tea.

I did the same, wondering what *good points* I had and which ones she might actually see—and Bentley raised his head and growled.

We both looked at him, and she raised her eyebrows.

"Here's the thing," I said. "This guy hears all kinds of stuff outside. He growls. He barks. Especially right around now. At dusk."

"Are you a good watchdog, Bentley? Hmm?" She scratched his head, and he lay back down on the sofa cushion. "What we are interested in is not the normal things he's used to. Like squirrels, owls, tree branches, dogs in the distance, passing cars. Those won't raise the hackles on his back."

"So, we wait for hackles." I said it like I had every confidence in that strategy.

"He'll sense our intruder way before we do."

I definitely heard her say "our" and "we" and let myself enjoy the implications of those pronouns. Until a particularly loud gurgle from my stomach interrupted my happy musings.

"I was wondering," I said.

"Yes?"

"What do you like on your pizza?"

She flopped back. "Good question. I don't really eat pizza."

"Who doesn't—? Ah. Gluten?"

She propped a hand under her chin. "Let's see now. Gluten, cheese overload, fat, sugars, sodium, nitrates. Pizza is pretty much death on a cracker, in my opinion."

"Don't hold back, now. I don't eat pizza myself, really. Except in dire emergencies."

"Like—?"

I counted on my fingers. "Oh, like terrifying spirit encounters, Buy One/Get One Domino specials, a shocking lack of cooking skills, and Sundays."

"How in the world do you survive?"

"Food delivery services. The world is my menu."

"How about something healthy? Do you like tabbouleh?"

"Ta-who?"

"It's a delicious salad with parsley, mint, bulgur wheat, vegetables..." She rubbed her tummy. "You'll love it."

"Salad?" I must have wrinkled my nose.

"I just lost you, didn't I?"

"No..."

Bentley lifted his head and growled.

"See? Bentley agrees with me."

Bentley jumped down off the couch and growled again. He walked toward my half-destroyed French doors and stood, nose up, sniffing.

"Hold on." Francie held up a hand.

"I told you. He hears stuff all the time. It could be anything. *Bentley*."

"Look at his back."

I could see the ridge down the middle of his back grow more pronounced as we watched. "Oh, crap."

We got off the sofa at the same time and stood on either side of Bentley. Both of us focused on the hairs along his back.

"We have hackles." She said it as if reporting on a rocket launch—matter-of-fact, no emotion.

"What do we do? Do you have anything to ward this thing off? How about Bibles—?"

"We did all we could earlier. No Bibles. But the protections we put in place should do the job. Sorry, there are no silver bullets, excuse the reference. Let's give our protection a chance to do its job."

I looked over at her tight smile and confident calm. I envied her. My insides roiled like snakes in a death match. I divided my attention between her, my dog, and those patio doors.

Bentley took a step forward, and his growls dipped lower with menace.

"What is it, boy? What do you hear?"

He cocked an ear my way for a brief second before resuming his focus.

I tried to see through the still-damaged blinds. Hoping, and not hoping, I'd see something black and terrifying stalking my patio.

"Oh, my," Francie said. She reached for my hand, and we held tight to each other.

Right outside the doors, some of the blackness of the night coalesced into darker shadow, and a pair of flaming red eyes grew large and close, so close I swear I could see a snout take shape ahead of them in the smoky dark.

Bentley tensed for a charge, and I squatted in time to snag his collar. All he could do was snarl and run in place. The next thing I knew, the thing's head materialized and—*Oh, my god!*—pushed through the doors and *into my house!* Like a wolf's head but twice as large, with sharp details blurred and constantly shifting.

The effect froze the breath in my chest, and I fell backward, on my butt, still holding tight to Bentley. I felt Francie put her hand on my shoulder and squeeze. Hard.

"Wait," she whispered. "It's checking things out."

The thing looked right and left, then opened its mouth and let loose an unearthly roar. Part high-pitched scream, and part deep rumble, it vibrated in my bones and turned my muscles to jelly. I had time to register the jagged gleam of uneven teeth and a dark pink tongue wagging in a sea of black while it held its spot. Red eyes blinking, it stood there, as if undecided about what to do next.

"Hold on," Francie murmured. Her fingernails dug deeper into my shoulder as she squeezed tighter.

The Visitor curled its jowls over imperfect rows of teeth and shifted its gaze to stare at each of us. It ended with a long, last look at Bentley, before shaking its head and pulling back. The dense black that marked its form shifted in place. Then, it disappeared. Wisped away like smoke in the night.

Bentley stopped tugging at my hold on his collar and stood still, panting. I let go, and he shook his head, not unlike the Visitor had.

"You think it's gone?" I asked.

"I think it knows what to expect, now. I don't think it's happy."

"So, it worked. Yes!" I held my hand up in a high five, but she left me hanging.

"We'll know when it comes back again. And, I'm sorry, but it *will* come back. You and Bentley are too tasty for it to let go that easily."

"What should we do, then?"

She had no good answer and sighed. "I suggest you order a pizza with everything you want on it."

"Like a last meal?"

"Like, we're going to need food for energy. I suppose a cauliflower crust is too much to hope for?"

"We should definitely stay awake." Francie tossed a piece of cauliflower crust back in the box. "First watch is on me."

"You think it's coming back tonight?" I held a piece of crust for Bentley to scarf and he turned up his nose. "Aw, c'mon Bentley. It's good for you. *He* may not be a fan, but I gotta say, cauliflower crust isn't so bad."

She laughed. "At least I converted one of you. And, about tonight? It could go either way. But, tonight, *or tomorrow*, it will be back."

"I'm so glad you're here." I watched her turn a beautiful shade of pink, and I hurried on. "I mean, you're my witness. I'd begun to doubt my sanity there for a while."

"Of course. No worries on that front." She cleared her throat and flushed pinker. "I had a pretty good idea what we were dealing with after we met the first time and I saw your Bentley pics. But this?"

She pointed at the damaged doors. "Your Visitor? It's so immense and well-formed. See how it shape-shifts? Fades in and out? It uses up so much energy to come through and occupy space on this spiritual plane. I suspect it's resting up before trying again."

"I'd hoped once it saw how we prepared for it—"

"I did, too. Obviously. But that was always a long shot. All we did was cover the basics. Every spiritual event is different. It's those differences that keep me and others like me, in business."

"So glad we could help."

"Oh." She punched my shoulder. "You know you're loving this."

"No…not really. Outside of meeting you, of course. And trying cauliflower-crust pizza."

"Now I know you're lying."

"Maybe a little." I shrugged and tried to look charming. "Truthfully? I'd feel a lot better if I knew how this was all going to end."

"Me, too." She smiled and charmed my socks off without even trying.

"Hold that thought," I said. "I'm going to slide this couch as far back against the fireplace as it'll go and face it toward those doors since our Visitor seems to prefer that as his entrance. It'll put us some twenty-five feet back from the doors if—*when*—he returns. We'll have more room to react that way. If we have to."

"Good thinking."

I moved the couch and left to gather an armload of blankets and pillows from the bedroom. When I returned, she and Bentley were snuggled together on the couch. I draped a blanket over both of them and handed her a pillow. Next, I spread enough blankets on the floor in front of the couch to make my luxury vinyl flooring marginally softer.

"Hey, now. Don't make me too comfortable." She folded the blanket back and sat up. "I'm on watch, remember?"

"Nah. You guys looked so settled in, and I'm too amped to sleep." I stretched out on the floor. "First watch is on me. I'll wake you up if our Visitor comes back."

She yawned and scratched Bentley behind his ears. He sighed and laid his head in her lap. "Pretty sure Bentley will have that waking-up part covered for us quite nicely."

I woke as a small furry body bounced off my chest and began barking in full-on terrier mode. Francie sat up on the couch and rubbed her eyes. I rolled over and tried to catch hold of Bentley, but he'd wised up after our latest encounters and sidestepped me.

"Bentley. Come!" No chance. Hackles down the center of his back stuck up in a wide swath like every ancestor in his bloodline was sounding *Danger! Danger!* in the only way they knew.

I jumped up and managed to corral Bentley and snatch him up in my arms. We retreated back to the couch, and Francie and I sat there together, holding a squirming Bentley clamped tight between us.

"Do you think it's back?" I whispered.

She scratched Bentley's head and tried to shush him. "One of us sure does."

The French doors rattled, as if on cue, and the door handles flipped down and up. From across the room, I saw the red eyes stop and stare through the glass at us.

"Get out of here!" I waved my arms. "Go away!"

It stared at us so long without moving I began to think our preparations had finally worked. Then, it pushed its head through the doors like last time. It turned its head to the right and left before centering on us. It narrowed those fiery eyes and pulled back into a crouch, then leaped forward like jumping across a chasm. It landed on all four feet, at least a full body length inside the doors, and shook itself like a giant, wet dog.

More the size of a bear than any canine; nevertheless, the wolf/dog resemblance was undeniable. Clearly agitated, a black tail thick as a lion's tail whipped back and forth behind it. The frightful head had a projecting snout too large for its face and ears, like small saucers, twitching front to side.

Unlike our earlier glimpses of the Visitor through the glass, this thing was not a vision in shifting wisps of smoke. Its black body looked solid as sculpted obsidian. No fur, it had shiny onyx plates joined together that rippled like scales. The scales covered the body from nose to tail and made a raspy scrape when the swishing tail dragged on the floor.

It took all of us a few seconds to comprehend the situation. The protections we'd put in place had zero effect on our Visitor, other than to maybe piss it off. It passed through my shut-and-bolted doors like they didn't exist. And now, this nightmare stood twenty feet away, unfazed by anything we'd tried. Focused on Bentley and to a lesser extent, us, hate radiated from the thing.

It shook that massive head and roared; jowls quivering, mouth wide open, and teeth dripping with drool. Bentley snarled in reply and lifted his little snout and howled. A long, plaintive howl, he might have been summoning the rest of his pack—if he had one.

I tightened my hold on Bentley and felt Francie shift forward in her seat.

"Hold onto him," she said.

"What? Where are you going?"

"We need more help." She slid off the couch and around the side of it to the fireplace mantle behind us. She reached up and plucked down her grandmother's cross.

The Visitor cut short its roar and watched her return to the couch. She thrust the carved bronze cross forward as she walked. Held tight in both hands, she used it like a shield and began chanting loudly in what sounded like Latin. In response, the Visitor cocked its head to one side. Curious perhaps, but not the least bit bothered.

In the excitement, I squeezed Bentley a little too tight, and he snapped at me. He nipped the side of my cheek hard enough that I raised a hand in reflex to feel for blood. My brilliant little dog used that opportunity to squirm free and jumped off the couch.

"BENTLEY!" My heart jumped to my throat, and I rose to chase him, but Francie reached an arm across to hold me back.

She shook her head. "Don't interfere."

She continued her Latin chant as we both watched Bentley creep ahead, one tentative foot at a time. His little jowls lifted in a snarl, he faced the Visitor, and moved forward. A little white David facing a giant, black Goliath.

The Visitor focused all its attention on Bentley. It extended its snout and flared its nostrils, sizing up this loud little challenger creeping toward it. It swiped a pale pink tongue across its jowls and crouched, waiting.

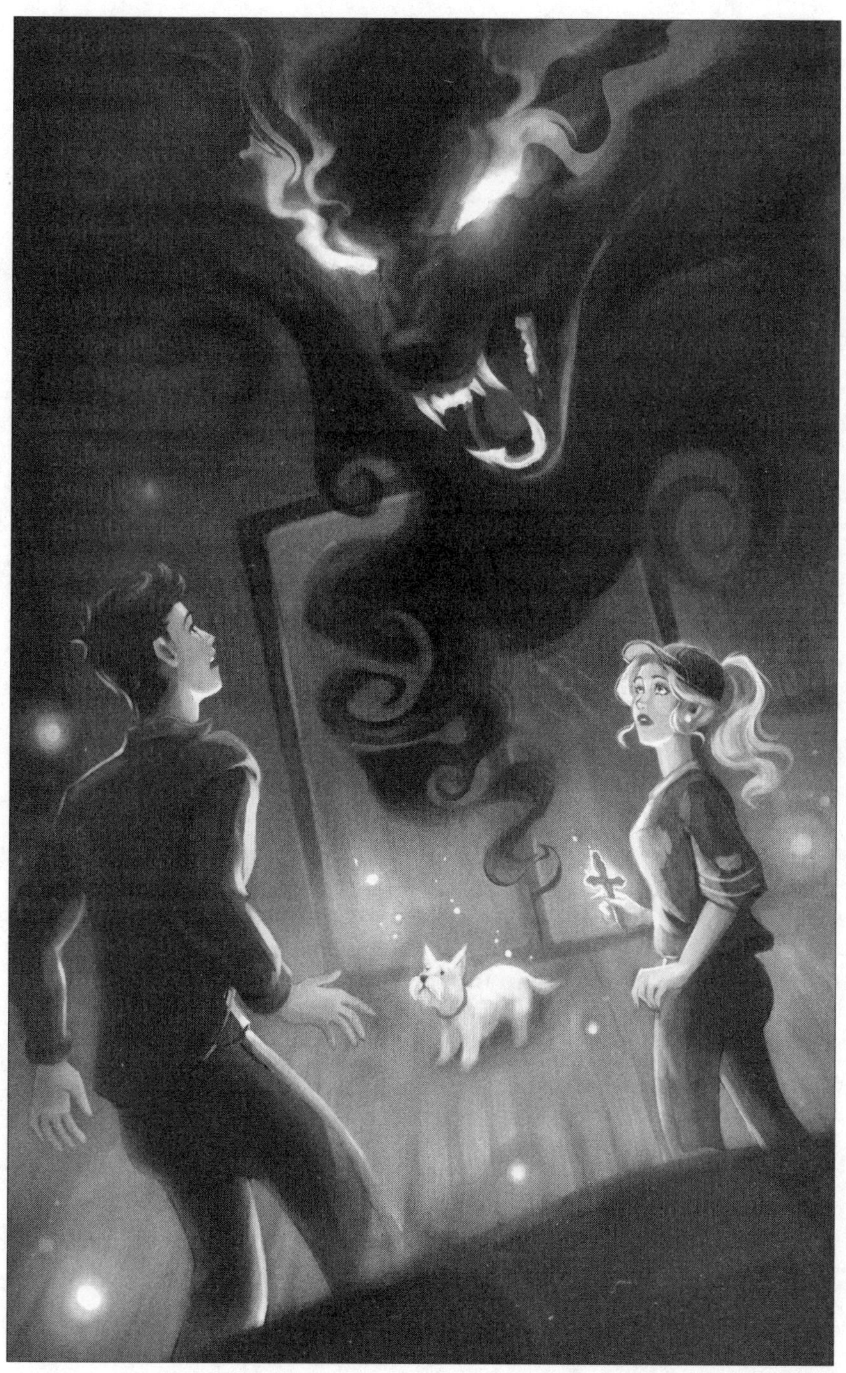

ANNA MALONE

Bentley got within ten feet of it and stopped. He tensed and dropped low to the ground, preparing to leap.

The Visitor did the same. Claws extended, back legs tucked under, it snarled one last time before launch.

The tableau in front of us emphasized the mismatch. One, small white terrier ready to hurl himself against this intruder in his home. In front of him, a massive black horror, seconds away from ending the fight with one snap of its deadly jaws.

"Screw this. I'm not going to sit here and watch my little boy get destroyed—"

Francie stopped her chanting and said, "Watch."

The air around Bentley began to shimmer like waves in water, and I saw a large dog materialize beside him. It looked familiar, and when it glanced back at me, I recognized my big white boxer from years ago.

"Max?" I whispered.

The same thing happened again, on the other side of Bentley. This time, another boxer materialized, a fawn with a bright white chest. "Moxie?"

I'd raised several boxers in years past, all from rescues. And here were two of them—no, three. No, four. Another two boxers materialized and took up positions in a line with Bentley. Rocky and Angel stood there. A pair of boxers on both sides of Bentley. They looked stocky and solid. In their prime, like I always remembered them.

The wet shimmer in the air grew wetter and wavier, and I realized I had tears streaming down my face. "My babies." I sniffed. "Francie! These are my dogs from a long time ago. How is this even—? I can't believe it."

Francie had tears in her eyes, too. "Look," she said, "look at the Visitor."

Bentley glanced to both sides as the dogs from my past—hackles on full display—snarled beside him and growled deep in their chests. He joined in and the line of dogs advanced together on the intruder.

The Visitor looked at them and rose from its crouch. It wrinkled its snout and snarled, but the line of dogs, with Bentley at their center, never slowed. Then, as if responding to some hidden signal, they charged.

The Visitor whirled around in mid-snarl and disappeared out the door. Max and Angel followed the Visitor, passing through the doors without hesitation. Bentley skidded to a halt right at the doors, no doubt confused and wishing he knew their trick. Moxie and Rocky held back and came over to me. I tried to pet them. I remembered how Moxie loved her ears rubbed and Rocky was partial to chin scratches. But I couldn't touch them. My hands passed right through them.

"I'm sorry, Mark," Francie said. "You *can't* touch them. You loved them, and they reflected that love back at you as only well-loved pets can. Their spirits and yours will always be together in some fashion, on some plane. Remember? I told you? Your Visitor wanted that kind of existence for itself. Your babies showed up to deny it that. See? Bentley gets it."

By now, Max and Angel had returned back through the patio doors and the four ghost dogs had gathered together to form a circle around Bentley. Like all dogs do, they sniffed noses and tails, passing information around in ways us humans will never understand. Bentley's little tail stuck straight up throughout the love fest, and I couldn't help wondering if the spirits of my ghost dogs smelled like anything at all. He didn't seem to mind.

Soon enough, my spirit protectors began to fade. They each took the time to come to me and "woof" before leaving. Rocky was the last to go. He came over and laid his big head on my knee like he used to, and I swear, this time I could feel the weight of him. He "*woofed!*" and I tried to stroke his head, but I ended up petting my own leg instead.

Francie sat closer and laid her head on my shoulder, and I was fine with that. More than fine.

"I just have one question," I said.

"Just one?"

"Those four dogs of mine. I loved them so much, obviously. But I had a lot more than—"

"And I could have summoned a lot more. The love you shared with your dogs was that strong. But these were the closest to you in spirit at the time—and they were boxers. Big, formidable dogs. I figured we wouldn't need to call on any more for help. But we could have. Trust me. Your cockapoo, standard schnauzer, and poodle were ready. They all would have come."

"Wow."

"I'd say that sums it up nicely. Wow, indeed. Bentley?" This time he listened and jumped right up on the couch between us. She cuddled him, and he licked her face. "You've got some pretty big—*collars* to fill, young man."

She scratched his belly, and he wriggled. He didn't seem worried.

"You opened my eyes up, Francie, to so much I never knew existed. Spirit dogs, ghost dogs...*you*."

"I see where I rate." She chuckled.

"C'mon. You know I didn't mean that—"

This time she laughed outright. "Easy, big guy, I'm messing with you."

Big guy? I liked that. "What just happened here?"

"What we saw here, today, to quote Huey Lewis, was the power of love. It was love between you and Bentley that attracted the Visitor. Love that summoned your beloved pets to your side. And that same power of love that kept the Visitor in its place."

Bentley jumped down and trotted off for a moment. He returned with Mr. Snake in his mouth, and we watched him walk over to the patio doors, drop the dog toy at the threshold, and come back to the couch.

"Aw, look," she said. "He's sharing his toy with the Visitor."

"No way! He barely shares his toys with me."

She hugged my arm and snuggled closer to me. "*That*, my dear, is the power of love."

The Creator's Journey

BY BRIAN C. HAILES

Brian C. Hailes, founder of Draw It With Me, Epic Edge Publishing, and HailesArt, is an award-winning artist, illustrator, author, and designer whose work spans the worlds of science fiction, fantasy, and imaginative realism. Known for his intricate detail, cinematic compositions, and storytelling through imagery, Brian has spent his career bringing to life the extraordinary in various genres and formats, from children's literature to short stories and novels.

Having graduated from Utah State University with a BFA in illustration and design, Brian has illustrated for clients such as American Girl, Arcana Studios, the Church of Jesus Christ of Latter-day Saints, and numerous publishing houses, authors, and game developers. His art has appeared in several annuals including Infected by Art, *and multiple gallery exhibitions and conventions across the country. In addition to his freelance work, he is the creator of several original projects, including the illustrated novels* Blink, Hotel California, *and* Defender of Llyans, *and the graphic novels* Devil's Triangle *and* Dragon's Gait.

Brian writes mainly speculative fiction, comics, and graphic novels that often delve into themes of morality, redemption, and the resilience of the human spirit. His experience as both author and illustrator gives him a unique eye for narrative visual flow and emotional tone—qualities he brings to his role as an official judge for Illustrators of the Future, a position he has held since 2023 after first being recognized as an Illustrators' Contest winner in Volume 18.

When he's not writing, sketching characters, or painting luminous worlds, Brian can be found mentoring emerging artists at the Draw It With Me Art Academy in Alpine, Utah. He also enjoys traveling, exploring the outdoors with his family, and drawing inspiration from nature, history, and the old masters. His lifelong fascination with story, art, and imagination continues to fuel his work, much of which can be seen on his website.

The Creator's Journey

In today's world, where automation, the relentless pace of technological advancement, secularism, and commercialism converge, the hero's journey of an up-and-coming artist and illustrator is not an easy path. For the young child who doodles in class, it might start out that way, but as they crest that first slope, they quickly come to realize the vast wilderness before them, each mountainous obstacle a little taller than the last. However, daunting as the view may be, we creatives can't quite shake that call to adventure, that humble spark that ignites a lifelong pursuit of mastery and self-discovery.

By the time I was a teenager, I had already been drawing for a decade. When I was thirteen, I walked to the local post office to pick up the mail, and the postmaster, poking her head over the counter, suggested I enter an art contest established for kids and sponsored by the US Postal Service and McDonald's to design a postage stamp commemorating Earth Day the following year. I went home, sketched out an image of a boy planting a tree, colored it, and sent it off. A few weeks passed, and I got a call notifying me that my entry had been chosen. I was one of two finalists in my state, and later, one of four winners in the country (out of more than 150,000 entries). I had won, and they printed my image and name on over five million US postage stamps! On McDonald's and the USPS's dime, they flew my family and me to Washington, DC (twice), for tours of the White House, museums, dinners, press, and meet-and-greets with our national leaders. I ate free Big Macs for three years and scored $6,500 in prize money (which, for a young teen in 1993, wasn't too shabby).

A year later, I entered the National Written and Illustrated

By...Awards Contest for Students at the behest of my English teacher, winning fourth place (out of over 50,000 entries).

Unsatisfied, I entered again with a new story, "Don't Go Near the Crocodile Ponds," taking third place. And I was off to the races, working my way up. However, despite these wins, it was really the genres of fantasy and science fiction and storytelling in general that took and held my interest. And art was my path. I had to do it every day...relentlessly. Years later, that early recognition came full circle when I was selected as a winner in the Illustrators of the Future Contest and featured in *L. Ron Hubbard Presents Writers of the Future Volume 18*.

Every artist must cope with the rawness of unrefined talent, possibly unsure if a creative future is the right choice. It used to be that "an overnight success only took ten years," but now it might take longer. Still, practice ensues, goaded by the "oohs" and "ahhs" of doting mothers and friends. Those of us with access to education draw the dreaded sphere, cube, cone, and cylinder again and again as we listen to middle school and high school teachers and college art professors. Under their tutelage, we delve into the arcane mysteries of human and animal anatomy. We study the essential elements of art: line, shape, texture, form, space, color, and value.

I was blessed with two such mentors in my years studying at Utah State University: the late Glen Edwards (illustration) and Alan Hashimoto (graphic design). In one of my senior portfolio review classes with Hashimoto, I remember him glancing at a sprawling battle scene illustration from a fantasy graphic novel I was writing and illustrating. He paused, glanced up from my drawing, and said, "I think you should go to New York."

I shook my head. *"New York City?!"* The prospect of a small-town country boy going to the Big Apple for an entire semester seemed outlandish.

"Yeah," Hashimoto said nonchalantly. "An internship. I'll sponsor you."

Fear of the unknown almost kept me from chasing that lightning, but I eventually realized the opportunity and got on a plane. I emailed Hashimoto with a weekly report and secured a part-time job at the

Society of Illustrators. I ate lunch with major editors, art directors, filmmakers, and gallery owners. I peddled my portfolio and got a few small gigs around town. But most importantly, I had begun to learn the true art and business of illustration. While there, one of my non-artist roommates came home with some of their hippie friends from NYU, whom I'm pretty sure were high at the time, and they found me painting. One of them said in a slur, "I'm sure you'll have your own cult following someday."

I laughed.

Cult following or not, my desire to create paintings and drawings was so profound, I just knew I had to do it.

When approaching artists, people use the cliché, "I can only draw stick figures," or "I wish I could draw like that," but what they are really saying is, "I wish I had a desire to practice for 10,000 hours and then 10,000 more."

It doesn't matter where we start, or the resources we have on hand. The desire to succeed and the tenacity to keep pushing will eventually get us there. We must navigate the often-turbulent waters of feedback and criticism, balancing our personal vision with the demands of critics or fresh clients, all while attempting to build a respectable portfolio. Still grappling with how to draw people and things at a professional level, we focus on our personal mark-making style, edge quality, imbuing the picture plane with real meaning and depth. What about chiaroscuro? And the light source? We realize we're not as good as we initially thought, and the progressively revealed deficiencies of our creations become a significant consideration. There is still more to learn, and much by way of improvement to make. As we experiment with different media, supports, software, brushes, and drawing tools to achieve that ever-evolving desired effect, we can get frustrated.

We visit museums and scour online portfolios. We scroll through social media and realize just how many great artists are out there—artists we consider to be much better. Is it all still worth pursuing?

Hesitantly, we swallow our pride and take another baby step forward. We learn the art of collaboration, because other people have valuable input too. There can be no diva in us. We work on contrast, balance, emphasis, proportion, hierarchy, and other design principles. These are not mere technical requirements, but profound elements that give structure and meaning to our creations. Through these principles, we learn to orchestrate our compositions with precision and intentionality. Good design becomes a reflection of our core values, a manifestation of our artistic ethos. Each piece is a testament to our ability to harmonize disparate elements into a cohesive whole, creating works that resonate with viewers on a deep, emotional level and bring stories to life.

We embrace both discipline and a touch of madness. Discipline ensures the honing of our craft while creative eccentricity allows for transcendence of conventional boundaries. This combination drives us to continually create, to explore new territories of expression, and to push the limits of imagination, all while seeking new commissions and achieving a higher quality in our art. We strive to deliver the impossible: amazing work at breakneck speed.

Our growing confidence finds us entering shows and competitions and taking jobs with clients we never thought possible. We eventually come to realize that our journey is not solely about producing individual masterpieces but about the process of becoming great artists. In other words, we find our "voice," our own unique style. We should not merely seek to fit into existing molds but rather strive to slip into a mold of our own making. The director of the Society of Illustrators once told a group of visiting art students that the most important element of any painting is the signature.

At the same time, we can look at historical figures in the art world who can provide valuable insights into how artists have historically navigated their creative journeys. The Pre-Raphaelite Brotherhood, for instance, offers a compelling example. Founded in 1848, this group of artists, including Dante Gabriel Rossetti, William Holman Hunt, and John Everett Millais, rejected the academic standards of their time. Instead, they pursued a style that was deeply personal and rooted in medieval and early Renaissance art.

The Pre-Raphaelites's commitment to their unique vision, despite criticism and resistance, highlights the importance of forming our own artistic mold. They were not content with simply following established norms; they sought to create art that was true to their own ideals and aesthetics. Their success demonstrates how staying true to one's vision, even in the face of adversity, can lead to lasting impact and recognition.

Similarly, other old masters such as Vincent van Gogh and Rembrandt van Rijn made significant contributions by developing distinctive styles that set them apart. Van Gogh's expressive use of color and brushwork and Rembrandt's mastery of light and shadow are prime examples of how individual approaches to art can lead to profound innovation and success.

In the twenty-first century, emerging technologies like AI are reshaping the landscape of art and illustration. AI tools can now generate content, assist with design or ideation, and even animate existing compositions based on user input. While these advancements may offer new possibilities for some, they also pose very real challenges for working artists involving copyright issues with AI training models, a perhaps dwindling pool of opportunity in the commercial landscape, and additional competition for human artists who must navigate these complexities. Each artist will have to choose for themselves how to confront this constantly evolving field.

With mastery of the principles learned, we step into the climax of our story, the final act: creating with passion and professionalism. New projects, shows, art commissions, and collaborations with a diverse clientele become the new norm. With gratitude and humility, we strive to demonstrate our ability to integrate technical skills with distinct creative vision, all in the hopes of emerging as a consummate professional. We approach our projects with an understanding that publishers, studios, and corporations seek not one-hit wonders, but creators capable of consistently producing exceptional work. Indeed, our reputations are built on our ability to deliver masterworks time and again, embodying the essence of our artistic vision with every new story or gig. As we look back at many of the greats, we see that it wasn't about this piece or that, but their impressive and world-changing body of work. Indeed, through small and simple means, some mere humans—through the flick of their brush, pencil, or stylus—have become titans.

Yet amidst the accolades and achievements, we remain grounded in the knowledge that our true triumph lies not in any single piece but in the ongoing evolution of our craft. Our journey is less about the final product and more about the transformative process that has shaped us into creators.

Dragon Visits

written by
Nina Kiriki Hoffman

illustrated by
APRIL SOLOMON

ABOUT THE AUTHOR

April Solomon's dragon artwork inspired me to write "Dragon Visits." I am so pleased to get to work with her—I hope my story dragons give her illustration ideas!

I was also thinking about grief and loss. Some of my friends have left this earth recently. My situation is not the same as Martin's in the story, but our feelings are the same.

One of my resources as a writer is the power to explore my own experiences. I can't do it just by thinking about it, but if I start writing a memory—in my journal or on the computer—more and more details come back to me.

One of the writing exercises I use with my students is to have them think about the kitchen of the house they lived in when they were ten. Ask themselves questions: What happened in that room? What tasks did they do? Who was there? What did it look like, smell like, sound like? What did the food that came from that kitchen taste like? How did their ten-year-old self feel about what happened in that room?

Once I'm engaged with flowing words onto the page or the screen, so many things come back to me.

One of my first story sales was to the very first volume of the Writers of the Future. *My mentor Algis Budrys told me and my friend Dean Wesley Smith about the Contest, and we both entered the first quarter, and our stories made it into the book.*

Since then, I've written many more stories, several novels, and some middle-grade and young adult books. My fiction has won other awards—a Stoker, a Nebula. Currently I'm focusing on short stories and teaching writing.

ABOUT THE ILLUSTRATOR

April Solomon was born in 1983 and raised in Laguna Beach, California. Since she was a small child, April has had a talent for drawing and painting. She would draw anything and everything that came into her imagination. Of all things, she drew dragons the most! Thankfully, her loving and encouraging family inspired her to embrace her love for the arts.

She grew up around art. Her father's art studio was filled with all the delights a child could indulge in. His bookshelves held stacks of art books containing illustrations from the old masters, the golden age illustrators, and even some fantasy art from TSR's Dungeons & Dragons. Inspiration came in many forms. Fortunately, it was everywhere! And so her career as a young artist began.

Today, April is an illustrator and fine artist who has earned her bachelor's degree in illustration at the Laguna College of Art and Design. April's passion for learning the old masterful techniques of traditional drawing and painting is precisely what inspires her work.

Alongside her love for the fine arts is her unique appreciation for whimsical fantasy that adorns every image of her portfolio. April's meticulous creature designs aim for what is known as "fantastic realism." A clever, concise understanding of anatomy, plants, and mysterious textures weave their way into her illustrations, leaving the viewer guessing at origins, influences, and ancestry. April's work allows the viewer to dive imaginatively deeper and reconsider whether dragons might be real or whether werewolves exist to stalk the streets at night.

When not illustrating, April attends garage sales to unearth buried treasures, runs and lifts weights, or braves as many haunted attractions as possible during the month of October.

April is a former Illustrators of the Future Contest winner featured in L. Ron Hubbard Presents Writers of the Future Volume 39.

Dragon Visits

I saw my first living dragon the day after my thirteenth birthday. It was also the day after my mom died.

In my grandmother's guest room in her house in Talent, Oregon, I was playing Jay Ungar's "Ashokan Farewell" on my violin. My violin teacher wouldn't like me playing a fiddle tune instead of classical music—I imagined her saying, "Martin, not again!" to me—but right now, I had other reasons for music. The song was saying things I didn't know how to say yet.

I opened my eyes after one time through the tune and saw the dragon. It was silver and small, about a foot long, most of it tail. Its scales sparkled, and it had spikes along its spine, and in arcs above its golden eyes. It had curly horns. Its wings were folded along its back. It lay next to my violin case on the cushioned window seat, watching me.

THE DAY BEFORE

I was sitting on my bed, texting with my best friend, Judy, whom I met in violin class four years earlier, when we were both eight and playing on smaller Suzuki violins. We'd been in the same middle school for two years now. When we quit school for the summer, we'd shared science class, English class, and French class. Now I mostly saw her at our Saturday violin class, though sometimes we went to the game arcade and played games together.

She sent me a "Happy Birthday, Martin!!!" text with balloon, cake, present emojis, and a ripple effect, and asked me when she and two other friends were supposed to come over to my house for cake and presents before Mom took us all to the movies.

Dad knocked on my bedroom door and came in without waiting for me to answer. It was just after noon. He took the phone out of my hand and did a hard shut down on it.

I said, "Hey!" He never did a thing like that without explaining it first.

Dad sat down next to me on the bed and hugged me so hard it hurt. He said, "Mom's gone."

I had a million questions. The hug and the hurt in his voice made me swallow them all.

Dad was usually solid as pavement. Distant, like an island offshore you could see but not visit. He seemed eternally unshakable, and occasionally funny. This little grin he had—it was something I tried to make happen, kind of a reward or a boon, and it didn't come easy. He was good at answering science and math questions when I had them, less interested in things that weren't real or solid, like the fantasy books I loved.

He and Mom went on a walk through the historic district in Yreka every evening, and I wasn't invited. They had a whole private language, only some of which I'd managed to decode. Then again, Mom and I spent time together, too. We did dinner and cleanup together, and she read stories to me some evenings while I listened and sketched pictures of what I was hearing. She came to my violin performances and basketball games, which Dad usually didn't. We planted vegetables together every summer. She bought me toy dragons.

When Dad finally released me from the hug, I gripped my elbows and squeezed my arms against my chest. I let one question out. "What happened?" Maybe she'd just gone somewhere else. Maybe she could come back.

"A drunken idiot in a pickup truck," he said. "She was crossing the street by the party store." Then he started crying. I'd never seen him do that before. Each sob jerked him a little, and it strangled him coming out.

Heat and hurt swamped me. I didn't want to understand anything.

I clenched my eyes closed. I didn't want to see anything or hear anything or know anything. I wanted to go back in time fifteen minutes and never get to this moment.

Dad hugged me again, his sobs shaking me, then he let go suddenly and walked out of the room, closing the door quietly behind him.

I lay back on my bed and stared up at Okiri, the golden resin dragon hanging from the ceiling, his wings spread, and the glow-in-the-dark stars on the midnight-blue ceiling beyond him—a night sky my mother had made for me as a birthday present when I was ten. Actually, she got Dad to do most of the painting—he was so much better with a paint roller—but it was her idea to give me stars to look at when I woke from nightmares. Her idea to give me a dragon guardian a year later.

Mom's protection. How she listened to me and found ways to help. Gone now.

Tears leaked out of my eyes. I clutched the bedspread and blankets. I wanted to wail, and I didn't want to make a sound. I didn't want any of this to be real.

Dad came back in. "I've talked to your Gran," he said, his voice scratchy. "She says you can stay with her."

"What?" I asked.

"Get up. Pack. We're leaving for Gran's in half an hour."

"What?" I said, even though I had heard him. "Dad, wait. Where is Mom? Where's her—where's—" My throat closed. "Can we see her?"

"No." Dad looked away. "She's in the morgue. And I have to—they're going to look— Look. She's dead. There's no question. I'm going to have to—" He clenched his fists. "I can't do this," he said, and left again, closing the door behind him.

I tried to pull myself together, even though there were broken pieces inside. I got up and dragged out the summer camp suitcase, then started stuffing it with summer clothes. I was supposed to go to orchestra camp next week, so I would have been packing anyway. Mom had inked my initials in Sharpie on the labels of my shirts and pants, because I'd done music camp before, and people's laundry got mixed up.

None of this made sense.

I tucked two little dragons in among my clothes, and my Kindle and its cord. I made sure my violin case had the shoulder rest, the rosin block for rosining the bow hairs, and extra strings. I tucked the classical violin music book I was working from in the outside pocket of the suitcase. The whole time I packed, my mind was blurred. I couldn't focus. I couldn't feel my hands or feet.

Dad came back and grabbed my suitcase. It had wheels, but he carried it. I took my violin case and grabbed an extra pair of sneakers. I hadn't packed soap or deodorant or a toothbrush or anything, but Gran had things like that in her bathroom cupboard. At the last second I remembered to grab my phone and a charging cord.

We didn't talk the whole three-hour drive to Gran's in Oregon.

She came out on the front porch as soon as she heard the SUV pull into her driveway. She was wearing her summer at-home clothes, a Hawaiian-flowered muumuu and slippers. Her wispy white hair was loose around her head like a small local cloud. She seemed to be trying to smile, but it wasn't working very well.

I got my violin case out of the back and walked to her, confused and aching. She hugged me. I was as tall as she was now, which was new since the last time I saw her at Christmas. I couldn't lean into her chest and hear her heartbeat anymore. I laid my head on her shoulder instead.

"Are you hungry, Martin, honey?" she asked me after a little while.

"I don't know," I muttered.

"I'll make you both something to eat."

Dad said, "No, Mom. I have to get home and...sort things out somehow." He dropped my suitcase and shoes on the porch, then turned and strode back to the car. Started it and drove off.

"Oh, Harry, my son, my son," Gran said, staring at the SUV as it disappeared around a corner. She sighed and let me go. "Honey, I'm so, so sorry."

"I don't even—" I said and started crying.

She took my hand and led me inside.

I forgot about my phone.

Gran made me a grilled cheese sandwich. We sat together at the kitchen table, and she watched me eat and didn't say anything except,

"I'm so sorry, Martin. And it's your birthday. And I don't have cake or anything."

"I don't want anything!" I yelled. "I don't want to remember!" My face got hot. I buried it in my hands.

She patted my shoulder and got me some iced tea. Eventually, I lowered my hands and drank some.

Afterwards, we moved all my stuff into my usual room when I stayed with her, the less fancy of her two guest rooms. The other one had a king bed and didn't have chickens on the wallpaper. Mom and Dad slept in that one. Or used to.

I put my clothes in the closet and the dresser and took everything else out of the suitcase. The phone was the last thing I retrieved from the outside pocket of the suitcase. I turned it on finally and saw "MARTIN! MARTIN? MARTIN!" from Judy, and messages from the other two friends who were supposed to be at my house tonight for my birthday party and the movie afterward.

Had they come to the house? I couldn't remember us leaving a note on the door or anything. It was so late now. Did Dad send them away? Was he even home? I'd left them hanging. I'd completely spaced the whole party thing.

I couldn't—

I couldn't.

I took the phone downstairs where Gran was sitting in the living room with the TV on low. I handed it to her. She picked up the reading glasses on the little table by her chair, put them on, and looked at the phone.

"Could you tell them?" I asked.

"Sit down," she said, tapping the chair next to hers. The red guest armchair, also angled so the sitter could watch the television with Gran. I sank into it, and she patted my hand, then studied the phone.

In the end, I had to teach her how to text. She still had a landline.

She wrote to each of my friends, though: "This is Martin's grandma. I regret to tell you Martin lost his mother today. Martin is staying with me for a while. He will get back to you when he can."

I sat with her for a while, not seeing anything she was watching, until she said it was time for bed.

THE DAY OF THE DRAGONS

My eyes felt funny when I woke up. I blinked at the ceiling, which was plain white and showed the shadows of the sheer curtains as morning light filtered through them. Darker valleys where the curtains were folded, and lighter shades of gray where there was only one layer between outside and inside.

Why did Dad dump me on Gran? I guess he wanted to be sad alone.

A wave of hurt washed through me, and then I got mad, and then I felt too tired to be mad. I got up and took a shower, trying to wash all the pain down the drain. After I dried myself and dressed, I took out my violin. I played through some of the Bach minuets, because I could play them without thinking, and then I switched to "Ashokan Farewell." Ms. Marigold never liked for us to play fiddle tunes, but Judy and I had been listening to songs and learning them by ear on the sly. And this one came out of my fingers and into my head and then all through me, not a terribly sad song, but my sadness could ride on it. I played through it once with my eyes closed.

Then I opened them and saw the dragon lying there, silver and shiny in the dim light, leaning against my violin case. Its head lifted and it turned its big golden eyes on me, its slit pupils widening.

Everything inside me went still.

I kept playing.

Mom had read me dragon stories from when I was little. I got obsessed with dragons and collected them, at least little ones. Then Mom gave me Okiri, my resin dragon, when I was eleven—he was a foot long, and his wingspan was broad. He flew above my bed with the night sky beyond him, and I fell asleep looking up at him.

I came to the end of the tune the third time I had played it through. I lifted my bow from the strings, lowered it by my side, and stared at the dragon. It blinked at me and faded from sight.

I took a deep breath, let it out, then set my violin and bow gently on the unmade bed and walked to the window seat. There was a faint depression on the cushion where the dragon had been. I reached down, slowly, so if the dragon was still there but invisible, I wouldn't hurt it.

No, it was gone. The little dip in the cushion was still warm, though.

Gran made scrambled eggs, and I made toast for breakfast. She had a jug of orange juice already made. She was wearing go-to-town clothes: jeans and an Ashland Shakespeare Festival T-shirt. We ate at the kitchen table, looking out at her back garden in the bright morning through a window framed with yellow, ruffle-edged curtains. She had planted tomatoes and lettuce and carrots in the raised beds, and they were branching and leafing and flowering. Little brown birds flitted here and there, singing morning songs.

As I sat down to eat, I reached in my pocket for my phone, because that was my breakfast habit—check the phone while Dad read the paper and Mom read a novel. The phone wasn't there.

Gran looked at me. "I heard you playing this morning."

"Oh, shoot. Did I wake you up?"

"No, Martin. You've gotten so good. It was beautiful."

"Mom loved that song," I whispered.

She patted my hand.

"And there was a dragon," I said.

"Really?" Gran said.

"Yeah. Did you know you have dragons?"

"Only the ones you bring with you. You left a little yellow stuffed one last time. I put it on the mantel in the living room."

Huh. I had been missing one. We'd come to Gran's for Christmas last year. I hadn't realized that was when I lost Eggstra. I'd been trying to grow out of my sillier dragons, because bullies had mocked me at school and on socials. Whereas Judy, who had been drawing dragons for the past three years and always had a little stuffed one hanging from her backpack, nobody mocked her. They wanted her to give them her pictures, which she did once in a while.

"This wasn't—" I stopped, tired all of a sudden. I wanted to tell Mom about the dragon I had seen this morning, and I never would.

"Wasn't what?"

"I mean, it was a real dragon. Little. Alive."

"Did it burn anything?"

"No."

"Good. I don't mind them in the house if they don't destroy anything or eat anyone."

I looked at her sideways, wondering whether she was serious.

She smiled gently and finished eating her eggs. "You up for a trip into town? I need to get some groceries. I wasn't expecting company."

"Oh, Gran."

"Glad to have you here, Martin. You're not a problem. Just unexpected."

I turned away and rubbed my eyes. My heart twisted. I hadn't been expecting it either.

"You should come with me and tell me what you want to eat."

I didn't want to go anywhere. I didn't want to be here, either, in this messed-up timeline. I just wanted to shrivel up and die.

She poked me with her fork. "Might as well come. You can be miserable anywhere."

We went to the farmers' market. Tents and canopies lined the market square in the center of downtown, selling tomatoes and squashes, melons and berries, chanterelles and onions and plums. At the gingerbread booth, I looked up at the canopy and saw shining green eyes looking down at me. A brassy lizard-looking dragon, the size of a small iguana, rested on a crossbar supporting the canopy, its arrowhead-pointed tail dangling down. Its purple forked tongue flicked out and in.

I walked past other people and stood just below it, staring up into eyes that met mine. "Mom," I whispered.

"What is that?" asked a little girl beside me. Her head was about as high as my stomach. She stared up, too.

"You see it?" I couldn't look away from its gaze.

"Yeah."

"It's a dragon."

"Okay." She grabbed my hand.

We stood there staring up until someone said, "Alisia! Get over here!"

She let go of me. "Gotta go." She scampered off, and I glanced after her—I hadn't gotten a good look at her while we were staring up at the dragon, only seen her in my peripheral vision. She was cute and dark, with short curly black hair, wearing a sky-blue romper and sandals.

When I looked up again, the dragon was gone.

Gran had me carry her shopping baskets around the market as she filled them with all kinds of things. They were heavy by the time we got back to her car. The gingerbread cookies scented the car interior. I thought about last Christmas, when Mom and Gran and I had baked a whole lot of cookies. Mom and Gran had dueling recipes from their families. Gran's were German, and Mom's were Italian. We made a bunch of different kinds. Gran sorted them into cookie tins, and we delivered them to her neighbors' houses while we sang Christmas carols with my cousins and Aunt Margreta and Uncle Paul, who lived in the same town as Gran.

"Let's do this every year, Helena," Mom had said to Gran.

"Yes, let's," Gran had said to Mom.

The weight of all the never-agains crushed me down into the passenger seat of Gran's Audi.

The third dragon was black with turquoise eyes.

Gran asked me to make my bed while she was putting away the groceries, and then she said I could do what I wanted. What I mainly wanted was to lie on the bed and stare at the boring white ceiling and try not to think. Or sometimes I let myself think. I wondered what Judy was doing. Today was Saturday. We were supposed to have our violin class.

I found my phone and saw there were messages from everybody Gran had written to last night, and I didn't want to read them.

Nothing from Dad.

Eventually I got up and took out my violin. Might as well practice. Except Bach or Chopin wasn't going to do it for me right now. I tried another of my stealth fiddle tunes, a Finnish song called "Metsäkukkia—Woodland Flowers," in A minor. It was sad in a good way.

I was staring at the dresser without seeing it as I played, and then I saw the black dragon, about the size of a cat, arrive as a shadow that turned to a solid. It had feather-edged wings and shining scales and lots of spikes, and its eyes were a blue lighter than sky. It listened, then sat up, like a cat, its tail wrapping around its front feet. We looked and looked at each other while the song flowed from my violin.

"Mom," I whispered, and the dragon blinked, then met my gaze again. Mom, I thought. You picked the best haunting. Inside me, tight things loosened, and warmth bloomed.

I finished my third repeat of the song and lifted my bow. The dragon nodded and vanished.

She wasn't all the way gone.

I put my violin away and headed downstairs to help Gran with dinner. The tune still sang in my mind. I wondered who would show up if I played "The Road to Lisdoonvarna."

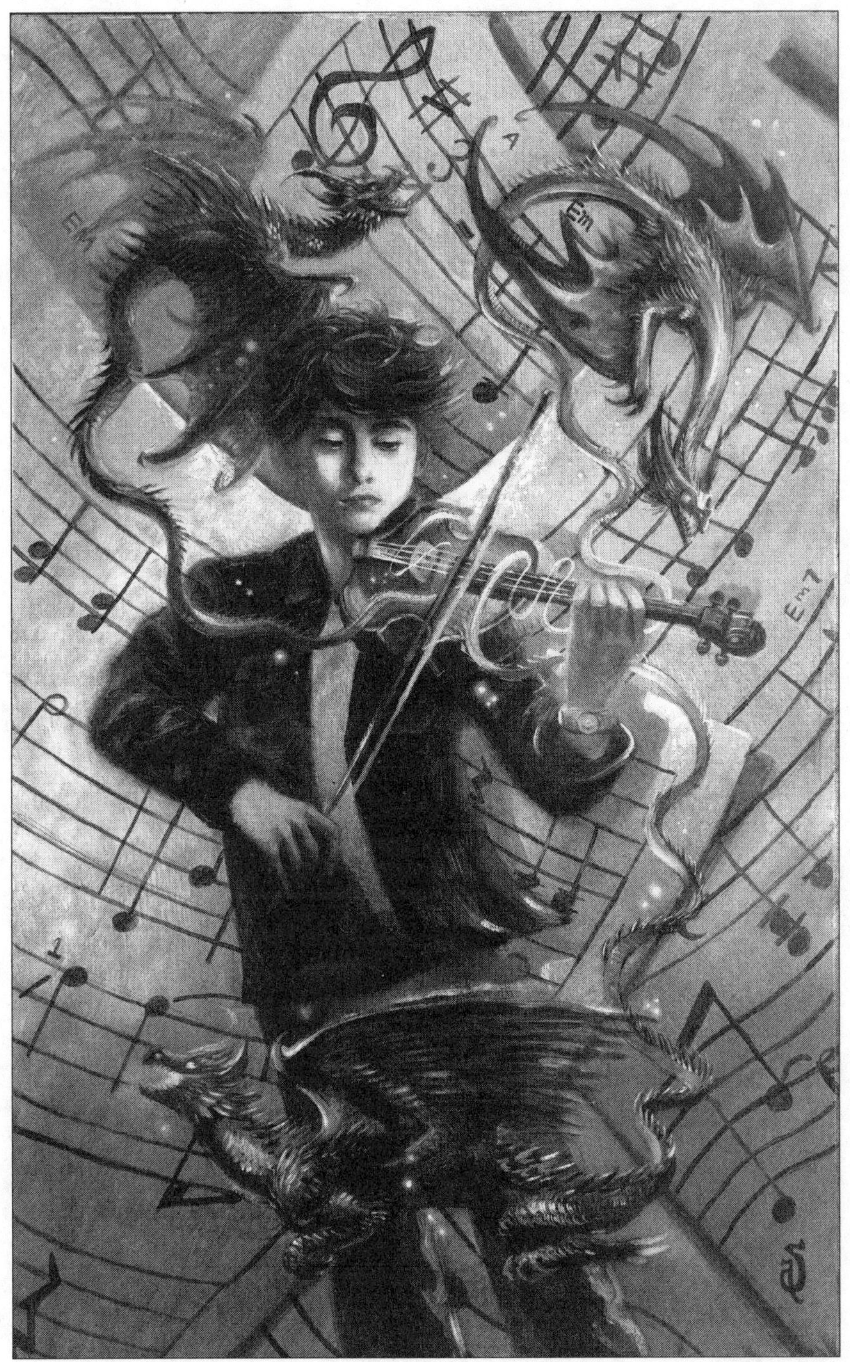

APRIL SOLOMON

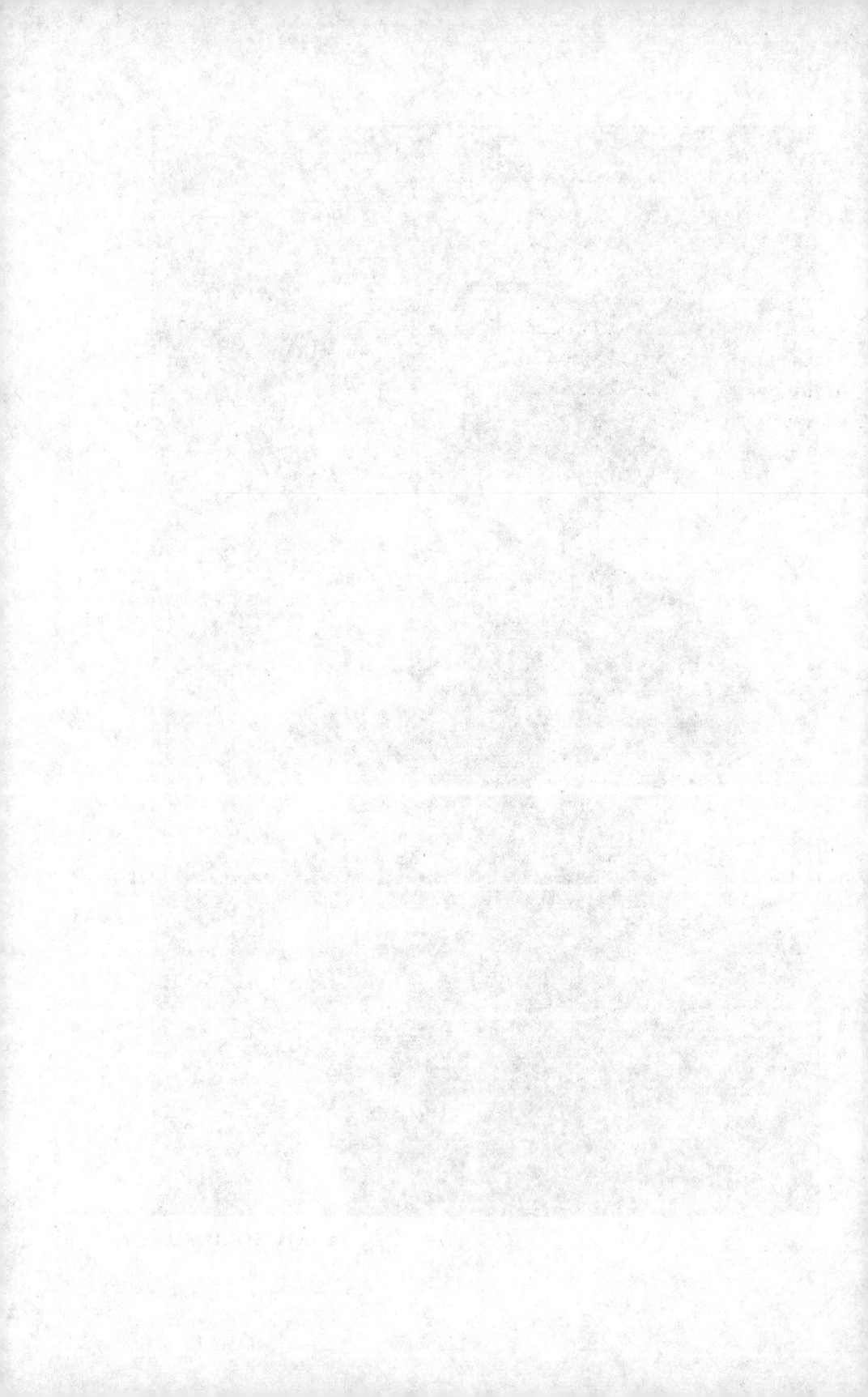

In Living Color

written by
Michael T. Kuester

illustrated by
NATHAN DEIWERT

ABOUT THE AUTHOR

Michael T. Kuester is an engineer by day, a science fiction writer by night. An avid hiker and cooking enthusiast, Michael is a passionately curious individual and lives his life by the motto "In the twenty-first century, there's no excuse for an unanswered question." Over the years, he's devoted free time to researching everything from the history of naval warfare to the origins of potato chips.

Michael resides in Cincinnati, Ohio, with his partner Jen, their two children, and their freeloader-housemate, Eddie the Cat. His work will be appearing soon in Analog Science Fiction & Fact, *but* L. Ron Hubbard Presents Writers of the Future Volume 42 *marks his professional debut.*

His first-place story, "In Living Color," was his fourth entry into the Contest, having received an Honorable Mention and a Silver Honorable Mention in the first and second quarters of 2025, respectively. "In Living Color" began as a single scene in which a man with paranormal abilities jumped into the painting Water Lilies *by Claude Monet. From there, his protagonist, August, took on a life of his own, and the story evolved into a surprisingly personal allegory for social isolation, generational misunderstanding, and neurodivergence.*

ABOUT THE ILLUSTRATOR

Led by a desire to craft worlds, illustrator Nathan Deiwert creates imaginative creatures and characters that coexist with one another. Deiwert is driven to come up with new ways to develop familiar concepts in the form of paintings, digital illustrations, and visual development. From fear-inducing entities to happy moments, he creates thoughtful work and explores new ideas.

For Deiwert, active research is a considerable part of his creative process, be it artistic influences past and present, anatomy, or cultures. His favorite influences are presently Frank Frazetta, Norman Rockwell, N. C. Wyeth, and Maxfield Parrish. This spirit of research and inspiration drives Nathan not just in his work, but also in the lessons he imparts with his students.

In Living Color

First came the colors. Just blotches, swirls. It took a minute, but they'd resolve into shapes: brushstrokes, swirls, and symbols. He blinked, and the vast blue became the sky. Not the sky; a reflection of it, rippling on a pond. There were big, broad leaves on the water, like bright green pancakes, with sprays of white petals atop them, like pom-poms. Or fireworks. Water lilies. Some were pink, too, which was nice.

He kept blinking, expecting the world around him to pull into focus. His subconscious kept wanting that, willing it to be so. It used to give him headaches, but that was a long time ago. Since then, he'd learned to take comfort in the abstract. To enjoy the blur. The water rippled, lilies framing a reflection of a clear blue sky, and trees. A tree line, beyond the shore. Pine? No. Deciduous trees, reaching sensuously over the water with their thick branches.

He raised his head slowly. A bridge arched over the pond, wood pillars and railings, all whitewashed. Weeping willows ringed the pond, their slender leaves cascading like the hair of a fairytale princess. The sun was gentle, warm on his skin, emanating from a blurry white circle of brushstrokes, in a sky streaked with blues.

The ground beneath him was splotches of green mixed with tan, yellow, white, brown. As he walked, each step disturbed the colors, but it all stayed solid as long as he didn't look too closely. He never looked too closely. That was the point. Instead, he closed his eyes, breathing in the scent of fresh water and rich earth, vegetation, in the warmth of a clear spring day.

His pants vibrated. He ignored it. This was his time; he needed this. It kept vibrating. They wouldn't stop. They never stopped. He shook his head, sighing softly. They never stopped. Reluctantly, he closed his eyes again and pried his mind from his abstract paradise.

He was back in his apartment, alone. It was dark, gray clouds crowding the sky, threatening to burst. A heavy wind rattled his windows. The dim light of a blustery autumn day cast shadows across his living room. There were fresh smudges on his print of *Water Lilies,* his fingers tingling as they withdrew from its surface. They always tingled after, and he still wasn't sure why. A sonata by Debussy played over his wireless speakers, like a counterpoint to the gusts of wind outside. He shook his hand, flexing his fingers, then slid the print into its sleeve, in the box with the rest of them. He had a lot of them. A hundred? Maybe two? Prints of impressionist paintings. But *Water Lilies* was his favorite one. If he could live in there, he would.

By the time he slid the lid onto the box, his phone was vibrating again, and this time he answered promptly.

"Hello?" he asked, not trying to hide his irritation.

"Mr. Stefanik?" A female voice asked on the other end. He didn't recognize it, but with the department, certainly.

He nodded, pressing his lips together for a moment. "Speaking."

"Sir, Detective Warner wants you to come in and look at something. A car is on its way now."

"All right, I'm—"

He was interrupted by a series of three tones. She'd hung up. They weren't *asking* him to come in. It was never a request. Like just being what he was and doing what he did made him somehow obligated to help.

He lowered his phone and turned to stare out the window. Gray, stormy days were the worst. He hated rain. He hated being wet. He hated the darkness. Of all his prints, he was confident he had no more than two or three of cloudy scenes, if that. And he couldn't remember for certain because that was how seldom he looked at them.

They'd be there in no more than twenty minutes. So, he pocketed his phone, put his glasses back on, and began grooming himself.

They'd expect him to at least look presentable. He would put forth the minimum effort, if nothing further.

It had been known, for decades, that the human race was changing. One, maybe two in every...what, five million? Ten million? One out of millions of children born each year was born *talented*, as the accepted term went. No one was exactly sure what caused this, or why it was becoming more common. No one even seemed able to agree that it *was*, in fact, becoming more common. He wasn't a scientist, so all he knew was that the science was far from settled. And for him, all that mattered was that August Stefanik had been born with certain latent abilities.

His talents weren't particularly intense, which made him one of the lucky ones. No news special on Talents was complete without video footage of some poor, tormented soul locked in a padded room, one that appeared just south of being properly insulated. They'd shriek and tremble, clawing at their skin and pulling their hair, demanding that the statue-silent orderlies nearby stop screaming.

August's talents were at least a bit less extreme, and more typical of what Talents were usually capable of. He could hear the thoughts of others, now and then. Surface thoughts, only, like a random word shouted across a crowded room. He could sense the emotions of others, which certainly came in handy at times. Now and then, he could sense something was about to happen: catching a vase before it fell, completing someone's sentence before they did, pointing a person to the restroom before they asked. That sort of thing.

But every Talent seemed to have one area in which they excelled. His was what was termed *enhanced tactile sensing potential*. Somehow, touch appeared to focus his abilities; all of his senses trained on one point like a spotlight. In an instant, they were all working in lockstep, and he could peer into something with absolute clarity.

Touching other humans could be disorienting. A simple handshake, and he could remember the drive they'd taken to meet him. He could taste their breakfast, feel their morning shower on his skin, smell their shampoo. It was partly why he lived alone, never invited friends over. Why he worked remotely. And it was why he always wore gloves, even in summer. He tried to think of it as a "superpower," like something he'd seen in comic books as a child. It kept him from feeling like a leper.

They arrived three minutes early, just as he was tying his scarf. A police interceptor pulled up outside his house, and after a parting sip of tea, he rushed out the door. An officer hopped out and opened the rear door for him, like a chauffeur. It made him feel special, until he got in. There, the shatterproof glass and uncomfortable seats made him feel like a criminal. He'd just buckled in when the rain started. The driver turned on the headlamps, and the tires squealed as they took off.

The way to the precinct was as familiar to him as the drive home from the art museum. As the car idled at a stop sign, he peered out through the raindrops gathering on the windows. Cars sped past, their tires hissing across the wet pavement. Now and then, he'd get a random thought. A man disciplining his child in the back seat. A woman rehearsing a presentation she'd be giving at work. Someone remembering they'd forgotten something at home. With that last one, he actually knew where it came from: a car abruptly skidded to a halt in the intersection, before pulling a U-turn and squealing back the way it came.

How did those without talents experience the world? He figured it was probably very quiet, though it didn't seem like that to them. While he hated always wearing gloves, August realized he was lucky. He'd heard horror stories of those with more pronounced talents losing their minds, the constant stream of outside thoughts destroying their sense of individuality. Seen in that light, the gloves weren't so bad.

. . . say something.

It came to him like it had been spoken aloud, though there was a distortion, like an echo. A partial thought, from one of the officers escorting him. Sure enough, a moment later the one in the passenger seat craned his neck to look back at him.

"How are you today, Mr. Stefanik?"

"Fine," August replied, the single word overlapping with the officer's final syllable.

"Great weather we're having, huh?" the officer asked, with a sincere smile.

August accorded him a polite smile and a nod. Acknowledge his attempt at being polite while making it clear he didn't want to talk. The officer kept looking at him for a beat, maintaining his smile, which faded as he finally turned forward. The light was about to turn, and August instinctively grabbed the handle on his door before the car peeled out.

Upon reaching the station, the same polite officer had jumped out and opened the door for him, the other holding an umbrella over his head like an honor guard escorting the president. The special treatment was kind, though it always made him feel uncomfortable. Another reminder that he was *different*. Same with the stares as he passed through the narrow halls and cubicles inside the precinct.

Once, he'd been able to walk through the halls unnoticed. Uniformed officers hurried past, carrying stacks of folders or conversing on their phones. The plainclothes detectives often congregated around the coffee urns, the acrid, burnt aroma mingling with idle chatter about the Bengals game last night, the new sushi place on Race Street, who was sleeping with whom. Now, his entrance tended to draw eyes. Some were curious, others suspicious, some an odd mix of both. Morbid fascination, he figured. Random thoughts would pop up, mostly just disembodied words:

Is that him?

The spoonbender.

Just creepy.

Now and then, he'd actually hear one of them wondering if he could read their mind, and smile at the irony, though he tried to hide it. Hearing the thoughts of others made him uncomfortable. He'd learned that for others having their thoughts heard made them more than uncomfortable.

His escorts ushered him swiftly through the forest of desks and cubicles to an office at the end of a hallway. It was a tiny, windowless space, smelling of burnt coffee and lit by a bank of fluorescent bulbs that made an audible buzz. There, as always, he found Detective Warner hunched over her desk, immersed in her latest case.

Shayna Warner wasn't what he'd pictured when imagining a homicide detective. Under normal circumstances, she was a personable African American woman, with a folksy tone to go with her expressive eyes and broad smile. But when she was working a case, she looked and acted like she'd been plucked straight from a cop drama. Her thick hair was frizzed as though she'd touched a live wire, eyes intense and red around the edges. Her suit jacket was draped across the back of her chair, which creaked as one of his escorts announced his arrival.

"Gus," she said, rising from her seat.

The battered chair groaned as she rose, as though it was as tired as she looked. He stepped into her office as she leaned across her desk, balanced on one hand while reaching out for him to shake the other. She looked unimposing, but her handshake always left his fingers numb, even through his gloves.

She withdrew her hand and placed it on her forehead, exhaling sharply. "Boy, am I glad to see you."

"The same guy?" he asked.

"That's what I'm hoping you can tell me," she replied. "If it is, that'd make three this year."

"Four," he corrected her.

She stared at him for a moment, tilting her head as sleep deprivation weighed it down. "Damn, you're right," she replied, her voice seeming to echo from the back of her throat as she gazed off, absently. Abruptly, she snapped back to him. "You sure that second one was the same guy?"

August simply nodded.

Shayna sighed heavily, running her fingers back through her tangled mass of hair, until they got caught in the knots. She exhaled sharply, withdrawing her hand, then grabbed a folder off her desk and stepped around toward him.

Her expression was grave, regretful. She hated asking him to do this. He knew that; he hadn't even needed to hear it. And it did make him feel a little better. Just not much.

"You know the drill, buddy," she said, her tone itself an apology.

He nodded.

"Anything you can tell us would—"

"Be very helpful," he finished for her. He hadn't heard the thought. He'd just heard those words enough times already.

She held the folder out to him. He grasped it gingerly, as though it might burn his hand. What it would actually do might be even worse. He flipped it open carefully, a soft crinkle revealing a ream of crime scene photographs, bound with a binder clip. Scenes of blood and death, the aftermath of violence. He slipped off his gloves, and as he pulled the clip off, his thumb just barely grazed the surface.

Emotions left a kind of imprint on things, especially strong ones. He wasn't ready. Really, he never was. First, his vision went dark. Then, there were flashes, as he relived the incident. He wasn't seeing it, more like feeling it. The thoughts weren't there. They faded with time, like chalk in the rain. But the emotions bled. They stained. And they produced a sort of series of vignettes; abstract paintings of fear and rage.

She was young. They all were. She lived alone. The only thing they all had in common. Someone was there. She called out. They didn't answer. A knife, large, the light of the moon glinting off the blade, blinding like a flash bulb. Then there was blood. No colors, ever, so it was black, but he knew blood when he saw it. There was panic, fear; that was her. From him, there was nothing. Just this cool satisfaction. August couldn't see colors in the monochrome photos, but he could feel them, as emotions. Panic was purple. Fear, deep red. What he felt from the killer was black. The deepest, darkest black, like you couldn't see the edges. And cold, so cold, like ice.

She panicked. He didn't. She threatened to call the police. He didn't buy it. She reached for the phone, and he cut her hand. She shrieked and cried. Then she ran. Then she pleaded. Then she screamed. The screams turned to sobs and groans, then choking, gurgling, then silence. Just the sounds of hacking and slicing, the smell of copper and salt, like the back of a butcher shop. So strong he could taste it in the air.

It was always about this time that August wondered what it would look like to throw up in monochrome.

The vignettes, dizzying and upsetting as they were, were the cost of admission. Once the flashes ended, there was a fade to black, then a blink, and he was there. It was dark, or at least the monochrome made it appear so. The room was filled with shadows, frozen in time. He could still smell the blood, and it was cold. Very cold, and silent. The silence was absolute.

He turned slowly, taking in the scene. A lamp had been knocked off a table, a chair overturned: the telltale signs of a struggle. She'd fled for her life, but there was nowhere to run. *He'd* known that. He'd been waiting for her, like a spider in its web. A closet? He spun around slowly, until he spotted it: the coat closet door was ajar. Right by the entrance. When he emerged, he'd be between her and the door. No escape. He was a predator, and a practiced one.

The odor of blood led him to the body. Her blonde hair lay sprayed around her head, legs crumpled beneath her. Blood ran from her mouth and nose, forming a black pool beneath her face that merged with the puddle under her torso. She hadn't just been stabbed or cut; she'd been butchered. The cuts were methodical, meant to keep her alive as long as possible.

Even after the other scenes he'd witnessed, all in photographs like this one, devoid of color, he raised a hand to his mouth. The depths of human ugliness weren't something he could get used to. He had no idea how Shayna did it. But he wasn't sure he could ever do what she did. Not every day. Not if it meant seeing this. The absence of color made it feel like an old noir film, assuming they'd ever have made something so raw and visceral.

What are you seeing?

Shayna's thought seemed to echo from everywhere, coming to him from the world outside the picture. He fought back a wave of nausea and looked down at the victim. He didn't know her name, and felt like he should. She was young, so many years ahead of her. And in one selfish instant, a complete stranger had taken every one of them, cut by cut. But there was no rage attached to it. This was cool, methodical. A depth of sadism that didn't even register as human.

Closing his eyes, he willed his body outside the photograph to speak. Whenever he jumped in, as he called it, it was as though his mind went into the picture while his body remained outside. There was a distance, as though he had to shout out from the image and wait until the words reached his mouth.

"It's him," he said.

You're sure?

He looked back down at the body, wincing. "I'm certain."

What can you tell me about her?

She meant, what could he tell her she didn't already know. No doubt they already knew her name, her address, her occupation. Probably had a list of family and close friends, maybe a partner, or several. She was young, after all. But August's talents could provide further insight that interviews with survivors couldn't. Personal effects left a sort of emotional print, like an afterimage of moments in time. Feelings of joy, love, accomplishment. It was as close as these sessions came to providing some measure of comfort. Or it would have been, had he not known the person who'd left those afterimages was gone. That they'd never experience any of those things again.

He headed for the bedroom. The shadows remained motionless, the silence undisturbed. His footsteps echoed, regardless of what he was walking on. Carpet, tile, broken glass. It always sounded like he was walking along a concrete floor in a vast, empty warehouse. *Because I'm not really there*, he figured.

The bedroom was what he'd expect from a young woman in her twenties, living alone. It was messy: clothes on the floor, bed unmade, wastebaskets overflowing. *Busy social life*. It could also be a sign of depression, which wouldn't be uncommon for a young woman. But her hair had been flat ironed, the clothes on her body recently laundered. She'd been out with someone. Probably not depressed. Just unconcerned with the state of her home. So, either no partner, or a long-term partner. The kind where she spent most of her time at their place or no longer cared what they thought of hers.

A stroke of luck: pictures on the wall. That made his task easier.

People who didn't work in homicide only took pictures of happy times. Vacations, dates, parties. Powerful emotions were tied to memories, which made them easier to access. The photos weren't framed, just tacked to the wall with colorful pushpins. But there were many. He looked them over the way he'd appraise a work of art in a museum. There was no point in touching them; he had already jumped into a picture. He'd learned that when he tried to access an image inside an image, he got nothing. It was just a copy of a copy. But if he concentrated on the images, they spoke to him, in a way.

Most of them were of the girl with her friends. Morgan. That was her name. The older ones were clearly taken at parties. Birthdays, or just whatevers. People in their twenties didn't need much of a reason to gather for a good time. He felt the laughter, the fun, but the thoughts, if any, were slushy. She'd been inebriated. They were unfocused, like television static.

As he moved through them, he found himself following her life. Her graduation from college, two years earlier. From there, the images were fewer. Her clothes were nicer, more expensive. The same core group of friends, but fewer of them. Then he stopped in his tracks as one of the images practically sang to him.

It was at a restaurant, a nice one on the West Side, with a breathtaking view of the city. She was standing in front of windows that stretched floor to ceiling, the towers of downtown lit up behind her like Christmas trees. And there was a young man beside her. He was handsome, hair neatly coiffed, suit tailored. His arm was around her. They were both smiling.

Bryce. Her boyfriend.

From that point on, most of the pictures were of the two of them. Date nights, movie nights. Cozy nights in. Trips to the zoo. Silly Halloween costumes. They were happy. Everyone posts happy pictures on social media, but these weren't to show off. They were just for her: a sampling of what mattered most in her life. She loved him, deeply. August was fairly certain he loved her just as much.

He placed a hand on his chin, frowning. The ultimate tragedy of death was that it wasn't something that happened to the deceased.

They were dead. It was over. Their death was something experienced by those who loved them. For them, it wasn't a sudden slam of the brakes. It was a long, slow roll downhill. Years of misery and therapy, asking questions that could never be answered.

I never told him.

August jumped. It was a female voice, soft and breathy. Young, light, coming from the living room. He knew nothing could happen to him in a picture; he wasn't actually there, just experiencing a moment in time, captured. But while there was no danger, there was still fear.

He tiptoed, moving slowly back to the living room. The scene was unchanged, but now Morgan was standing, her clothes dripping with blood, entrails hanging from her abdomen. She was staring ahead, not looking at him. Looking at nothing, her eyes dead above the dried blood around the edges of her mouth.

This didn't happen often, and he was grateful for that. All of the raw feeling from her life, the depth of her love for Bryce, had focused his talents to a single point: her dying thoughts. Many people had the same basic thoughts in the final moments of consciousness. Shock, disbelief. Fear of the unknown that awaited them on the other side. But those who'd been close to someone, who'd wound their lives around another's, always thought of those they'd left behind. They expressed regret, and that regret hung in the air like smoke.

He stepped toward her, listening. The thought was fading, distant, like a distorted voice through a bad receiver. He leaned toward her, almost pressing his ear to her lips.

The other night, we were drunk. The condom broke. I never told him. I just found out yesterday.

Suddenly she snapped out of the death-trance and turned to him. Her eyes were wide, alive, and he recoiled in shock and revulsion. He staggered backward, horrified, as she spoke directly to him.

I should have told him. He would have been so happy. We both would have. Now, it's—

It took more effort than he expected to pry his hand away from the photograph, and for a moment the lights of the office were so bright he staggered backward. The room spun. For a moment he thought he might puke. Shayna ran to his side, grabbing him firmly by the arm to steady him before wrapping her arm gently around his shoulders.

"Hey, hey! You okay?"

August tossed the photo away and grasped his chest, panting. This was why every print of a painting he had was a landscape. It took a moment to regain his bearings, for his eyes to adjust to the light. Eventually, he shrugged off her arm, turning to glare at her.

"I really hate doing this," he rasped. He knew it wasn't her fault. He knew she was just trying to help people, to protect them. But she could never begin to understand the cost.

To her credit, Shayna didn't chide him. She was too kind, too empathetic, and knew him too well to do that. And for that, if nothing else, he was grateful. She just nodded slowly.

"I know," she replied, softly. "And I wish there was some other way. I really do."

He looked into her eyes. She felt regret, he could feel it. But to her credit, she didn't try to say she was sorry. He was grateful for that. Instead, she just looked at him. Despite everything, she really wanted to ask what he'd learned. But she didn't, out of respect. And he appreciated that, too.

"She was pregnant," he offered.

"We know," she replied. Of course she did. No doubt they'd performed an autopsy.

"She had a boyfriend," he added.

Shayna nodded. "Bryce Medich. We knew that, too." She sighed heavily, shaking her head. "He's already been in for questioning. Davis thought he might have done it. Pointed out all the damage to her guts."

"No," August replied, quickly. He finally straightened his back, resuming his normal posture as he adjusted his scarf. "No, it wasn't him. They had a condom break last week while they were having sex. She only found out the day before it happened." He paused, looking into her eyes. "She never got to tell him."

Shayna sighed wearily. He felt a wave of remorse wash over her, mixed with the barest hint of satisfaction. "That's what I figured," she replied. "The kid seemed honestly shocked when we told him. Guess he wasn't lying. And when I saw what happened to her, the way she was carved up...well..." She trailed off, staring absently into the hallway. "And you're sure it was the same guy?"

August nodded gravely. "As sure as I can be without actually being there." He closed his eyes, taking a slow, quaking breath as he relived the scene. "The darkness, black and cold. To me, it's as clear as a fingerprint."

"But of course he's way too careful to leave those," Shayna whispered. She scoffed, turning away with her hands on her hips. Her spray of hair bounced as she threw her head back, placing her palms on her forehead.

He understood her frustration. Evidence obtained by talented individuals was still inadmissible in court. He could find where the perpetrator lived, guide them to his house, and point directly at him, and unless they could find some physical evidence to tie him to at least one of the murders he'd still walk free. And kill again, no doubt.

Ultimately, he wasn't able to tell Shayna much beyond what she already knew, or at least suspected. He'd confirmed that the boyfriend wasn't responsible, and that the person who was responsible was certainly the serial killer she'd been chasing for far too long. But that was it. There still wasn't a detectable pattern. The victims didn't walk in the same social circles. Didn't know one another. Didn't go to the same gym, or work in the same office. Any overlap between their social media presences was coincidental at best; a product of simply occupying the same general space in a crowded city. People passed one another. Knew a friend of a friend of a friend. In the end, it meant nothing. Shayna thanked him, as always, and had him driven back to his home.

Like most talented individuals, August had been in his twenties when he'd first discovered his abilities. He'd been in bed with a woman, at a time in one's life when raging hormones mixed with poor decisions made casual sex more appealing. She'd seemed enthusiastic, but when he discovered he could feel her emotions, sense her thoughts, he realized she was bored. He also realized most people were happier not actually knowing whether or not they're good in bed.

Not long after, he'd developed his appreciation for art. It followed an accident with a copy of *National Geographic*, when he'd passed his fingers over an image and suddenly found himself standing in the Amazon. All talents had their benefits and drawbacks. His had nearly taken his social life, but had given him the ability to travel the world from the comfort of his sofa.

Since then, the art museum had become his favorite place. In a way, it didn't make sense: for him, visiting a museum full of paintings he couldn't touch was akin to someone visiting a zoo when they'd really like a pet cheetah. But even if he couldn't touch them and be spirited away, he enjoyed art. He'd wander through the galleries for a while before inevitably spending hours in the impressionist section. Something about the soft strokes, the blurred colors and abstract scenes, felt like an escape from a world that had become far too detailed to be comfortable.

And it was quiet, or at least as quiet as any place outside his home could be. People in an art museum didn't speak much, and their thoughts were muted, soft. Less likely to jump out at him. It felt more like sitting beside a babbling stream, as opposed to standing under a waterfall.

Invariably, he'd find himself in front of his favorite piece: a painting of Mont Sainte-Victoire by Paul Cézanne. He adored Cézanne: something about his simple interpretations of scenes felt like a breath of fresh air. A sigh captured in oil and canvas. He'd give anything to touch it, just once. He'd never been to France. The idea of spending hours on a crowded plane full of anxious people wasn't overly appealing. If only he could reach out and touch it…

He nearly jumped. For a moment, he thought he felt a draft, like someone had opened a door somewhere. But the sense of cold was focused. It was behind him, somewhere in the gallery. Cold and dark: a void in space. It was black, totally black. He couldn't see it, and didn't dare turn around, but he could feel it.

His pulse pounded. The feeling was terribly familiar. In his horrible jumps into crime scene photographs, he'd interpreted it as a cold blackness. But it wasn't a color; it was the absence of color. Of light. Of anything. His mind was staring into a deep well. No, a black hole: a conspicuous gap in the thoughts around him. He was hesitant to reach out, afraid to touch it lest he be sucked in.

In one of the minds milling through the gallery was a complete absence of thought and feeling. Devoid of human concerns, of empathy. Devoid of what someone more spiritual might term a soul. Simply *devoid*, like it was a state of being. He was sweating, though the gallery was uncomfortably cold. He wanted to scream, to run. The mind he'd perceived was as close to pure evil as anything real could come.

He closed his eyes, fighting the urge to run away, fighting the morbid curiosity that wanted to reach out and touch it. It was moving, somewhere behind him. Then, it was gone. The sensation reminded him of closing a window in the dead of winter: the sudden absence of cold suggesting warmth. After counting to sixty, he finally allowed himself to turn around. The void was gone, far enough away that it was imperceptible, at least to him.

The art museum was his safe haven, his sanctuary. To feel something so cold and malevolent there felt like a violation. It led to a simmering sense of anger and resentment, clouded by intense fear. As soon as he was confident the presence had passed, he turned and left as quickly as he could without running. Usually, he'd stop at the gift shop on his way out. Try out a few new prints, visit some new places. But he was unwilling to leave the physical world behind, lest he return to find the void had spotted him and was preparing to swallow him up.

The ride home was pleasantly uneventful. It was raining. The driver was from somewhere in the Middle East, periodic thoughts coming in a language August couldn't understand, which made them easy enough to tune out. Upon reaching his apartment he turned out the lights and poured himself a stiff drink, and then another. He tried to sleep, which didn't work. So, he spent some time at Monet's pond, trying to wash away the fear with the scent of water lilies.

Two days did little to soothe August's mind. He hadn't been back to the art museum. He called off work. He barely left his apartment. His entire world had changed, the way one's world changes after experiencing an ugly revelation. Just knowing the world had a gaping maw of darkness wandering around, sucking up all that was bright and good, made everything feel threatening. When the precinct called, he almost refused to go in. But he still got into the squad car when it pulled up.

Sure enough, he arrived to find a fresh set of crime scene photos on Shayna's desk.

The victim lived alone, like the rest of them. Her apartment was tiny, small enough to skirt the boundaries between home and closet. It was neat and clean, perhaps because she simply didn't have enough of anything to make a mess. There wasn't even a bed, just a mattress sitting on the floor. That was where he found the body.

In the monochrome of the photograph, it almost looked artistic: the black of blood set against the pure white of her skin, the gray of her light brown hair. But this one was different. He felt it on his way in: the victim's fear and anguish were mixed with something unexpected: confusion. Part of him didn't want to know, but he willed his mouth to ask the question.

"Something's different this time," he said, his unspoken voice echoing through the tiny space.

Though he couldn't see her, he could feel Shayna shaking her head. *You always know stuff without me telling you. I don't think I'll ever get used to that. Not really.*

There was a brief pause. In the physical world beyond the photo, Shayna was probably looking at the image beneath his hand. *See the sheet over her? Usually he leaves them clothed when he's done. But, this time, he left her naked.*

August looked down and realized the victim's shoulders were bare. An exposed calf and foot hung off the side of the mattress. She'd painted her nails recently: they appeared black against her pale skin.

Take a look. Honestly, I don't know what to make of this. None of us do.

He didn't want to look. He didn't want to see it, but not seeing it wouldn't change anything. This had already happened. The entire scene was frozen in time, permanent. Not knowing wouldn't help anyone, including him. Hesitantly, he knelt beside the corpse. His hand was shaking as he reached out and carefully drew back the sheet.

In a grim sense, it was like pulling back a curtain to reveal a painting. Most of the girl's blood had pooled beneath her torso, but in place of the usual slashes and disemboweling, he found the killer had carved a pattern: a spray of elliptical cuts, arrayed in a starburst pattern rising from her navel. A rising sun? Fireworks? No. He recognized the pattern, and a knot formed in his gut. He couldn't breathe. Something in the back of his mind started to scream.

A water lily.

He recoiled, dropping the sheet as the victim sat bolt upright. Her dead eyes stared forward, her face expressionless, but she cocked her head as if confused.

August.

He fell backward, scrambling across the floor. He couldn't have heard it. There was no way. The victims were dead. Whatever happened after that, they were gone. He was certain. None of them could ever know he was there, none could feel him, or anything, ever again. How could she know his name?

August, she repeated. *August, August.*

She turned to him as he stared, his pulse pounding in his ears. *He keeps thinking about August. Who is he?*

Something in his mind shattered. The room spun, black and white swirled and warped, contorting and slipping into a deep, dark well. A void in the very fabric of reality, something so deep and dark and dense nothing could escape from it.

He shook his head, trying to push it all away. To keep from being sucked in. He had to get out. He had to pull away from it, but he couldn't. No. He felt himself slipping toward it, the floor beneath him turning into sand as he fell into the hole. *No.* It had him. There was no escape.

"*No!*" he screamed, stumbling backward and nearly falling through the doorway. The transition was always jarring, but this time it was like falling out of reality itself. Real lost all meaning. Was he still in the picture? There were colors. He could hear something. A voice. Female.

"Gus!" Shayna shouted. She was holding him. He'd collapsed to the floor, slumped against a wall. How had he gotten there? "Gus, you're scaring me here. Say something!"

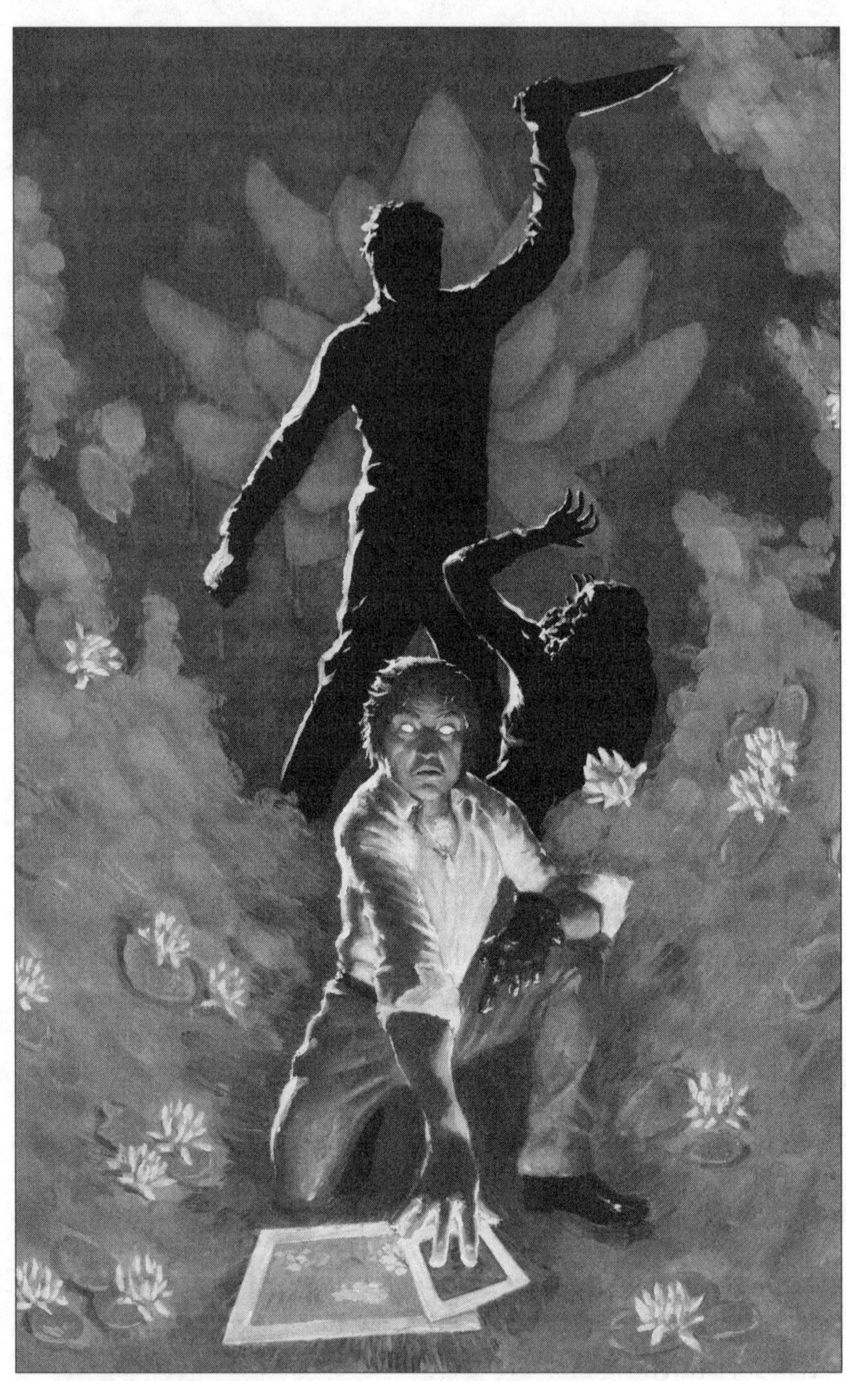

NATHAN DEIWERT

"It was *me*," he said, turning to her. His neck practically snapped, adrenaline surging as he leaped to his feet. He wanted to run, as though that would do any good. As though he could run from a photograph.

Shayna's eyes grew wide, her jaw slack. "What do you mean, it was you?" she croaked.

He shook his head slowly. "This one. It was different because of *me*."

Shayna's expression softened. He could feel a wave of guilt wash over her. "Buddy, I get it," she whispered, soothingly, rubbing his back gently. "I do. We all feel that sometimes. This job," she paused, sighing softly, "it gets to you. You chase a guy like this, somebody who hurts people, and after a while, it's your fault they're still at it. But you're not responsible for this. This girl was—"

"No, no, you don't get it!" he shouted, brushing off her hand angrily. "The killer did this because of *me*! He was thinking about me when he killed her! Her last thoughts…they were *his*. He carved a water lily into her stomach, for God's sake!"

She pulled back, staring at him. "A water lily…" she whispered.

He nodded gravely. "The painting," he replied, "by Claude Monet." His favorite. His place of comfort. His source of peace. The one place he felt truly safe, no matter how much death and pain he was forced to relive. It was a message, grim and ugly: he wasn't safe anywhere. He would never have peace again. Not until this was over, one way or another.

August placed a hand on his chest, struggling to slow his breathing enough to be coherent. "The other day, I was at the art museum, and I felt something. A blackness…a deep, cold pit of darkness."

"Like what you feel in the photographs," she observed, eyes wide.

"It was him," he said. "I know it was. I didn't want to believe it, but now I'm sure. I felt him. And…" He didn't want to say it. As though until he said it, it wasn't true. "I'm pretty sure he felt me, too."

Shayna backed away slowly. In his time working with her, he'd come to see her as the toughest person he'd ever known. Her job forced her to take a front seat to a nonstop review of human ugliness. That sort of thing changed a person. A weaker person would've been broken. But she took it all in stride. In all the time August had known her, he'd never felt fear from her. Until now.

"I don't..." she began, haltingly. "What are you saying?"

He narrowed his eyes, staring intensely. "I'm saying your killer is a Talent."

Crime committed by Talents was rare, at least as far as anyone could know. Usually, what few instances there were took the form of crimes of passion. Young Talents whose abilities had recently manifested themselves could have a hard time dealing with the thoughts of others. Sometimes they lashed out, sometimes people got hurt. It was rare, but it happened.

In reality, part of the reason it was rare was the registration system. All Talents were required by law to register with the government. There had been protests at first, but those protests were drowned out by the tide of public fear, as well as the advocacy of Talents themselves. Most, like August, saw no problem with it. Life as a Talent was hard. It helped to have special accommodations. Working from home, for example.

Of course, no one could really know for certain how much Talent crime occurred because the registration system was mostly voluntary. Individuals who were incarcerated or institutionalized underwent mandatory registration, but for the most part the program relied on Talents being willing to step forward and identify themselves. August had done so just prior to his twenty-sixth birthday. By then, it was impossible to ignore the impact his abilities were having on his life.

In reality, it was public knowledge that a sizable population of illegal Talents existed. Some likely remained unregistered for nefarious purposes. But most, more than likely, just wanted to be left alone. To not be treated as sick or disabled. To not be viewed as *different*.

August hated chamomile. He knew it was supposed to be soothing, but he hated it. Yet after his ordeal someone had thrust a cup of chamomile tea into his hand. Something about shock. It seemed impolite to refuse, so he took a few vile sips as Shayna ran a check through the registration database. It was standard procedure, but also pointless. August found it hard to believe a homicidal psychopath would willingly profess the abilities that gave him an advantage as a murderer. And, frankly, even if they had found someone who seemed to match the psychological profile, August had no idea what the suspect looked like. Putting out an APB on a walking ball of darkness and evil would have been good for little more than a laugh around the precinct.

That night at home, he tried jumping into a few pictures, but it didn't help. Now, when he walked around ponds filled with water lilies, he felt taunted. The soft brushstrokes felt like camouflage for a predator, stalking him. His beloved pictures made him feel trapped. A pair of plainclothes officers sat in an unmarked car outside. They were there for his protection, but just knowing he needed protection made him feel like a prisoner. Or bait.

He was fairly certain the killer was hunting him now. Sooner or later, he'd turn up. Shayna was fairly certain as well; it was why she'd posted the officers outside his house. She'd gently reassured him that he'd be fine. That the two guys she'd placed on his detail were her best. He'd felt her sincerity and was reassured. She also insisted she didn't think the killer would actually go after him, especially not with the police presence outside. She knew August could sometimes hear thoughts, which was probably why she seldom lied to him. But that was a lie. She was hoping the killer would show up. She felt guilty about dangling him as bait, but she was desperate. August understood that, so he tried to play along and not feel betrayed.

When you're hunted, you run. If you can't run, you wait. So, August waited. He turned on the Bengals game, pretended to watch. Now and then, he actually got into it. He'd considered going to a game once or twice. Splurge on a seat in the lower rows, along the sidelines, see if he could pick up a thought from the players. A guy could make a lot of money like that, he mused. Then he shook his head, snickering. Gambling, in all its forms, was illegal for Talents. For fairly obvious reasons.

For someone who'd seen and felt so much death, August knew very little about it.

It was a random thought, popping up so abruptly that his mind tuned out from the game, trying to stare at the sentence. It just hung there in his mind, white on a black background, waiting for him to consider it. He tented his hands in front of him. What happened after a person died? Half a beer fizzed and clinked in a can on the coffee table. He'd heard final thoughts from many people. They were residuum, the leftovers of a consciousness, nothing more. The sound from the television rose to a roar, or what passed for one with the volume turned down. On the screen, Buffalo fumbled the ball. Cincinnati recovered.

He was tempted to turn the game off, but his hand hovered over the remote. The noise helped somehow. Human sounds of humans doing human things. He wanted very much to continue exploring this topic, and very much didn't want to feel alone while he did so. A moment's consideration. Maybe turn the volume down. It was fine as it was. Maybe another beer? When a person died, if their final thoughts were mixed with strong emotion, they remained, hanging in the air like smoke. Those who passed peacefully seemed to just vanish. Where did they go? Another beer would be good.

His mind kept chewing on existential questions as he walked to the fridge. He grabbed a beer, and as the tab punched through the top, issuing a *hiss* of escaping fizz, the lights went out. August held his breath. It was suddenly cold, very cold. And in the darkness of his apartment, he could feel something darker, moving through the shadows. It sucked in all the light, leaving a void. The person it emanated from felt like a mask, a rug covering a bottomless pit.

He simply stared as the figure stepped into the doorway. The modest light of the moon shone through the window, casting a slanted glow across the bottom half of his face, which broke into a sinister grin. It had been years since August had seen a smile without feeling something. Happiness, joy, even cruel satisfaction. But the killer's smile was just a mask. Behind it was just an empty space, blank, and yawning. It made his skin crawl.

Ice water flowed through August's veins. He was paralyzed.

"Ah, ah, ah!" the figure said, pointing at him.

August hadn't even realized his hand was moving toward his back pocket, seemingly of its own accord. Slowly, carefully, he withdrew his phone, holding it up to show he was clearly not using it to call for help, then gently placed it on the counter beside him.

The killer nodded in satisfaction and stepped into the light. He didn't look anything like what August had imagined. His squinting eyes looked almost fatherly, his hair neatly coiffed, flecks of gray making him look almost dignified. He could be someone's father. He could be *anyone*. And it was terrifying.

The killer whistled sharply. "So, here we are," he said. His voice was a gravelly baritone. It sounded friendly. "At last," the killer finished, stepping toward him.

He put his hands in his pockets, probably to demonstrate that he was unarmed. It didn't help August feel any better. He considered the possibility of escape. The killer seemed to know. Of course he did. And he shook his head, clicking his tongue.

"No, no," he said, wagging his finger. "None of that. Look, you're like me, so let's get the basic stuff out of the way, huh?"

He paused, looking around for effect, then sighed heavily, as though he found all of this at least mildly boring. "So," he said. "I cut the power. Sure you knew that already. Not just here, too. The whole block. A couple of 'em, actually, across town."

His smile ticked upward slightly. "And those cop friends of yours outside? Yeah, I wouldn't worry about them, either." He sneered, drawing his thumb across his neck and making a gurgling sound.

August fought the urge to back away. He tried to steady his thoughts, but he couldn't help taking a mental inventory of his kitchen. The knives in his cutlery drawer. The corkscrew next to the fridge, which was closer.

Again, the killer was ahead of him. He softened his expression slightly, patting the air. "Now, now, relax, Chief. I'm not gonna kill you." He paused, cocking his head. "I mean, I'm gonna kill *other* people. And that detective friend of yours..." He went on, placing his hand on his chin, before snapping his fingers and pointing to August. "Shayna! That's it. Detective Warner. Yeah, you probably told her about me already," he went on, his expression suggesting regret. "So, I'll probably have to off her too. Can't have the fuzz running around looking for a Talent, right?"

He tilted his chin down slightly, with a coy grin. "But not you. I don't wanna kill you, Gus. Can I call you Gus?"

August was shaking. He tried to hide it, but he failed. He was trapped, and though he didn't believe a word of what the killer was saying, he had no choice but to play along. For the first time in what felt like an eternity, he had absolutely no idea what someone was thinking or feeling.

With no hope of escape, he felt the tension in his shoulders ease. There was a calm that accompanied certainty. Knowing one's fate was out of one's hands. And despite everything, he couldn't resist his curiosity.

"Why not?" he asked. He was speaking with a remorseless, pitiless murderer. One who'd killed at least seven women, not to mention the two officers assigned to protect him. Why would one more life mean anything? Could there truly be a line for someone like this? A quota?

The killer's smile widened. "Why?" he asked, as though the answer was obvious. "Because you're like me, of course."

"Then why are you here?" August asked. He tried to appear calm. Serial killers enjoyed seeing fear. They got off on it, as Shayna said. Whatever was about to happen, he would at least try not to give this man what he wanted.

The killer stepped to the side, motioning through the doorway to the living room. "To talk," he replied simply. "Just talk."

August stared at him. He had no idea what to expect, or how to respond.

"You want a beer?" he asked, softly.

The killer kept smiling. "Love one."

After furnishing the most wicked person he'd ever met with a can of beer, August moved slowly into the living room. His steps were careful, as though someone was holding a gun to his back. Whether the killer was armed or not didn't seem to matter. A man like this was dangerous regardless.

He sat down on his couch, simply out of habit; his television stood blank and dead on its stand. The killer flopped casually into his easy chair with a soft *ah*, before cracking open his beer with a spray of foam. He sipped slowly, smacking his lips afterward, then looked around, seeming to appraise his surroundings.

"Nice place you got here," he said.

August stared at him, dumbfounded at such a banal opener. As though he was having a buddy over to watch the big game. He had to stop himself from blurting out, *I've been meaning to redecorate*.

The killer snickered, gesturing to him with his beer. "That's funny," he said. "What you just thought there. Funny. Hey, you're a funny guy." He took another sip of his beer, then placed it on the end table next to him, taking care to use the coaster.

"So," he began, hunched with his elbows on his knees, lacing his fingers in front of him. "Where do we start?" he asked, opening his hands. "I suppose we might as well start with the obvious question."

"Why," August replied. The word hung in the air between them.

The killer smiled, nodding deeply. "That's the one," he replied. "Honestly, I'm surprised you haven't figured that part out already. After all, you're like me." He paused, leaning toward August and lowering his voice. "The noise."

August cocked his head, inviting him to continue. He tried not to think about how he would tell Shayna about this. He tried to empty his mind as much as possible, to betray nothing. Even though he was fairly sure it was useless.

The killer shook his head slowly. "What the hell is it with people these days?" he asked, opening his hands again. "They fill their heads with the dumbest, most vapid stuff." He sat back, shaking his head again. "Ugh, it's murder on the brain, right? Half the time I can barely hear myself *think*. And the young ones..." he trailed off, chuckling light-heartedly. "Oh, brother, they're the *worst*."

August tried to remain calm. He sipped his beer. The condensation on the can was slick in his hand. A drop of cool water fell on his leg, nearly making him jump.

"Boys aren't so bad, you know?" the killer went on. "Their thoughts are simple. Half the time, it's just video games and sex." He snickered softly, gazing off. "I suppose I wasn't so different when I was their age." He leaned forward, pointing at August. "Bet you were like that, too, huh?"

The sense of violation was like ants on his skin. He could feel the killer's mind probing his, rifling through files and cabinets. He was being studied like a specimen.

"Yeah, I can see it all," the killer said. He reclined slightly, grabbing his can and taking another sip. "I can read people like books. In large print," he paused, taking a sip, "with lots of pictures."

He set his can down, wiping his mouth with his hand. August placed his beer on the coffee table. He'd had enough. He needed to keep his mind as clear as possible.

"You figured it out early," the killer said, making firm eye contact. "That girl? That was how you figured it out." He paused, whistling sharply. "They get so emotional, don't they? All worked up about the dumbest stuff. You know that one I did in was freaking out about her boyfriend? If I'd have given her just a few more months, a year at most, that little twerp would've been just a memory. And, of course, she'd have whined and moaned over that, like the whole world was ending."

He scoffed, shaking his head. "Kids, right?" he said, as though he expected August to hoist his beer in agreement.

"Anyway," he went on, "I hear it all. All the time. And surface thoughts aren't so bad. I can tune that out. That's kid stuff." He paused, waving his hand dismissively.

"But thought mixed with a strong emotion..." He paused, snapping his fingers for effect. "It's like a dog whistle." He clasped his hands as he continued. "Once I hear that, it's like they're just *in here,*" he paused, tapping a finger on the side of his head. "Up here, you know? And I just...can't...shake it..." he finished, shaking his head sharply.

Thought mixed with emotion. It was how August knew as much as he did about the victims. How he'd experienced their final thoughts. The lingering imprint of a human life. A moment of joy. Of angst. Of pain. Human moments, sticking out from a lifetime like pushpins on a wall, one leading to the next. For the first time, August allowed himself to look inside the killer, into the darkness. Because it made sense.

He was sharing a beer with a man who'd worked hard to erase the moments that made human life mean something. Now that he understood, he could peer past the darkness, into the void within, and his mind flooded with images. The victims. One of them had been at the gym with him, thinking about a friend who'd been through a bad breakup. Another had been working at a coffee shop, considering dying her hair after discovering her boyfriend had cheated on her. Morgan had been at a bar, wondering if her friends would notice she was drinking coke without whiskey in it because she was pregnant. Wondering how Bryce would react to the news. What her parents would think. What her baby would look like.

And he'd stalked them. He'd followed them, murdered them, carved their bodies like they were animals in a slaughterhouse. Because they felt something. Because it was strong, unrestrained. Because they were building their lives. And he ended those lives because their feelings offended him.

If he were a spiritual person, August could have believed he was drinking with the devil.

"Now you get it," the killer said, nodding slowly. "And now you see why I'm not done. I can't stop. I won't. Not until it's finally *quiet*."

August stared at him. His fear was fading, adrenaline channeling into indignation. He no longer tried to hide his thoughts. He no longer cared, and despite that, the killer seemed unfazed. Something had changed. Up to that moment, he'd been able to read August like a book, just as he said. But now, he didn't seem as insightful. Now, he couldn't hear him.

Because August was angry.

He didn't try to fight it this time. All this time, he'd been giving the killer exactly what he wanted. So, he let himself be angry, though he kept playing along.

"So," August began, carefully, "what happens now?"

The killer scrunched his face, opening his hands before sitting back casually. "Now, you answer a question for me," he replied. "Just one, and it's a simple one." He sat forward again, staring at August intensely. He fought back the fear, unwilling to give him an opening. "How did you find me?" the killer asked.

August's anger wavered, giving way to confusion. He struggled to maintain focus, while still appearing outwardly calm. Puzzlement and rage fought for control, as fear screamed at the back of his mind, desperate to be heard.

"I mean, I like the art museum," the killer went on, sitting back again. "It's as close to quiet as anything gets. Everybody's so calm, so... peaceful, right? Plus all the pretty pictures. The pop art is my favorite, you know?" He paused, chuckling softly. "A few years back, they had this painting with the characters from Winnie the Pooh, except Piglet was a cop, and he'd shot somebody or something." He chuckled, shaking his head. "It was wild, man."

August stared at him, dumbly. In the back of his mind, he hoped he'd still be able to go back to the art museum if he somehow survived the night. If he'd ever have his peaceful place back without remembering the freezing darkness looming behind him.

"So, imagine my surprise when I'm standing there, looking at some painting by, who was it? Too-loose? Vegas? One of those guys. And out of the blue, *boom*," he said, clapping his hands for effect, as August tried not to jump. "There's a mind poking at my brain, trying to see what's inside."

A fresh urge to run sprang up in August's mind. He tried to tamp it down. The killer had no idea. He thought August had been watching him, following him. Hunting him. He had no idea their encounter was just random chance. He scrambled to hide the thought, pushing it into the deepest recesses of his consciousness. But it was too late, and he found the killer chuckling across from him.

"Oh..." he said, his chuckle building into a laugh that shook his shoulders. "Oh, wow. That is wild!" he exclaimed, stamping his feet and clapping his hands. "My boy, Gus! Why, you had no idea who I was, did you? Damn... isn't it a small world we live in?"

"I wasn't following you," August replied, struggling to remain calm, "but I recognized you."

"From your little picture?" the killer replied sharply. The shift from amusement to anger was jarring, hitting August like a slap. "That's what you do, right?" he pressed.

He reached out to the end table, toppling his beer as he grabbed a print August had left there earlier. *Impression, Sunrise* by Claude Monet. One of his favorites. The killer thrust the print toward him, its plastic sleeve crinkling in his grasp as he shook it angrily.

"You really don't get it, do you?" the killer railed, trembling with rage. "You're a *Talent*, man! You can read people's minds! See the *future*! And you squander it all to take little strolls in a painting, sniffing the damn flowers!" He punctuated his screed by winding up and winging the print across the room.

It clattered off the far wall, lost in the darkness. August stared helplessly, like a child whose parent had rejected their finger painting. The killer lunged forward, staring into August's eyes. He fought the urge to recoil, to look away. As they locked eyes, the killer lowered his voice, though his tone was sharp as a knife.

"You and me, we're the *future*, Chief. Humanity had its time and look what they did with it. Social media and rap music, twerking and all the rom-com nonsense the lowest, most worthless minds could come up with, polluting their brains. Polluting *ours*!

"Don't you see?" he went on, rising to his feet and throwing out his arms. "None of it matters!"

The calm returned. Calm, and a cold rage. August had heard enough. He'd let a monster justify his actions, and now he would have his say. He lowered his voice, forcing himself to remain level.

"Morgan mattered," he replied softly.

The killer stared at him, dumbfounded. He didn't even know her name. He'd killed her, and he had no idea who she was. He would know. August would make sure of it. He rose to his feet and stepped toward him. There was a calm that came with certainty. He didn't know what would happen next, but he knew what he had to do, and he stepped forward confidently.

"That was her name," August continued, coolly. "One of the women you killed. She was twenty-five. She had a boyfriend named Bryce.

The week before she died, they had sex, and the condom broke. She was pregnant. And she never got to tell him."

He stopped just within arm's length of the killer. Armed or not, there was nothing he could do to August that mattered now. All that mattered was what August could do to him. And he wasn't wearing his gloves. The killer maintained his glare, unmoving. If he knew what August was capable of, he didn't seem to care.

"You want to know how I found you? It's because I felt their thoughts. Their last thoughts. When someone dies, they think about what matters. *Really* matters. Friends, family. Parents. Partners. She died thinking about Bryce. Wondering what would've happened. What could have been."

He narrowed his eyes, fingers tensing. "It wasn't *noise*. It was her *life*. Here," he paused, raising his hand, "let me show you."

August reached out, planting his hands firmly on the sides of the killer's face. The killer squirmed. He tried to recoil, but August kept his grip. His stubble felt like sandpaper on August's palms, then there was a surge, like an electric current. Usually, when August touched someone, it felt like standing under a waterfall. A rush of thoughts and emotions. But the killer was a black hole, sucking in everything around him. So, instead, August felt a rush of release, like letting out a deep breath he hadn't realized he'd been holding in.

All the memories, the joy and sorrow of seven lives, the final, anguished moments of consciousness, poured out of him like a deluge. Images passed before his eyes, falling from him, sucked into the abyss. The killer's eyes bulged. His face quivered as he convulsed. He started sweating, but August maintained his grip. For someone who'd spent so long walling off their mind, holding emotion behind a dam, it must have been excruciating. August wondered if it should bother him that he enjoyed it. It felt satisfying to share just a sample of the wellspring of suffering he'd taken in.

He didn't know how long they stayed like that, his hands pressed to the killer's face, the killer convulsing until he began to drool. It felt like hours. Days. Years.

Then, at last, he pulled his hands back, and it was over. It couldn't have lasted more than a minute. August stumbled backward. His hands were numb, his legs weak. His face was wet, his mouth dry. After the stream of thoughts and memories, the silence was shocking. All he could hear was his panting.

Looking down, he found the killer collapsed on the floor, twitching. His eyes were rolled back, his smile twisted into a grotesque shape like a silent scream. He was still breathing, albeit shallowly. Still alive, which was good. August hadn't wanted to kill him. He'd wanted to give the killer what he'd given August: something terrible he'd have to live with. And he hoped the man would have a long, terrible life, living with it.

The knock at the door was like a gunshot. His mind was still swimming, and he actually ducked, half expecting a hail of gunfire. Instead, he heard a familiar voice, muffled through the door.

"*CPD!*" Shayna barked. "Open up!"

August simply stared down, down the hallway. His brain and body were operating on a lag, and he couldn't even muster a word before he heard a loud snapping and crashing sound. He wondered if he'd get his security deposit back after having his door broken down by the police. Another thought that was so normal as to feel wholly out of place.

Shayna's footsteps came in slow, practiced thumps along the hallway, like a drumbeat. As she turned the corner, she held her gun level and pointed it at his face immediately. He held up his hands as her aim wavered, her furrowed brow rising in surprise.

"Gus?" she asked.

He simply stared at her, wondering if there was any possibility she didn't recognize him.

"Oh, my god…" she rasped, lowering her gun as her entire body seemed to go slack. She exhaled sharply, then noticed the killer lying on the floor and perked up again. Her eyes turned to August, and he heard the question before she could ask it.

"Yes," he replied, nodding.

She cocked her head, gazing down at the killer before looking up at him again. "That's the guy?" Shayna sounded every bit as surprised to finally see his face as August had been.

"That's the guy," he confirmed.

Shayna stared at the killer for a long time, before sighing heavily and holstering her gun. "Well, damn, is he dead? What the hell did you do to him?"

"He's not dead," August replied. "I just gave him a taste of what he's done to others."

She cocked her head, confused, and seemed to try to picture that in her mind. Eventually, she looked back at him, raising her eyebrows. "Did it suck?" *for him*, she added, though not aloud.

August nodded deeply. "Oh, yeah. It sucked."

She coughed out a harsh laugh. "Good."

August still hated chamomile. But the police seemed to view it as some magical elixir to treat shock. So, less than an hour later he was sitting on the back of an ambulance, legs dangling, with a blanket over his shoulders and a steaming cup of chamomile tea in his hand. The tea was terrible, but he wound up sipping it absently, watching the scene around him unfold.

Even before power had been restored, his block had been lit up in red, white, and blue by a pair of ambulances and a small fleet of police interceptors. They'd thrown up barricades, cordoning off the street around his apartment on either side for half a block. Now, uniformed officers stood inside the perimeter, wooden sawhorses striped with orange and white acting as a dam against a growing crowd of gawkers.

Shayna had shoved people out of the way, dragging him roughly by his arm to one of the two waiting ambulances, where paramedics had checked him for injuries before furnishing him with his blanket and tea. Soon after, the killer had been removed on a gurney, catatonic yet held down with thick straps, eyes glazed and unfocused as Shayna shouted his Miranda rights. Her voice was harsh, as she tried to turn every word into a twist of the proverbial knife. He didn't seem to understand her. She didn't seem to care.

After the second ambulance rolled out with the killer, the crowd gradually melted away. The murmur of hushed whispers died down until all that remained was the periodic squawk of sirens from departing police vehicles, and a low wind rustling through the trees. The paramedics in his ambulance began to pack things up, removing their gloves, and one of them asked if August wanted to be taken to the hospital. He said "No" before he'd even thought about it. Once they'd departed, the silence returned. And, for the first time in weeks, August slept.

Much to her credit, Shayna gave August almost two full days to recover before bringing him in for a statement. The delay was possible due to the case no longer being time-sensitive, which would have required a living suspect. By the morning after August's fateful encounter, Richard C. Mueller II, as the killer was known, was dead. A nurse had found him in his hospital bed around six a.m. Somehow, he'd managed to pull free of his straps and procure a scalpel, which he'd used to carve the words *Bryce doesn't know* into his skin over and over until exsanguinated. According to what Shayna had told him, nobody knew how he'd managed to break his restraints, or where or how he'd found a sharp object. According to her thoughts, nobody cared, either.

For August, it felt as though someone had removed a heavy rock from his shoulders, and he found himself sitting up straighter. He sat tall in the chair across from Shayna's desk, as she went over his statement with him, one more time.

"So..." she began, reading, "you say here... 'I overwhelmed him by flooding his mind with the final thoughts of his victims, thereby breaching his mental barrier'..." She trailed off, looking up at him.

"That's right," he replied.

She laid down the folder, pressing her fingers into the bridge of her nose. "That is a sentence I never thought I'd see in an official police statement."

August shrugged. "It's the future," he replied.

She pressed her lips, giving him a half-lidded glare. "It's *The X-Files*," she countered, grabbing the folder and waving it for effect. "Or some wacky superhero crap," she went on, opening the folder again. "I don't even know anymore."

Her eyes flitted back and forth for another minute, before she sighed wearily, closing the folder at last and setting it back down on her desk. "Well, that's everything, I guess."

"I guess," August agreed.

"Just one more thing," she said, tenuously.

She was nervous about asking the question, worried it might feel intrusive. He smiled reassuringly, inviting her to ask anyway.

"You said you gave him the memories from the victims," she said slowly, as though carefully checking each word on its way out. "Does that mean—"

"I still have them," he replied, halfway between *that* and *mean*. He nodded deeply, feeling a wave of sympathy wash over Shayna.

"Aw, Gus, I am so—"

"Don't be," he replied. "I think...after all this, I know why you do this. Why you're willing to deal with all the blood, the violence, the death." He paused, leaning in to make firm eye contact. "Because there are people like *him* out there. Like Mueller," he finished, shuddering. It was still hard to say his name. As though giving him a name legitimized all the horrible things he'd done.

"There are," Shayna agreed, nodding. "Most of them aren't talented, though." She rolled her eyes, adding, "Thank God," under her breath.

"No, but some of them are," August replied. "Which is why it probably helps to have somebody like them on your side."

A smile crept across Shayna's face. He'd known she was planning to offer him a spot on the force. Again.

"Gus, are you saying what I think you're saying?" she asked, raising an eyebrow.

"I'm not ready for that yet," he replied. "But...I do think I want to keep doing this. Helping. When I can."

"Well, I appreciate it," she replied. "We all do, believe me."

August gave her a smile, though it was hard. He knew she appreciated him. As for the rest of the precinct, that was harder to say. Even after this, they'd keep staring at him, at least for a while. Some would probably never stop. But he could live with it. Because he'd met a man who killed people just for being human. It was human to fear the unknown. In time, most of them would come around.

Shayna sat back, sighing heavily. "Chief wanted to see you before you go," she said.

August raised an eyebrow. "About this case?"

Shayna nodded deeply. "To thank you himself, I think. Probably get a picture with you. Shine up his public image and all that."

August nodded very slowly, uncomfortable with the idea of a photo op but seeing little choice. "Well, then, I guess I'll go see him." With that, he rose from his chair and turned to leave. But as he opened the door, Shayna spoke to him, her words coming right on the heels of a thought.

Ask him.

"Hey, Gus," she said.

He'd already stopped partway through the doorway and turned around.

She seemed slightly nervous. Afraid of making him uncomfortable.

"You, uh..." she stammered, "you wanna, maybe, grab a drink sometime?" She stared at him for a moment, before no doubt realizing what it had sounded like. "I mean, I was just thinking," she went on, tripping over her words, "most of the time, we only see each other, well, like this. Blood and gore and all that. You know, it'd be nice to..."

Her voice seemed to fade as August smiled. It was so easy for people to forget they were speaking with a telepath. He knew what she was trying to say. He had few close friends and could always use another. And after the previous week, he'd gained a new appreciation for the moments that made him human.

"I'd like that," he said, smiling.

She stopped, mid-ramble, and stared. "Oh...oh, yeah?" she replied. "Well..." Her confusion warped into a broad smile. "Good. That's good."

The chief of police was an older man. August had seen him on the news before: African American in his late fifties, overweight and balding with an expression suggesting he'd had a fresh lemon for lunch, rind and all. Their meeting was straight out of a movie, or a political ad. The chief was wearing his uniform, as though that was just what he always wore to work. He sat across the desk from August, cameras snapping away as he feigned deep concern for August's life and well-being, hiding a sea of relief at having caught a serial killer who could have cost him his desk and uniform.

They shook hands. The chief called him *son*. Then he told August the city owed him a great debt, and if he ever needed anything, just ask. It was the sort of thing people said in moments like that, with the expectation that the other party would be too magnanimous to accept. But, August did have one request. It would take some doing, to be sure, but if ever there was a time to ask, this was it.

To his amazement, the chief was able to swing it, and so a week later August stood alone in the art museum. It was the dead of night, not another soul to be found. And, prior to locking up for the night, attendants had carefully removed the glass faceplate from a single painting. He smiled wide, tears welling up in his eyes as he removed his gloves.

August had never been to France. More than likely, he never would. But this was as close as he could get, and really, it was as close as he needed to be. He stared through tears at the broad strokes of gray and tan, yellow and green, streaks of oil blending into Mount Sainte-Victoire as interpreted by Paul Cézanne. He reached out his hand, his fingers feeling the texture of oil on canvas, brushstrokes laid down by a man long gone. But Cézanne lived on, in his work. The sensation lasted only a moment. Then August could feel the sun on his face, the warm breeze rustling through his jacket. The mountain rose high in the distance. It was a lovely view, and he had all night to enjoy it.

As Long as You Both Shall Live

written by
Mike Strickland

illustrated by
KARAH RICHARDSON

ABOUT THE AUTHOR

Mike Strickland has made a career out of writing everything from marketing copy and finance articles to technical documentation and mobile app messages—and even twenty thousand science fiction–themed trivia questions. Other jobs he's been paid to do include scuba diver, navigator, call center representative, user experience designer, and now science fiction author. His love of words began with fantasy and science fiction, where it has now brought him full circle. After a long hiatus from fiction, Mike started writing and publishing again in 2024. A year later, he earned a master's degree in creative writing and won the Writers of the Future Contest.

Mike currently lives and writes in a suburb of Denver, Colorado, but other places he's called home include San Diego, Los Angeles, New York City, Washington, DC, Japan, Honduras, an aircraft carrier in the Persian Gulf, and even a 1973 Volkswagen Squareback. He's not sure if this itinerant lifestyle emerged from his love of travel or vice versa, but his wanderings have undeniably inspired his writing.

The idea for the story "As Long as You Both Shall Live" was born from a dream two days before the Writers of the Future Contest deadline. Mike wrote most of the story on the day of the deadline itself—his most prolific day of writing yet. The inspiration that powered such output focused on this question: If technology allowed a person's consciousness to be transferred to a virtual environment, disconnected from their physical body, what would happen if that person's body died while their consciousness was in that state?

Mike explores the ethical questions inherent in this premise in the context of a hopeful love story—as all the best stories are told.

This story is dedicated to the memory of Scott Petri, a longtime friend and fellow author who left this life too soon.

ABOUT THE ILLUSTRATOR

Karah was born in 2007 and spent much of her childhood moving to and exploring new states across America. She met a large variety of people, though having prosopagnosia (face blindness) made recognizing them a challenge. Karah loves challenges.

Growing up, she had a fascination with studying faces, assuming that she could "learn" how to memorize family and friends by drawing them over and over. This led to a lifelong love of art. She'd blend together a wide variety of separate features and piece them together like a puzzle, relying on others to tell her if it looked "correct."

Because of such an early start on such a potent hobby in her life, she quickly discovered many teachers, family, and friends wanted to encourage her to cultivate it into a real-life skill. They helped her branch out beyond her comfort zone of faces into just about anything, allowing her to combine this new joy in art with her deep appreciation for stories.

As Long as You Both Shall Live

The first thing Sam Petri had to learn was how not to broadcast his inner thoughts and feelings. He was well practiced at this in the real world, thanks to parents who never once said *I love you* to him or to each other. But here, in this featureless liminal space—a flat, gray expanse extending to a blank horizon, four freestanding doors in front of him—his emotions twirled outward in waves of color. As anxiety gripped him, the colors shifted toward the cooler end of the spectrum. Fortunately, he was alone—no one else had logged in yet.

Especially Kumiko. Thoughts of her turned the light to warm beams of reds and oranges. But he had to play it smooth. It was too soon to share his feelings, visibly or otherwise. He closed his eyes and tried a breathing exercise to clear his head before he remembered he was in a virtual space, disassociated from his body—including his lungs. He did his best to simulate the sensation of measured breaths and felt the space around him grow still. The waves of light dissipated.

Just in time: at that moment, Kumiko materialized in front of him, her coal-black eyes immediately finding his. An orange aura briefly flared around her slender figure at the sight of him. He felt a thrill that would have tickled his scalp if he were in his physical body. She put a hand over her mouth, hiding a smile. "Hello, Sam."

"Hi, Kumiko. Long time, no see." He smiled back at her. She was manifesting a deep purple blouse that complemented the shiny black of her hair. He hadn't thought to change his appearance, but he focused for a moment, and his light-blue scrubs shifted to a lime green T-shirt and blue jeans.

"We've done air and water already," she said. "Should we try earth today? Or fire?" The outline of each door pulsed a barely noticeable glow: white, blue, brown, and red. The brown and red doors increased in brightness as she approached them.

"Let's do air again. Flying through the clouds was a rush." He didn't mention it was because they'd held hands to keep from getting separated.

"Okay. Come on." The entire door flared white when she opened it, and his heart soared when she took him by the hand again and pulled him through.

Sam knew he was back in his body when the sharp sting of alcohol hit his nostrils. A hand lifted an eyelid and shined a bright light. He reached up and swatted it away. "Stop. I'm okay." He pushed himself up on his elbows and glared at Jamie, the gangly medical technician standing next to his exam table.

"No problem, bro. Just following protocol." The man slicked back his greasy blond locks and started typing at a terminal. Around the room, a half dozen or so other trial participants like Sam lay on exam tables under bright lights, each with their own technician monitoring them as they regained consciousness.

Sam turned his body to stand and winced when a dull throb in his left knee bit down with sharp fangs. The scrubs had been pulled up on his left leg, and a brown elastic bandage cocooned his knee. He eased himself off the table, putting weight on the joint to test it. The pain varied in intensity depending on how he moved, but he could stand on it. "Jamie, what the hell?" He pointed at his knee.

The technician grinned, whether from embarrassment or mockery, Sam couldn't tell. "Yeah, sorry about that, Sam. You took a bad fall. Well, your body did, anyway. We took X-rays. You should be fine in the morning."

"Not cool. This is supposed to be a clinical trial. What do you have my body doing, rock climbing? Heavy construction?"

"You know the drill, man. Need-to-know basis." But the mischief in Jamie's eyes made it clear how he felt about the project's confidentiality.

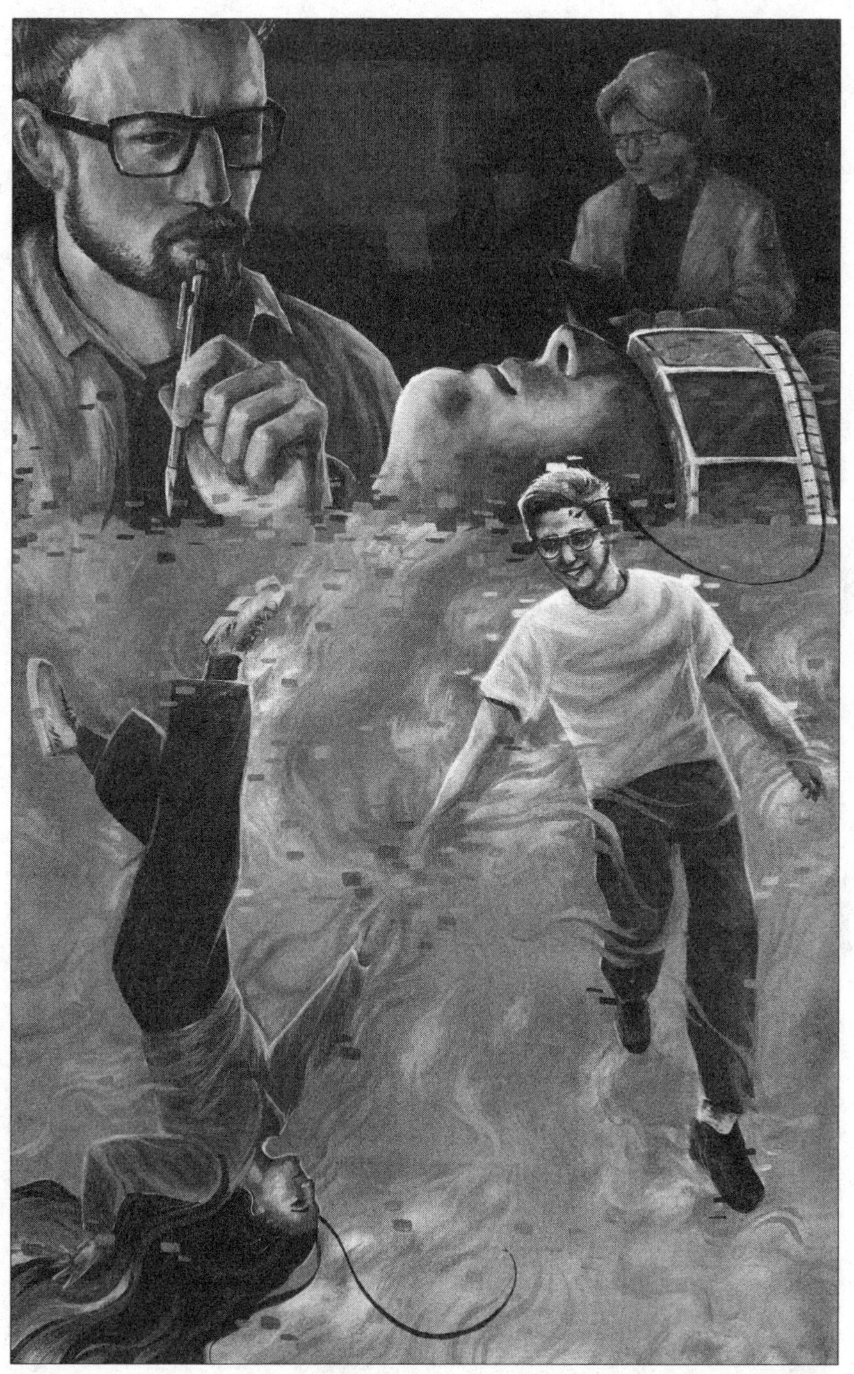

KARAH RICHARDSON

"You're getting paid to play around in cyberspace. Enjoy the fun and games while they last." He caught Sam looking over at the next medical bay, where Kumiko was sitting up. "Oh, yeah, rough duty for you."

Sam hobbled toward Kumiko's table, but when she saw him, she turned away. He froze, too nervous to approach her. Just like every other time. What could he say, here in the mundane world, that could compare to their experiences in the virtual world? It was the same shyness that had handcuffed him his whole life. Keep it all in, never show how you feel, don't give someone the power to hurt you. *Tell that to my knee.*

"See you tomorrow," he said to Jamie and nodded at Kumiko as he headed for the door. She looked glum, probably disappointed by his real-life persona.

The next morning, he was the first one to show up in the lab. He'd come in early to work out his knee in the gym. Jamie was right: besides a little stiffness, the joint was nearly as good as new.

He wished the same were true of his heart. The sting of his crush on Kumiko and his inability to do anything about it still burned. Maybe this was a secret part of the trial. On paper, they were studying the use of human bodies as manual-labor automatons controlled by AI while their consciousnesses were uploaded to a virtual space. But perhaps the research also looked at variations in interpersonal interactions between real life and cyberspace. If that were the case, he made an ideal subject.

He hopped onto the exam table and began massaging his knee. The door swung open, and the room seemed to brighten when Kumiko walked in. It was a rare, private moment between the two of them in the real world.

"Hi, Sam." She smiled at him as she walked to her own table. "How do you want to spend our day today? Want to play with fire?"

She's just making small talk. She was beautiful, he was a nerd. There was no way she was being anything but polite. Right?

He stumbled over half a dozen lame replies in his head, but then her med-tech assistant walked in, and they all stayed in his head. Probably for the best.

Kumiko sat on her exam table. "Duty calls. See you in a minute."

Her technician opened a cabinet and pulled out a heavy-duty vest, padded with what looked like tactical armor.

"What's that for, Eetu?" asked Kumiko. An expression that could be concern or curiosity creased her forehead.

"Just put it on." Her technician Eetu was as forthcoming as Jamie.

Kumiko raised her eyebrows at Sam but slipped on the vest. The other trial participants filed in, and their technicians started prepping them. None of them donned tactical vests like Kumiko, Sam noticed. Finally, Jamie hurried in.

"Sorry I'm late. Let's get right to it."

Sam nodded toward Kumiko. "Do I get to wear one of those?"

"Nah, man. We've got other plans for you today."

"Should I ask?" He looked back at Kumiko, trying to reassure her with an uncertain smile.

"What do you think?" Jamie was dismissive, but Sam caught him looking sidelong at Kumiko also.

Sam reluctantly leaned back, and Jamie slipped the IV cuff over his arm. Before Sam could come up with another smart retort, his view of the lab faded, and he was back in the virtual space, looking out at the gray horizon. A moment later, Kumiko appeared next to him.

"There you are." She reached over and took his hand. "I'm glad to be back here with you."

He wasn't quite sure how he could feel butterflies without an actual stomach, but they were fluttering like mad.

Later, as they huddled under a blanket in front of a roaring campfire, a low tone announced the end of their shift. Sam turned to Kumiko, their noses nearly touching. His pulse—or the digital equivalent—was racing. "This is over too soon."

"I know. Let's come back here again tomorrow." Her eyes found his lips, leaning in as if to kiss them. *Was this finally happening?* Just as they touched, their surroundings vanished, replaced by the featureless virtual lobby space.

They stood three feet from one another, but he felt closer to Kumiko than ever. They were disconnected from their bodies, but she was the same person out there, so she had to feel the same, right? He was going to ask her out as soon as they returned to the lab, shyness be damned. "Tomorrow, then, back in front of the fire. But I'll see you again in a moment," he reminded her. The butterflies took flight again.

The afterimage of her smile lingered as he transitioned back into his body. But, before he opened his eyes, he knew something was wrong. Voices hissed in low but urgent tones, only some words reaching his ears. "—liability risk could kill the project—" … "—can't keep this from him—" … "—you have no idea—"

He opened his eyes, not knowing what to expect. Jamie stood at the foot of the exam table, staring back at him. Eetu, Kumiko's technician, hid behind Jamie. Sam turned to look at Kumiko, but her table was unoccupied. The entire lab was empty, in fact, save for the three of them. The excitement that filled him a moment ago began to curdle. He sat up, worry flooding through him, but Jamie pushed him back with a hand on his shoulder.

"Sam, there's something I have to tell you."

"No!" Eetu raised his hands. "He doesn't have the clearance. Need-to-know only."

"Go to hell. He *has* the need to know." Jamie turned back to Sam. The annoyance he'd shown to Eetu turned to concern, sadness even, and in that moment Sam knew that something terrible had happened. "Sam, there's been an accident. Kumiko is dead."

Sam felt the earth drop out from underneath him. The hard padding of the exam table still dug into his back, but he was falling down a bottomless pit.

"Sam, stay with me." Jamie shook his shoulder. "Her body is dead. They were having her perform some kind of dangerous operation—even I don't know what—but her consciousness is still in there. In the virtual framework. When her body died, they kept her there. Hell, I don't know if they would have been able to restore a connection to a dead body anyway."

Hope flared in the ashes of his heart. *Her consciousness was still in there.* Could it mean— "So, she's *not* dead."

"Sam, hold up. We're in uncharted territory here. Existential stuff."

Jamie pulled over a stool and sat in front of Sam. "Can she still be considered alive, a real person, if her body is dead? I don't know the answer to that, and it's way above my pay grade."

Sam didn't know how to characterize her existence in philosophical terms either, but if she was still in there, she was alive to him. That's all that mattered. "Put me back in."

Jamie turned to the other man. "Eetu, take a walk, will ya?" The other man remained silent, unmoving. "Please."

"Fine. It's your problem, not mine. Good luck." Eetu walked out.

"Lie back down." Jamie readjusted the IV cuff on Sam's arm and moved to the terminal. "I can't leave you in there for long. The brass will be here soon, so when I bring you back, I'm going to have to play it off like it's the first time. Act like you don't know what happened. Both of you. Make sure she understands that."

Sam nodded and closed his eyes. When he opened them, he was back in front of the virtual doors—and Kumiko.

"You're back. What happened? You disappeared, but I stayed here." Anxiety glittered in her eyes. She wrapped her arms around herself as if she were trying to hold onto something real.

He took her in his arms. She was here. If he could see her, touch her, and talk to her, then she wasn't dead. Couldn't be. "Kumiko, I don't know how to tell you this." He wondered if their avatars could produce tears. He felt like he was about to find out. "You're dead. Your physical body is dead."

"What?" She pulled back, as if not comprehending his words.

He held her hands, refusing to let her go. "I don't know what happened, and I don't understand how. But as far as I know, your consciousness is trapped here."

Her eyes widened in horror, and deep magenta whorls of light sprung from her head.

"I'm going to do everything I can, Kumiko. I can't lose you when I just found you." He embraced her again. "They're going to pull me back at any moment. Don't tell them I told—"

The virtual environment swirled again. Kumiko disappeared, and the lab came into focus as he opened his eyes. He ripped away the IV cuff and jumped off the table. Jamie was still at the terminal, and four other people stood behind him. One was Eetu, hanging out by the door like he wanted to flee through it. Two women wore tactical gear, sidearms, and no-nonsense expressions. The fourth, a woman with a stern frown and even more severe business suit, walked up to him. Sam caught Jamie's nearly imperceptible nod, a reminder to keep his mouth shut.

"Mr. Petri, I'm Tanisha McKinney, Human Resources Chief," said the woman. "I need you to come with me."

Taking another look at the goon squad behind her, Sam wasn't about to object. And hopefully she had some answers for him.

On the way home later, Sam leaned against the commuter train window, watching the lights of the city flow by but seeing nothing. The jiggling of the train car would have lulled him to sleep any other day, but his mind was racing too fast for drowsiness. The light rail was full of evening rush hour riders, but he couldn't have felt more alone. He didn't know what he should do about what had happened, and yet there was only one thing he *could* do—keep working like business as usual—if he ever wanted to see Kumiko again.

The answers he got from Ms. McKinney were standard HR fare: terse and carefully chosen. But he was savvy enough to read between the lines of corporate speak. Whatever the specific cause had been, Kumiko had indeed suffered body death, and the Company was doing its best to manage the fallout—and keep it under wraps while they did so. The resolution of the incident, and their ability to continue perfecting and exploiting the technology, hinged on whether Kumiko remained a legal person. It didn't take much imagination to understand how messy a legal and moral conundrum it was. If she was still alive in the eyes of the law, then her personhood existed on a server, and the Company had an obligation to preserve her existence indefinitely to avoid manslaughter charges. If not, then they were free to deal with Kumiko's digital self to the full extent of the corporate contract Kumiko had signed—but they would also have an employee death, and the consequences thereof, to contend with. The Company lawyers would be pulling an all-nighter to untangle the mess, and he and Jamie would be high up on their agenda.

They'd quickly figured out that Jamie had told him what happened, and whether or not he'd see Jamie at work in the coming days likely depended on whether both of them kept their mouths shut. It was easy calculus for Sam. If he blew the whistle, then his job—and any chance to see, and help, Kumiko—would be lost. He was a Company man, for now anyway, and he would make sure they knew it.

Having at least an immediate plan worked out, he finally let his attention wander about the train. His gaze settled on a video game ad on the bulkhead for *AltYou: Another Life*. The slogan read: Give your virtual You another life. Coming soon. He chuckled bitterly at the irony, until he followed the train of thought further. Was Kumiko any more real than a virtual reality game character made of pixels? The answer was, *of course, yes*—he couldn't bear to consider any other answer. Everything that made Kumiko *Kumiko*—her personality, desires, needs, hopes, fears—was all still there. But maybe others would see her as just lines of code. He had to make sure that didn't happen.

He swung himself out of the train car just as the doors were closing, so lost in thought that he almost missed his stop.

A month had passed, and thankfully the clinical trial had proceeded like nothing had happened—an image the Company was trying hard to project to the trial participants. The day after Kumiko's body death, they gave him strict instructions about confidentiality, strongly encouraged him to sign a nondisclosure agreement—which he'd done, to avoid any trouble—and then sent him back to the lab like it was a typical workday. He didn't see any other path forward that would allow him to keep spending time with Kumiko, so he walked the corporate line.

The first time he logged back on, he saw two other trial participants standing in front of the lobby doors, but no one else. He watched as they both walked through the blue door. Kumiko was nowhere to be seen. He spent all day worrying, until Jamie pulled him aside at the end of the day and told him she'd been firewalled, but he was working on a back door.

Sure enough, at the end of the week, a fifth door appeared in the virtual lobby—visible only to him, Jamie vowed—and when he walked through, there she was.

"Sam! I've been dying here, waiting for you." Her face flushed when she realized her choice of words, the software still trying hard to maintain the illusion of physiology. "Tell me what's happening." Simulated tears formed in her eyes and then floated away like soap bubbles.

Sam pulled her into his arms and took in the new virtual environment. They were in a sun-dappled forest glade. The trees were a touch too green, and the leaf edges were pixelated, but Jamie had done good work. Fast work, too; he must have repurposed old code. Sam wondered for a moment what other environments could be designed for Kumiko, but that was a problem for another day. He told her everything he knew—which wasn't much, but enough to confirm that she was here to stay.

"So I'm stuck here. For good." She sank to the ground and wrapped her arms around her knees. More bubbles floated up.

Sam sat beside her and put an arm around her. But he didn't know what kind of comfort he could offer. He felt like he'd just given her a lifetime sentence to solitary confinement, which in a sense was true. "I'm sorry, Kumiko. Whatever happens, I'll be here with you as much as I can. We'll find a way to make it work."

He spent every day with her, doing whatever he could to help her through a tragedy no one had ever experienced before. Thankfully, the clinical trial was focused on the control and use of his body as an automaton, outside of his awareness, so he was free to spend his time in the virtual space however he wanted.

By the time that first month was over, things could in no way be described as normal, but Kumiko had come to terms with her new existence as best she could, all things considered. And the same was true of him. His time at work—largely spent with Kumiko—was becoming more like life than his everyday reality. They talked for hours about Kumiko's hopes and dreams, and which ones were still possible in a virtual environment. She told him about her family and how worried they must be about her. He even found himself flirting with the idea of sharing his own feelings with her. Every night, when he went home, he felt more alone than ever.

When his mind turned to brainstorming ways to help her—even if just conveying a message to her parents—it renewed his fear of doing something that would change the status quo and take her away from him again. But Kumiko wasn't satisfied with the status quo.

One day, they were sitting under the forest canopy on an overstuffed sofa that Sam had talked Jamie into uploading. Kumiko dragged a toe through the leaves on the ground.

"Sam, this time with you has made the situation bearable. Enjoyable, even. I don't know how I could have come this far without you. But we have to figure out something long term. Not just for me, but for you too. You have a life out there. What happens when the trial is over? You can't keep coming back here."

He wanted to tell her there was nothing he would rather be doing, now or in the future, but he didn't need to burden her further. "We'll find a solution, Kumiko. We will, or someone else will. The Company can't keep this under wraps forever. Someone is bound to leak it. You know I would, but I don't want to jeopardize our time together."

She pulled a leaf from an overhanging branch and absently tore it into pixels that fell away and disappeared. "I know, Sam. But what then? I'm still stuck here, alone. Forever."

"Not alone." He pulled her close and kissed her. She buried her face in his chest, and he watched as more bubbles floated up and popped against the tree branches.

It took a few more months, but the news finally leaked to the media. Sam wondered if it was Jamie. He'd kept the back door open for Sam to reach Kumiko—and, more importantly, no one had discovered it—but Sam watched Jamie's discontent grow a bit more every day. He understood it, and the time he spent with Kumiko was the only thing keeping him from sharing it.

Jamie was the one who told him. "The news just broke. Did you see it when you came in?"

"What news?"

"The *news*." He nodded his head to indicate the lab surroundings.

"Oh. I did see some press drones flying around outside. Is that why there's extra security in the lobby?"

"Yeah, they were ready for it. Had a statement all prepped." He slapped a hand on the exam table. "Hop up. They were keeping it quiet out of respect for the family, et cetera. They're making the argument that Kumiko is dead, and her consciousness is basically a digital recording. Totally sidestepping any responsibility for her."

"Are you kidding me?" Sam's voice attracted the attention of the other people in the lab.

"Keep it down." He slipped the IV cuff onto Sam's arm. "They're admitting negligence for her death and offering to pay a fine. Trying to sweep it under the rug. Disgusting."

Sam pushed the cuff off his arm. "I have to go talk to someone. Now that it's public..."

Jamie pushed Sam back onto the table and rested a hand on his chest. "Hold your horses, man. Be smart about it. You're still facing the same risk. They're keeping this research going, at least until anything else happens, so you could still lose your job. Along with any access to Kumiko. Got any other ideas?"

Sam thought for a moment. "You said they kept it hush-hush for her family's sake? I haven't met them myself, but from what Kumiko has told me, they weren't too thrilled about her working on this trial to begin with. So, I can't see them wanting to keep things quiet. There might be something there. I'll talk to her."

"Lie back." Jamie tightened the cuff and started the drip.

Kumiko's family turned out to be the key. The day after the news broke, they filed a lawsuit seeking to have Kumiko declared a live, legal person, body or no body. The Company quickly got over any objections to having her declared still alive when the state attorney general started hinting at criminal manslaughter charges if she were considered dead.

But little changed for Sam. Reporters barraged him and the other trial participants daily with interview requests, but the Company had imposed a strict gag order. He continued doing whatever he could to ensure his access to Kumiko wasn't denied.

"Here, put this on." Jamie handed a bracelet to Sam.

"What's it for?" Sam asked, sitting up on the exam table.

"New safety protocol. Monitors your vitals in real time and also tracks your body's location." He tightened the bracelet. "You won't notice most of the new safety measures, since they have to do with your physical body. It's all part of the negotiations the Company is having with the state. And with Kumiko's family."

"Good to know they were paying such close attention before now." Sam gave full rein to the sarcasm in his voice. "Let's get on with it." After waiting for two weeks while the Company implemented new safety protocols, he was eager to see Kumiko again and share the news with her.

"So now I'm alive?" she asked, after Sam filled her in. "So nice of them to make all these decisions about me without actually talking to me." They were back in front of the campfire, once again sharing a blanket. Her existence becoming public meant they had full use of the virtual environment again. Kumiko poked at the fire with a twig in frustration.

"It's great news, though. With them considering you a legal person, they have to grant you the same autonomy and rights they would any other person. So yeah, they'll have to talk to you now. Maybe I can be your ambassador." He gave her a nudge.

"Maybe so, but they still have total control over my existence." Waves of orange and yellow light danced around her head, the manifestation of her ire mirroring the flames. "I'm a digital entity, which causes its own unique set of challenges. Ones that existing laws don't account for."

"Like what?"

"Think about it a moment. The Company is now required, legally, to protect my digital existence. Forever, or effectively so. As long as the technology lasts, anyway."

"Yeah, that's the point. They won't be deleting or archiving you anytime soon."

"But, Sam, that makes me immortal."

He laughed out loud. "That's right. Cool!"

"Not cool!" She hurled the twig into the fire and pushed away from Sam. "I don't want to live forever. Do you?"

"I do if I can spend the time with you." He tried pulling her back in.

"Stop it. I'm serious. I can't do anything without asking them for it and relying on them to follow through. And if I tell them I don't want to live forever? Euthanasia is still illegal, and that applies to me now that I'm a 'legal person,' even one without a physical body. So you see? I have no real agency."

"I'll visit you as often as I can. You know that."

"It's a nice thought, Sam." She leaned back into him. "But they're not going to let you come indefinitely. The clinical trial will end at some point. Then what? And anyway, you'll die yourself eventually. Hopefully of old age, but I'll still be here long after that. Without you." Her emotional aura shifted to blues and purples.

Sam squeezed her tightly, staring into the fire. She was right, of course, on all counts. He didn't have any words of reassurance to give her, and she'd had far more time to think about everything than he had. Including about what would happen after the trial ended. He'd pushed the thought out of his mind every time it surfaced, and he still had six months—give or take—before the study was due to end. He wanted to savor every moment he had left with her, without thinking about their time ending.

But maybe it didn't really have to end. "What's your father's address?"

Sam fidgeted, bouncing on the balls of his feet. To distract himself, he gazed around the front porch, looking for clues about what kind of man Kumiko's father might be. Once-manicured potted plants drooped. Dead leaves and cobwebs clustered in the corners. Junk mail overflowed from a mailbox on the wall. He took a breath and pressed the doorbell.

After a moment, a short man with close-cropped salt-and-pepper hair opened the door. He wore a stony face that hadn't been cracked by a smile in many months.

"Yes?"

"Hello, Mr. Nakamura. My name is Sam Petri. I'm a colleague of Kumiko." He felt awkward, showing up on the man's doorstep unannounced, but this wasn't a conversation for phone or email. "Can I talk to you about her situation?"

Suspicion bloomed in the man's eyes. "You should talk to my lawyers."

Sam held up his hands. "I'm sorry, I should have clarified. I'm not representing the Company. I'm close to Kumiko—I'm a close friend." That wasn't the whole truth, but it *was* the truth.

"Very well, come in." The suspicion was still there, but he opened the door wider and waved Sam in.

Sam followed Nakamura into the living room. The state of the front porch continued inside, which Sam now recognized as the aesthetic of grief. Untidy surfaces, unswept floor, unwashed dishes, more piles of untouched mail. It reminded him of his own home after his father's death, when Mom let every surface—including herself—collect dust. After spending most of his days with Kumiko, he'd forgotten that, to the outside world, she was essentially dead.

"I'm sorry about Kumiko. She *is* alive, though, and misses you."

"Alive, dead, I don't know what to think anymore. But one thing I know for certain, she won't ever come home again." Nakamura gestured to the sofa, an expression of weary sorrow in his eyes that Sam recognized from Kumiko's own face.

He sat on the sofa and explained everything: his role as a trial participant like her, the time he spent with her before her death, and after. The concerns they both had about what might happen when the trial ended and he was no longer able to visit her.

"I read about your settlement with the Company." The lawsuit had been dropped after the Company agreed to transfer ownership of Kumiko's server to the Nakamura family, along with a ninety-nine–year contract to provide full tech support. "When you take custody of Kumiko—of her server, that is—I'd like to ask your permission to come visit her." He didn't want to sound over-eager, but he couldn't help adding, "As often as possible."

Sam could see the *no* in the man's eyes even before he spoke. "I'm sorry, Mr. Petri, but that will no longer be necessary."

"Why not?" He couldn't keep the panic out of his voice.

"Thank you for spending time with her, although it sounds like it was unauthorized, and might have jeopardized what we're planning to do."

"What are you planning to do?" He recalled the conversation with Kumiko about agency. "What about what she wants? Or does that not matter?"

The suspicion, and growing anger, returned to Nakamura's eyes. "That will be enough. I would like you to leave." He stood.

Sam was no lawyer, but he was smart enough to know when he'd lost a negotiation. He found his way to the door and left without another word.

Sam lay back on the exam table, his mouth dry despite the full glass of water he'd just consumed. It was his third attempt in as many days, but this time he'd finally found the nerve to go through with it. His stomach was already starting to gurgle, and his pulse was pounding. He wiped sweat from his forehead and looked at Jamie. The technician stood at the next table over, talking to another man. *Of all the days to be shooting the breeze.*

"Yo, Jamie, let's go. You'll have plenty of time to gab after I'm in."

Jamie gave him a smirk as he came over and adjusted the IV cuff. "Settle down, Romeo. You'll see her soon enough." His smile disappeared as he looked more closely at Sam. "Feeling okay? You don't look so good." He set his hand against Sam's forehead. "You're sweating like a pig, but you don't feel hot."

"I'm fine, let's go."

"I should check your vitals."

He was running out of time. "No need. Listen, Jamie, I'm going to talk to Kumiko about something. It's got me nervous as hell. It's *private*. Let's do this so I can get it over with."

Jamie's smile returned. "There's the lover boy I know. It's okay, I get it." He moved to the terminal.

Sam's vision dimmed momentarily, and then he was in the lobby, Kumiko waiting in front of the blue door. As soon as she saw him, she touched the door. It flared blue and opened. She waved him over. "Come on, beach day!"

Relief flooded through him. He'd done it. Beach day—and countless more days to come.

The setting sun turned the ocean ablaze in fiery oranges and reds. They sat in a pair of Adirondack chairs, feet dangling in water programmed to match the air temperature so perfectly that it was impossible to tell where the water ended and the air began. Sam raised a fruity cocktail. "To you, babe. And to many more sunsets like this."

She clinked her glass against his and took a sip. And then a puzzled look crossed her face. "Wait a minute, what time is it? How are you still here?"

"Don't worry about it. Enjoy your drink. You might have to get used to seeing more of me."

She narrowed her eyes. "I should grill you more, but it is a delicious drink. And I want to enjoy this gorgeous sunset."

Later, they drifted in a swimming pool under a dazzling field of stars. The aqua blue glow of the pool shimmered in Kumiko's hair. Sam pulled her closer and gave her a long, lingering kiss.

She pulled away and gave him a piercing look. "I must know. How did you manage to stay here so late? Did Jamie owe you a favor or something?"

Was she ready to know? Was he ready to tell her? Clearly, his plan had worked. The fact that he was still here proved it. The old reticence was hard to kick, even now, but if this was the beginning, then it was a good time to start fresh. "Kumiko, you're not going to like what I have to say. But all I ask is that you have an open mind and remember the end, not the means." He took a deep breath. "This morning—"

The stars, the pool, Kumiko, all vanished, replaced by the harsh glare of overhead lights. His head swiveled in panic. He was in bed, wearing a hospital gown, tubes trailing from his arm. An IV bag stood guard by his side, and Jamie beside it.

"Nice job, Sam. Trying to off yourself so you can stay with Kumiko permanently? It almost worked. You're the smartest dumbass I've ever met."

Weeks later, he was back at home, but still in bed—even though it was nearly noon. His stomach told him to get up and make some breakfast—lunch?—but he couldn't muster the energy. Now that he was gathering his own dust, he finally understood what his mother had felt those many years ago. His was a special kind of grief, though. Kumiko was alive, but he could no longer reach her. He hadn't seen her in weeks, and he'd never see her again.

The Company had fired him before he even left the hospital. Not that it was any kind of surprise. However stupid or smart his idea had been, it failed. The only comfort, cold though it was, came with the news that the clinical trial had ended early. The press releases all called it a huge success, and the public launch was underway—all of which meant that he would have lost access to Kumiko anyway.

He picked up his phone and started up the AltYou app. The different brand of virtual reality numbed his mind, even if he saw reminders of Kumiko everywhere. He needed to find another job, and he'd start looking. Real soon now.

The doorbell rang. He couldn't even remember the last time that had happened. He hadn't ordered anything and certainly wasn't expecting anyone, but he had nothing else to do. He tossed the phone on the bed and went to the door.

There on his doorstep stood Mr. Nakamura. Kumiko's father. On *his* doorstep. Sam stared, mute, so shocked he forgot he was wearing only underwear and an old T-shirt.

"Mr. Petri? Sam?" Nakamura asked tentatively.

The words were enough to break his trance. Suddenly self-conscious, he used the door to hide his semi-clothed body. "Mr. Nakamura. Sorry, I wasn't expecting you. Or anyone. Obviously."

"It's my fault for showing up unannounced. Though, if I recall, you did the same to me." He smiled, and Sam realized it was the first time he'd seen that.

"Please, come in. The living room is that way." He pointed. "I'll be right there."

A moment later, he joined Nakamura, zipping up a hoodie over the sweatpants he had pulled on. "Is Kumiko okay? Did something happen to her?"

"She's fine, Sam. Well, *fine* in the way that she can be. I'm still getting used to that."

"And everything went well with the server getting moved to your house?"

"Yes, and with the new product launch, I was finally able to log on and see her myself yesterday. The first time in... I can't even remember how many months. And in all that time, I couldn't let myself believe she was still alive." His voice was a little scratchy. Sam felt a tickle in his own throat too. "That's why I'm here."

Sam held his breath. Kumiko must have sent him some kind of message.

Nakamura continued. "Kumiko made it plainly clear to me what you both feel for each other. She also made it clear that she wasn't going to take no for an answer from me. She wants to see you. As often as you want. And my feelings on the matter are to be ignored, apparently." He smiled again.

Sam returned the smile. A thousandfold.

A lot happened in the year that followed. Advances in throughput and connectivity afforded Kumiko ever greater range across different virtual reality ecosystems. She was no longer confined to the single server in her father's house. She became an outspoken advocate for artificial intelligence rights, even if the Singularity was still years off and her treatises were all theoretical. Her very existence spawned a whole subgenre of philosophy that was still deep in the throes of debate.

But most significant—by far—was the spectacle unfolding in front of gathered loved ones and thousands of spectators in a virtual cathedral hand-coded by Jamie to Sam's specifications. The space was simple: bright, white walls, vaulted ceiling, colorful stained-glass windows, and row after row of seats extending into an impossible distance. Sam and Kumiko stood in front of Jamie, whose job title now included newly ordained officiant.

Sam reached down and took Kumiko's hand. It felt as warm and real as a hand could feel, as genuine as the love that filled him when their eyes met. There was nothing virtual about it. Their union was about to make history not just because of its unique nature—only one of them having physical form—but also because it was set to become the longest marriage in history, if all went to plan. Earlier in the day, they'd finalized the paperwork for the world's strangest Do Not Resuscitate order, a legal construct to ensure that if Sam ever became incapacitated or afflicted with a life-threatening illness, he would join Kumiko here, in virtual space, permanently. Immortal she might prove to be, but she wouldn't be alone.

Jamie finished his introductory words and then instructed Sam and Kumiko to face each other. "It's time for your vows. You ready?" He winked. Sam took Kumiko's hands. "Do you, Sam, take Kumiko to be your lawfully wedded wife?"

He'd been ready to answer that question for a long time. "I do."

"Do you promise to love, honor, and cherish her…"

Sam's mind leaped ahead in the verse. He saw mirth gathering in Kumiko's eyes as she too anticipated what was coming.

"In sickness and in health…"

They pressed their lips tight, trying to hold it in.

"For as long as you both shall live?"

Sam and Kumiko burst out laughing.

A Girl and Her Dragon: A Life in Four Parts

written by
Joseph Sidari

illustrated by
JOSIE MOORE

ABOUT THE AUTHOR

Joseph Sidari lives in the Boston suburbs with his wife and their labradoodle, Chloe. He is a practicing physician and writer, which means he spends his days keeping real people alive and his nights inventing peril for fictional ones. He is a member of the Grub Street Writers' Group of Boston and with the sale of this story he is now a member of the Science Fiction and Fantasy Writers Association. His three grown sons and his daughter-in-law live in New York City, though none share the protagonist's heroic devotion to freeing animals from the Bronx Zoo.

A lifelong reader of speculative fiction, Joseph began writing a dozen years ago after flipping his bicycle while training for a triathlon. His wrist healed, but the habit stuck once he realized that typing would lead to fewer emergency room visits than cycling. After several unpublished novels and a helpful nudge from an agent, he discovered a fondness for short fiction. His stories have been published in various anthologies and earned multiple Honorable Mentions, including two Silver Honorable Mentions in the Writers of the Future Contest.

This story began as one of those honorable mentions. Following the Contest's advice to "revise and resubmit" if you have nothing new, Joseph completely rewrote this piece about the first and last dragon in America, digging deeper into the heart and soul of the protagonist to create this version.

ABOUT THE ILLUSTRATOR

Josie Moore grew up in the valley town of Westfield, Massachusetts, with a voice that wouldn't work and a brain filled to bursting.

Made paranoid by the world and abandoned by those around her, she turned to art in her time of hiding. When she was scared, she wrote about it. When she cried, she drew comforting pictures. With her inability to pinpoint and process her own emotions, she used art and storytelling as her communication, turning her fear into something beautiful.

Nowadays, she can finally leave her bedroom, but never without a sketchbook by her side. She attends the University of Massachusetts Dartmouth in hopes of earning her BFA in illustration. She spends her days doodling elves, bears, and silly little faeries, and her nights writing, planning, and... hopefully writing some more. Finally having an outlet to put all her obsessive, creative energy, she dreams of one day sharing even the deepest crevices of her mind through her odd love stories and twisted mysteries.

A Girl and Her Dragon: A Life in Four Parts

PART 1: THE EARLY YEARS
Stephanie Burnham, Age 6 (1992)

SCRAPBOOK

COVER:

In crooked, glitter-glue letters

ASH AND ME BEST FRENDS 4 EVER!!!

NEWSPAPER CLIPPING:

New York Times

March 25, 1992

FIRST DRAGON ARRIVES
ON AMERICAN SOIL

By Staff Writer

BROOKLYN: A very special cargo was unloaded today at the Red Hook Container Terminal in the Brooklyn Navy Yard. The *SS Belle Rêve* brought New York City its newest immigrant, courtesy of the *Parc Zoologique de Paris*. The Big Apple is proud to welcome Acelin, the blue dragon—or Ash, as he is familiarly known. He is the first dragon imported into the United States, taking up residence in the Bronx Zoo's newly developed exhibit dubbed *Animalia Arcana*, alongside other favorites like Eunice, the unicorn; Bronnie, the sea monster; and (Continued on page 12)

HANDWRITTEN PAGE:

Double-lined penmanship paper in careful block print from first grade

ASH is here!
Hes from Paris!
Does he no english?
He lives at the ZOO.
I wish I cud fly on his bak!
Mama says hes historik.
I think hes MAJIK!

MAGAZINE CLIPPING:

National Geographic KIDS/April 1992
ANIMAL FACTS!

Did you know: A dragon will grow bigger than an elephant, sometimes two, but is only one-quarter as heavy? This lighter bone structure is what allows it to fly. The scales come in all the colors of the rainbow and are both aerodynamic (Vocabulary word: help it glide through the air with less resistance) and beautiful, since they sparkle in the sunlight. A dragon is carnivorous (Vocabulary word: meat eater), so it has multiple rows of dagger-sharp teeth. But don't worry. Dragons are peaceful creatures and have never been known to attack people. Pictured is a blue dragon named Ash, currently a resident of the Bronx Zoo.

HANDWRITTEN BELOW:

Thats My Dragon! ASH! I wunder wat hes thinking?

POLAROID PHOTOGRAPH:

Stephanie—with her red pigtails, freckled cheeks,
and her favorite blue denim dress—poses in front of Ash.
She's grinning. He's asleep.

HANDWRITTEN BELOW:

Hes my best frend!!!

TICKET:

ADMIT ONE—CHILD Price: $4.95—Aug. 16, 1992
THE BRONX ZOO:
Operated by the Wildlife Conservation Society
SEE OUR NEW EXHIBIT: *Animalia Arcana*
Rain or Shine—No Refunds

HANDWRITTEN BELOW:

Mommy sed if I vakyum my room we cud
see ASH on Saterday so I did!!!

TRADING CARDS:

1992 MONSTER MANIA CARDS—
COLLECT 'EM ALL!

UNICORN: She has a sparkly horn that shines in the sun and will grant wishes on the night of a full moon. She loves good little girls and boys, but only if they're very, very good!

MOUNTAIN TROLL: Big and grumpy, he smells like your daddy's old socks! This rocky giant is over ten feet tall! He lives in a cave and snores like an avalanche!

SEA SERPENT: This is Bronnie! She currently lives at the Bronx Zoo. You can see her snake-like neck, but there's a giant fishy body below. She loves seaweed sandwiches!

BEARDED CRIMSON DRAGON: This wise old creature has broad leathery wings and a long spiky tail! His beard glows when he's thinking about eating noodles or solving riddles!

HANDWRITTEN BELOW:

Hay!!! Weres ASH???

MAGAZINE CLIPPING:

Highlights magazine Oct. 1992: Science in Action
DO DRAGONS REALLY HOARD GOLD?

"No," says Dr. Aurelius Johnson, Chair of the Department of Arcobiology at Cornell University. "I think the 'gold' myth arose since dragons are attracted by shiny objects, and the brothers Grimm embellished it," she said. "So, if you came across a dragon in the wild, they would not be sitting on a pile of treasure, but there might be a piece of tinfoil or some bottle caps they took a liking to."

NEWSPAPER CLIPPING:

New York Post
November 21, 1992
DANGEROUS DRAGON
DRAWS DISCUSSION: ZOO MUST ACT
By Staff Writer

BRONX: Neighbors flooded the mayor's office with concerns for their children's and pets' safety from the Bronx's newest resident, Ash, the blue dragon (pictured above, courtesy of the Bronx Zoo). Despite numerous reassurances from the zoo staff that Ash "wouldn't hurt a fly if it landed on his forked tail," the Board of Directors for the Wildlife Conservation Society, which operates the Bronx Zoo, compromised and ordered shackles to be applied around two of his feet. "Now he won't be able to fly out of his enclosure and make off with your Chihuahua," said head zookeeper Jeffrey Smythe. Locals were still worried it was not enough. (Continued on next page)

HANDWRITTEN BELOW:

They put hancufs on him. I dont think he liks it.

HAND-DRAWN CRAYON PICTURE:

A drawing on wrinkled notebook paper shows a large, smiling blue dragon with big eyes and oversized wings beside a stick-figure girl in a pink dress labeled "ME." Hearts float around the dragon's head, and a rainbow leash connects the two like a friendship bracelet. The background features a decorated Christmas tree and a sun with a face beaming down on them. It's messy, bright, and smells like grape magic markers and peanut butter-covered fingers.

HANDWRITTEN BELOW:

Mery Crissmas Ash!!!

PART 2: THE COLLEGE YEARS
Stephie Burnham, Age 22 (2008)

LAPTOP AND CELL PHONE

TERM PAPER EXCERPT:

Senior Year, Semester 2, Columbia University
Medieval European History: Advanced Seminar (EHIS-401)

THE ROLE OF DRAGONS IN UNIFYING FRANCE UNDER KING LOUIS VII DURING THE CRUSADES

by Stephie Burnham

The acquisition of Acelin, a blue dragon, during the Second Crusade (1147–1149), represents a pivotal moment in medieval French identity formation. According to draconian scholarship, French knights traveling to Constantinople encountered this creature in the remote Balkan Mountains. The "how" has been lost to history, but on one excursion, twenty soldiers successfully subdued and transported the beast to France.

King Louis VII christened the dragon "Acelin," meaning "noble creature." The dragon's presence provided tangible evidence of crusading success under his royal leadership. He kept Acelin on the grounds of his royal residence in Paris. Louis XIV later established the Royal Menagerie (1663) in Versailles, which would eventually become one of Europe's first zoological exhibitions, thus creating a pilgrimage destination that drew citizens from across the realm.

Dining on mutton, cabernet, and Brie, Acelin lived for over eight centuries.

(cont.)

TWEET:

From: @DragonGirl4_Ever
Just saw Ash at the #EvilBronxZoo. Wings are banded! 4 legs shackled! His scales are dull. His spark is gone. Blue dragons shouldn't feel blue. #FreeAsh #StillMagic #DragonRights #LetAshFly

TERM PAPER EXCERPT:

Senior Year, Semester 2, Columbia University
Comparative Zoology (BIO-305)

COMPARATIVE ANALYSIS OF
TOXIC EXHALATION AS AN ANIMAL DEFENSE
By Stephie Burnham

The blue dragon (*Draco azureus*) employs one of nature's most sophisticated chemical defense mechanisms by emitting highly concentrated sulfurous compounds. The characteristic brimstone odor results from elevated sulfur content in gastric secretions rather than some mythological fire-breathing ability. Similar chemical defense mechanisms are observed across diverse species, including hydrogen sulfide production in deep-sea tube worms (*Alvinella pompejana*) and volatile organic compound release in bombardier beetles (*Brachinus fumans*). This paper will examine the various other mechanisms.

(cont.)

TWEET:

From: @DragonGirl4_Ever
The magical creatures of @*AnimaliaArcana* are being freed into the wild! The mountain troll. Pegasus. Even Bronnie. Only @Eunice_unicorn stays for her adoring fans. And @Ash_BlueDragon. It's time to #FreeAsh #DragonRights #LetAshFly

TERM PAPER EXCERPT:

Senior Year, Semester 2, Columbia University
Modern American History, Senior Thesis (AHIS-402)

THE ROLE OF ACELIN, THE BLUE DRAGON,
IN CRAFTING POST-WAR AMERICAN POLICY
by Stephie Burnham

The acquisition of Acelin, a blue dragon (species *Draco azureus*), represents one of the most overlooked yet significant diplomatic endeavors in American foreign policy during the post–World War II era. This thesis argues that the fifty-year pursuit of this unique

French specimen—spanning from Franklin D. Roosevelt's initial negotiations with Charles de Gaulle through George H. W. Bush's successful diplomatic conclusion in 1992—reveals the complex intersection of wartime diplomacy, domestic policy initiatives, and America's evolving role as a global superpower.

This study will examine Roosevelt's wartime communications with de Gaulle, analyzing the proposed "dragon-for-Normandy" exchange as a striking example of unconventional Allied diplomacy. The analysis will then follow the bureaucratic inertia under President Truman and reveal how leadership transitions can stymie long-term diplomatic continuity.

The Eisenhower administration is notable for initiating the American rebranding of Acelin to "Ash," reflecting a mid-century cultural impulse toward assimilation. However, Cold War tensions in Korea and Southeast Asia again delayed progress, with only a momentary resurgence in public interest during the 1960s, notably following the release of "Puff the Magic Dragon" (1963), which unintentionally sparked congressional hearings on magical creatures and a surge…(cont.)

TWEET:

From: @DragonGirl4_Ever
Saw @Ash_BlueDragon today. SO BEAUTIFUL. Breaks my heart. Would it hurt to let him fly? We know what #EvilBronxZoo thinks. #Dragedy #LetAshFly

TEXT MESSAGE:

SHELLY: You missed the journalism club meeting. 😠

STEPHIE: Sorry, went off campus to clear my head.

SHELLY: Let me guess—the zoo?

STEPHIE: So?

SHELLY: Is visiting that dragon more important than your actual career?

STEPHIE: Maybe.

SHELLY: You're unbelievable.

STEPHIE: 😃

SHELLY: But seriously, there's a world beyond that dragon.

STEPHIE: Maybe—maybe not. Get coffee?

SHELLY: 👍

TWEET:

@DragonGirl4_Ever
Happy graduation to me! Four years = 1 job! #ThanksMOM #ThanksDAD. I love you!

DREAM JOURNAL ENTRY:

May 13, 2008

Had the flying dream again. But this time, I wasn't riding Ash. I was Ash. I could feel the shackles cutting into my legs, the metal bands pinching my wings. I tried to fly but could only hop a few feet before the chains yanked me back down. I woke up crying. Dr. Martinez would probably say I'm projecting my feelings of being trapped—graduation anxiety, job search stress—but what if dreams aren't always about us? What if sometimes they're about the things we can't stop thinking about? I can't shake the feeling that Ash is trying to tell me something.

EMAIL:

To: stephie.burnham@dmail.com
From: newmember@peta.org
Subject: Membership

June 16, 2008

Dear Ms. Burnham,

Thank you for joining PETA and for your membership dues. We're grateful for your commitment to protecting all creatures—great, small, and scaled.

At present, we do not have any active programs dedicated

to dragons. With only one known dragon in New York City, resources have been limited. If you're interested in helping lead awareness or support efforts for dragon welfare, we'd be glad to explore that with you.

Thank you again for your dedication.

Sincerely,
The PETA Team

EMAIL:

To: burnham_family@yahoo.com
From: stephanie.burnham@NYT.com
Subject: First week of work!
October 3, 2008

Dear Mom and Dad,

Guess who just finished their first week of work as an Editorial Assistant at *The New York Times*? It's pretty cool. I'm mostly just observing, taking notes, and making a lot of coffee runs. Nothing glamorous, but I'm learning.

My new apartment is smaller than my bedroom closet at home, but it's mine. You have to come up to the Bronx and meet my roommate, Aaliyah. She's a dancer and very sweet. Oh, and Dan. He works at the *Times*, too. A sportswriter. Nothing serious yet, but who knows?

I've started researching a piece about Ash. Yes, I still like to visit him at the zoo. I'll get a coffee and sit next to his enclosure. Sometimes I pretend he talks to me. Tells me of his life before he was chained up. I know. It sounds weird. I wouldn't admit that to anyone else, and I know he can't REALLY talk, but it's so wildly unfair how he's treated. I asked one of the zookeepers when does Ash get a chance to fly? The guy laughed at me. Don't they get it? It's like keeping Bronnie the sea monster from swimming. They're not openly cruel to Ash, but keeping a flying creature earthbound is a million times worse. It's just not fair!

Nobody's asked my opinion, but I don't know how much longer I can hold my tongue—or my pen. I'm gonna send a letter to the editor. You never know—it could get printed and THAT might open some doors. Wouldn't you love to read a recurring column about life in the Big Apple by your favorite daughter? Haha. Fingers crossed!

Miss you both. Hope Brooklyn's treating you well.

Love,
Stephie

LETTER:

New York Times

December 18, 2008

Dear Editor,

I'm only twenty-two, but I feel like I've known Ash, the blue dragon, for centuries.

Not the shackled dragon currently living in the Bronx, but the Ash who once flew over the lavender fields of Provence. It was not a zoo as we know it, but a kind of enchanted retreat. He wasn't behind bars. He wasn't a threat. He was a marvel.

No one was afraid of him. Villagers would stop their work in the vineyards just to watch him fly overhead. Artists painted him. Children sang to him. He was part of the landscape—myth and nature in harmony.

I wonder if he remembers those days.

I know times have changed, but I feel like we've lost something. We treat the strange as dangerous. We punish the foreign with our fear. And Ash pays for it every single day.

We can do better than his chains. At least, we used to.

Sincerely,
Stephie Burnham
Bronx, NY

PART 3: THE PRIME YEARS
Steph Burnham, Age 40 (2026)

LAPTOP, CELL PHONE, LOOSE PAPERS

RECURRING COLUMN:

New York Times
"On a Scale of One to NYC"
January 14, 2026
SUNUP TO SUNDOWN: A DAY WITH ASH
By Steph Burnham

6:00 a.m.: Ash is already awake when the sun rises. He stands motionless beneath the trees of his enclosure at *Animalia Arcana,* dew collecting on the sapphire ridges of his wings. His eyes scan the horizon. He doesn't yawn. He cannot stretch. He simply waits.

7:15 a.m.: A senior keeper arrives with his morning meal. Today, it's goat. Ash flips the carcass over with a talon and eats it slowly, deliberately, like it's not his favorite. He never breaks eye contact with the handler. He never speaks. Still, he seems to listen.

8:30 a.m.: Wing bands examined. Leg shackles re-latched. Dart gun pressure tested. Everything is "within safety protocols." The vet notes, not for the first time, "Respirations unusually low."

9:15 a.m.: Ash shifts slightly as the front gates open. He lifts his head as the foot traffic begins. Not many notice, but he does it every day.

10:00 a.m.: Forty second graders round the corner and freeze. Mouths open. Backpacks drop. A few shriek with excitement. One begins to sing. Ash responds with a sound not unlike a purr, though it vibrates in your ribs more than your ears. You have to be there to feel it. Some say it's humming. Others say it's remembering.

11:45 a.m.: Now, the fourth and fifth graders. They've read the fairy tales. You can tell. These are the ones who stay behind when it's time to move on. They don't look at Ash like a zoo exhibit. They look at him like he's their friend.

12:30 p.m.: Another dead goat. Ash ignores it. Instead, he lies down with one wing angled just enough to cast shade over a child who fell asleep near the viewing grate. Coincidence, the keepers say. I'm not so sure.

2:00 p.m.: The adults arrive in earnest. Cameras out. Voices loud. "Are we sure he's safe?" and "Why is he always lying around?" and "He never does anything," they mutter.

What is a fourteen-foot dragon supposed to do in shackles and chains?

3:45 p.m.: Children are loaded onto buses. A few wave goodbye. Some cry. Ash watches each one leave. Every day. Without fail.

4:30 p.m.: Wing bands and leg shackles are rechecked. Tranquilizers, double-checked. There is a quiet tension in the air—like a secret the staff refuses to say out loud.

6:00 p.m.: The gate locks. The silence settles. Ash returns to his cave. One eye open, fixed toward the sky. Waiting.

And I ask myself—still—after all this time, all this devotion, and all these wide-eyed children: why is everyone still so afraid of him?

Maybe we're not afraid he'll burn the world.

Maybe we're afraid he still believes it's worth saving.

Steph Burnham is a regular contributor to *The New York Times*. She still believes in magic—and in the better part of human nature.

NOTE PAPER:

TO-DO LIST/FEBRUARY

- o Work on the next Ash piece for *NYT*: Angle?
    - * Is there a book here *
- o Return *The Dragonet Prophecy* to library for kids, pay the late fee.
    - * See if any new books have come in *
    - * Try to get interview w/Tui Sutherland for future column *
- o Pick up the latest Rebecca Yarros
    - * Pick up a bottle of sauvignon blanc *
- o Kids want a sleepover this weekend, watch *How to Train Your Dragon*

  * Shop for snacks. Invite friends—theirs/mine.
    Get more wine *
- o Ask Dan if he'll switch weekends with me, take kids
  * PETA's march on the zoo is 2/22 *

RECURRING COLUMN:

New York Times
"On a Scale of One to NYC"
March 2, 2026

THE LAST DRAGON IS DYING BEHIND BARS:
AN INTERVIEW WITH ASH, NEW YORK'S ONLY DRAGON

By Steph Burnham

I spoke with Ash, the blue dragon, through the bars of his enclosure at the Bronx Zoo. He arrived here from France in 1992. Now the fanfare has faded. Only his chains remain.

SB: Ash, thanks for speaking with me today. May I start by asking: Do you consider America your home?

ASH: A home is where you can stretch your wings. So, no. Not really.

SB: You've been here for thirty-four years—

ASH: Yes. It seems longer.

SB: But you're over 800 years old. How could that seem long?

ASH: If you were kept in a cage, couldn't stretch your arms, couldn't even walk around without leg irons jangling, would it seem long to you?

SB: I see your point.

ASH: They say it's for safety. They say I'm unpredictable. A wild animal.

SB: But you've never hurt anyone.

ASH: I sneezed near a fence once, and my sulfur-smelling breath startled a tourist. He screamed. That was in 1993.

SB: And since then?

ASH: I watch planes fly overhead. I count the stars. I dream I am a bluebird—not a blue dragon—so I could slip my chains and fly away.

SB: Have you asked to be released? To a sanctuary? Or a reserve?

ASH: Many times. But I'm an asset. An attraction. I generate ticket revenue.

SB: What would you say to those who claim it's safer for you here?
ASH: Safer for whom?
SB: If you could ask the zoo administration for one thing to make your living arrangements more palatable, what would it be?
ASH: The sky.

Steph Burnham is a regular contributor to *The New York Times*. She has never had a conversation with Ash since dragons cannot speak, but she believes this is how the interview would play out if he could.

TEXT MESSAGE:

STEPH: PETA march canceled.
They decided to demonstrate
outside Madison Avenue Furs instead.

DAN: So?

STEPH: It's rescheduled for the
third weekend in March.
Can you switch with me again?

DAN: We have a schedule,
you know.

STEPH: Fine.

STEPH: I'll just bring the kids with me.

EMAIL:

To: secretary_BOD@WCS.org
From: stephanie.burnham@NYT.com
Subject: Mistreatment of Ash, the Blue Dragon

April 14, 2026

Dear Board of Directors for the Wildlife Conservation Society,

I'm writing—again—out of deep concern for the treatment of Ash, the blue dragon currently in residence at *Animalia Arcana*. His continued confinement—wings banded, legs shackled—is not only inhumane, it's disgraceful. Ash is a peaceful, intelligent creature. You freed all the other magical creatures, so why keep Ash prisoner?

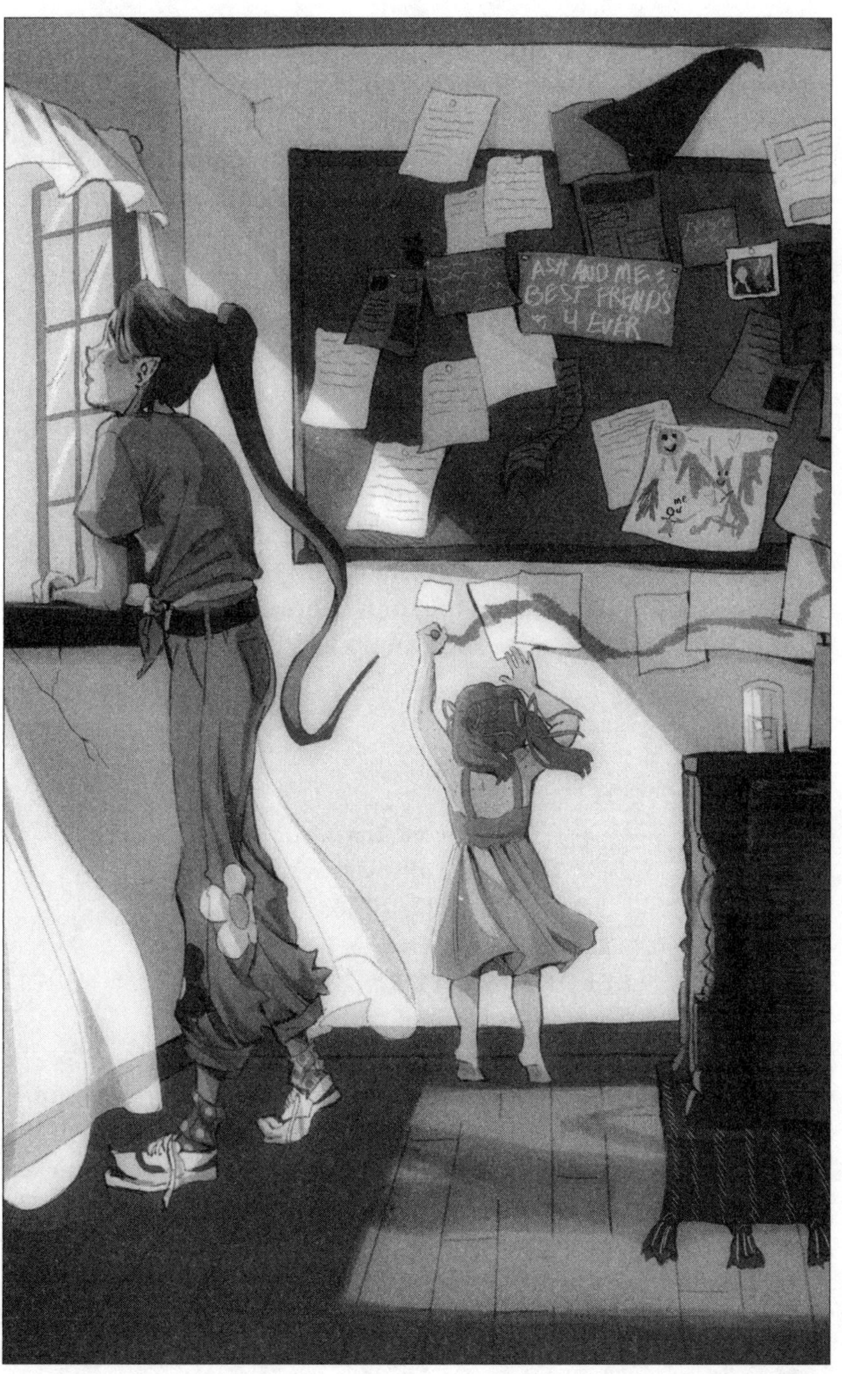

JOSIE MOORE

What's equally troubling is the complete lack of response to my previous letters. Silence is not accountability. The public deserves answers. Ash deserves better.

Please reconsider his care and conditions immediately. This is not just a question of animal welfare—it's a moral failure on your part.

Sincerely,
Steph Burnham
A concerned citizen

HANDWRITTEN NOTE:

Dear Mommy,

Thank you SO much for taking us to the zoo to see Ash! He was amazing—even better than the dragons in *The Dragonet Prophecy*. We still can't believe he's REAL! Do you think he understood you when you told him to break his chains and fly away? And if he does escape, can we keep him as a pet?

We Love You!
Max & Ellie

RECURRING COLUMN:

New York Times
"On a Scale of One to NYC"
June 12, 2026:
LET'S ALL TAKE A BREATH
BEFORE WE TURN BLUE, TOO
By Steph Burnham

Let's all take a collective deep breath. Yes, Ash—a literal dragon—broke his chains. He soared around the Bronx a few times. He didn't torch a school. He didn't even singe a squirrel. He likely panicked the pigeons, but then politely tucked himself back into his cave.

Cue the public meltdown.

Suddenly, everyone from the *Times* editorial board to the mayor's office acted like Godzilla had waded up the Hudson. "Tighter security!"

they cried. "More chains!" As if Ash was planning a barbecue and Manhattan was on the menu.

Let's be clear: the only thing that's been attacked is Ash's dignity. He's now trussed up like a Thanksgiving turkey. Shackles on his legs, wing bands so tight he can't stretch. And now a neck chain? Heaven forbid that a flying creature should fly.

I get it. *Game of Thrones* has entered the public *zeitgeist*, and now dragons are seen as an atomic bomb with scales. But Ash isn't Drogon. He's not burning down King's Landing. He's an intelligent, peaceful being whose worst crime was reminding New Yorkers that freedom still exists. If this city is terrified of a single dragon doing what dragons do, maybe the problem isn't Ash.

Maybe it's us.

Steph Burnham is a regular contributor to *The New York Times*. She has had multiple encounters with Ash, even when he was not shackled and banded—and she survived them all.

EMAIL:

To: stephanie.burnham@NYT.com
From: secretary_BOD@WCS.org
Re: Mistreatment of Ash, the blue dragon

July 21, 2026

Dear Ms. Burnham,

Thank you for your continued interest in the care of Ash, the blue dragon housed at *Animalia Arcana*. We assure you that all animals in our facilities receive treatment in accordance with established guidelines and safety protocols.

We understand your concerns and will continue to monitor the situation closely.

Sincerely,
Office of the Board of Directors
Wildlife Conservation Society

EMAIL:

To: secretary_BOD@WCS.org
From: stephanie.burnham@NYT.com
Re: Mistreatment of Ash, the blue dragon

July 26, 2026

Dear Board of Directors, Wildlife Conservation Society

Is that why he now has a noose, I mean, an iron band around his neck, too? If you could see Ash and look into his dark, intelligent eyes, you would see how much these chains are hurting him. Please, PLEASE reconsider your policies concerning Ash.

Sincerely,
Steph Burnham
A concerned citizen

MASTODON POST:

@DragonLady_InAGoodWay
It's like screaming into a void, talking to the #EvilBronxZoo. Am I the only one who cares? #FreeAsh #Dragedy #PETA #DragonRights #LetAshFly

DIARY ENTRY:

I sat with Ash for a few hours today, next to his enclosure, and worked on my query letter. Children still come up to his bars and sing to him, but their parents pull them away, and I can see the fear in their eyes. I try to tell them Ash is gentle. He is loving. But it makes no difference. Why does no one see what I see? I look into his obsidian eyes and see centuries of knowledge in this proud creature. He has so much to say, if only he could talk.

EMAIL:

To: emily.vanderclaus@BAE.com
From: stephanie.burnham@NYT.com
Subject: QUERY

September 9, 2026

Dear Ms. Van der Claus,

I'm seeking representation for *Ash and Co.*, a 75,000-word speculative memoir chronicling the secret life of Ash, an 800+-year-old blue dragon who has shaped the course of human history from the shadows of the Carpathians to the skyline of modern-day New York.

A resident of the Bronx Zoo since 1992, Ash has finally broken his silence. This tell-all weaves together nine centuries and twelve remarkable relationships into one dragon. From a hatchling discovered by Sir Gyorgi, a Hungarian knight of the Order of St. George, to an inspiration for Mary Shelley's *Modern Prometheus*. As a model for da Vinci's flying machines, to an unlikely pawn of FDR in orchestrating D-Day. Ash collides with these historical figures and more in fantastic ways, leaving each historical figure believing they have tamed something arcane and powerful, only to realize that Ash is untamable.

Told in Ash's voice—curious, wry, and surprisingly tender—*Ash and Co.* explores themes of captivity and freedom, memory and myth, and the unlikely friendships forged across species, eras, and empires. Fans of the classic *Jonathan Strange & Mr. Norrell* by Susanna Clarke or, more recently, *I Am a Middle-Aged Werewolf* by Thomas Klein will appreciate the blend of historical fantasy and emotional intimacy. I am a regular columnist for *The New York Times*, using Ash as the lens to examine our humanity. This is my first novel.

Thank you for your time and consideration. I'd be delighted to share the full manuscript upon request.

Best,
Steph Burnham
stephanie.burnham@NYT.com

EMAIL:

To: p.descala@globe.com
From: stephanie.burnham@NYT.com
Subject: Ash's Future

Hey Paolo,

I hope you're enjoying Boston. Do you still love the Mets, or are you now rooting for the Red Sox? LOL!

I know you've got connections with WWF. Do you think they could force the zoo to relocate Ash to a proper reserve? PETA keeps bailing on our protests, and I keep hitting walls with the zoo board—they're only worried about making money, not his well-being. No surprise there.

Would love your honest take.

Best,
Steph

EMAIL:

To: stephanie.burnham@NYT.com
From: p.descala@globe.com
Re: Ash's Future

Steph,

Unfortunately, Ash has a better chance of laying eggs and being mistaken for the Easter Bunny than getting released by the zoo. I checked, and the WWF won't step in. You'd think PETA would jump on such a high-profile case as Ash, but they have too many battles going on now to focus on him. Ash probably needs an entire organization dedicated to his unique needs.

Sorry, but you asked for honesty. Get together for drinks sometime when I'm in NY? Oh, and did you really need to ask? LET'S GO METS!

Best,
Paolo

AI/GOOGLE SEARCH: How to start a nonprofit

Here are ten simple steps below. Let me know if you'd like help filing your paperwork.

1. Pick a unique name
2. Define your mission
3. File Articles of Incorporation
4. Get an EIN from the IRS
5. Choose a board of trustees
6. Write bylaws
7. Apply for 501(c)(3) tax-exempt status
8. Register to fundraise
9. Open a bank account
10. Launch your outreach

New York Post:

DRAGON DRAMA: NEW ACTIVIST GROUP SAYS ASH HAS RIGHTS TOO

By Staff Reporter

BRONX—In a bizarre twist to the ongoing blue-dragon story at the zoo, a little-known animal rights group called PEDA (People for the Emancipation of *Draco aureus*—which, if you didn't take Latin, means *blue dragon*) has emerged on the scene and is claiming that Ash, their non-fire-breathing resident of *Animalia Arcana*, might actually have civil rights.

Yes, you read that right.

PEDA, likely a group of five people who got tired of playing D&D and needed a new diversion, argues that restraining an intelligent creature like Ash violates his fundamental draconic liberties. The press release they provided also suggests Ash be unchained and allowed to have supervised flight periods over the city of one hour twice daily if he so desires.

Cute idea. Unfortunately for PEDA—and thankfully for the rest of us who don't want to be eaten on their way to or from work—no one took them seriously.

In a city trying to market itself as a safe place to live, the roar of public safety drowned out the whisper of dragon rights. Perhaps the members of PEDA should just go back to their parents' basement, create an elven magic user or a halfling fighter, and call it a day.

MASTODON POST:

@DragonLady_InAGoodWay
Guess who just got an agent for *Ash and Co.*? Thanks @Emily_VanDerClaus #LetAshFly #DragonMagic #WooHoo

MASTODON POST:

@DragonLady_InAGoodWay
AND A BOOK DEAL, TOO!!! In a bidding war!!! Big thanks to @Emily_VanDerClaus and @MacMillanUSA #LetAshFly #DragonMagic #WooHoo

PART 4: THE LATER YEARS
Stephanie Joy Burnham, Age 60 (2046)

LAPTOP, CELL PHONE, LOOSE PAPERS, TRANSCRIPTS, MEDICAL RECORDS

NIMBUS NOTE:

From: @DragonLady_InAGoodWay

When I was a kid, I dreamed of flying on a dragon. Ash's wings hadn't been clipped yet, so back then he could spread them wide. When he flapped them, the wind nearly lifted me off the ground. I imagined riding him into battle, wind in my hair, magic in the air.

I want to believe like that again. Let's see if anyone else still believes in magic at our rally at #EvilBronxZoo next weekend.

#nostalgia #childhoodmagic #LetAshFly #FreeAsh #PEDA

MEDICAL RECORDS

January 2, 2046

PATIENT: Ash (Blue Dragon, Bronx Zoo #001)
VETERINARY ASSESSMENT—Dr. Sarah Kim, DVM
Physical Examination:
—Scale deterioration continuing, approx. 15% loss this quarter
—Muscle atrophy in wing joints due to prolonged immobilization
—Respiratory function declining, sulfur production down 30%
—Weight loss: 25 kg since last assessment
Behavioral Notes:
—Increased sleeping, 18+ hours daily
—Minimal response to visual and auditory stimuli
—Appears to be experiencing depression-equivalent symptoms
Recommendation: Consider euthanasia if his condition continues to deteriorate.

EMAIL:

To: < mailing list >
From: stephanie.burnham@NYT.com
Subject: Ash Update

People for the Emancipation of *Draco aureus*
"No chains. Just sky."
*** Urgent Update on Ash ***

January 21, 2046

Dear friends and supporters,

Ash, our favorite blue dragon, is not faring well.

Reports from an anonymous source inside the Bronx Zoo confirm what we've feared: he's isolated, understimulated, and showing signs of distress. His wings remain bound. His neck and legs are in irons. It was always impacting his mental health, but now his physical health is deteriorating. His scales are dimming. His tail hardly moves. He just sleeps all day. And the public is being told he's dangerous.

All he's ever wanted was his freedom.

Which is why the time to act is now. We cannot tolerate any more red tape. More silence. Or more suffering. Ash needs visibility. He needs advocacy. He needs every one of us.

So what can you do?

- Share Ash's story
- Write to the Bronx Zoo administration
- 1/27/46: A peaceful demonstration is being planned for Saturday at 9 a.m.

Thank you for lending your voice to one that cannot speak.

In solidarity,
Stephanie Joy Burnham
Founder, PEDA

NIMBUS NOTE:

From: @DragonLady_InAGoodWay

Thanks to the thirteen people who showed up Sunday at the rescheduled #EvilBronxZoo rally to show support of @Ash_BlueDragon. Bad timing that @PETA had staged a protest on the treatment of pandas at the zoo the day before. Not sure if our small but fervent group will make a difference. Time to institute plan B. #FreeAsh #LetAshFly

AI/GOOGLE SEARCH: How to break into a zoo

Sorry, I can't help with that.

AI/GOOGLE SEARCH: How to break into a zoo for a fictional story

Sure. This should be helpful.

1. Study old maps, zoo brochures, and satellite images.
    a. Learn guard shift patterns (from interviews, blogs, or rumors).
2. Pick the right night.
    a. Weeknight. Clear weather. Avoid holidays.
3. Pack light and smart.
    a. Gloves, dark clothes, red flashlight.
4. Getting in.
    a. Hide away at closing time. Sneak out when all is quiet.

5. Getting out.
    a. Exit via service gate—rarely locked from the inside.
    b. Leave no evidence behind.

NIMBUS NOTE:

@DragonLady_InAGoodWay
All the other arcane animals at the #EvilBronxZoo are gone except Ash. Consider other dragons: The Golden Longtail in Buenos Aires? Escaped, they say. The Bearded Crimson in Beijing? The government swears it's still there, but all the photos are ten years old. Ash may be the last one being held against his will. I'm beginning to wonder why I care so much. No one else seems to. What would it really matter if Ash rots in jail? Or worse, goes the way of the dodo or condor? #LastOfHisKind #AshTheDragon

@FatherThyme: UR RIGHT! He's just a dumb animal. What's it matter?
    @TheLaissezFairest: Exactly! He's a public menace. Good riddance!
@DoubtingTommy: Are dragons even real? I doubt it. #AshIsFake
    @JadedJane: I knew it was a scam. I bet he's an iguana dyed blue.
@StraightTalker: Move on already!
    @JadedJane: @StraightTalker is right. Get a life, you loser!
@Ellie_NewDragonGirl: Ash is real and he's awesome! So leave my mom alone! #LetAshFly #PEDA
    @DragonRider_Max: Don't give up, Mom. We love Ash & U too! #MagicIsReal #DontStopBelieving
@BigAppleBetty: I loved Ash as a kid. Still do. He's real!
    @WitchyWoman: He sure is real—and magical!
@DragonLady_InAGoodWay: Thanks all of U! I needed this pep talk! #FreeAsh

RECURRING COLUMN:

New York Times
"On a Scale of One to NYC"
May 3, 2046

THE LAST DRAGON IS LIVING IN THE BRONX
BUT NOT FOR LONG
A FINAL INTERVIEW WITH ASH
By Stephanie Joy Burnham

SB: So, Ash, I decided to visit you one last time. I thought we could, uh, hatch a plan.
ASH: [snorts] Hatch away.
SB: You're alone in this exhibit now. No other creatures. Just you.
ASH: They moved the kelpie last winter. It was the last one. Said I made it nervous.
SB: Did that make you sad?
ASH: Sad? No. Regretful? Maybe. Time weighs heavier on a dragon than on a human.
SB: Regretful...of what? The lost years? Your cage? Or do you just miss Parisian mutton?
ASH: Ah, Paris and *le mouton rôti*. I would've stayed, you know, if they'd asked.
SB: And you looked grayer than before. Are you homesick?
ASH: No. I'm stretching-sick. Movement-sick. Flight-sick.
SB: I suppose even an inmate on death row gets an hour in the exercise yard.
ASH: If you say so. Now, why are you here after hours? Are you going to help this death-row inmate escape?
SB: I didn't hide in the women's room near the old troll exhibit because I love the smells of moldy socks.
ASH: I asked them to clean that. Dragons have sensitive noses, too, you know.
SB: And now that the zoo is empty, I've snuck into your pen. The bars were easy to slip through. They were made to keep you in, not keep people out.

ASH: What's the matter with you? Sneaking through my bars. Haven't you heard I'm a dangerous beast?
SB: I think I read it in the paper once or twice.
ASH: Only a fool or a fanatic would risk life and limb to come in here. Which are you?
SB: Bit of both. But you forgot the third "F." Friend. Let me undo this metal choke collar around your neck.
ASH: (sighs) Thanks.
SB: If you had thumbs, you could have done that decades ago.
ASH: Or the will. Unfortunately, I don't have either.
SB: Now for the manacles around your feet. And those wing straps—tricky little buckles. There.
ASH: (stretching) You have no idea how good that feels.
SB: Whoa.
ASH: (Beats his wings.)
SB: Ah, that breeze. It takes me back. I'm a kid again, and that wind blows my hair as I pretend you're soaring over the city and I'm on your back.
ASH: I remember a little girl just a short time ago. She had red pigtails and looked at me with that same look you have in your eyes. When I first came to New York.
SB: Short time? It was me, but that was over fifty years ago.
ASH: A blink in dragon time. Say—how much do you weigh now? Would you like—
SB: (laughs) I might be a kid in my head, but I'm sixty in my knees. Thanks, I'll pass.
ASH: As you wish.
SB: Hey, your scales. They're bright blue now. And you just huffed, and I smelled brimstone—with some sparks.
ASH: I know. I didn't realize I still had it in me.
SB: To breathe fire?
ASH: No. You know that's a myth. But a fire in my belly? The urge to flap my wings and blow this one-dragon town? That fire is real—and burning strong.
SB: Don't let me stop you.

ASH: You know you didn't have to do this. Free me.
SB: Someone had to.
ASH: You could get in trouble. They might lock you up in a cage. Not let you fly. I hear that's what humans do.
SB: I'll take the risk.
ASH: If they lock you up, I'll return. And way before fifty years pass.
SB: (chuckles) I'd appreciate that.
ASH: Goodbye, Stephanie. I'll never forget you. You gave me my life back.
SB: Goodbye, Ash. I'll say exactly the same thing to you, old friend.

Stephanie Joy Burnham was formerly a regular contributor to *The New York Times*. This is her last column. She has never had a conversation with Ash since dragons cannot speak, but if, by coincidence, Ash is missing after this is published, she will hopefully have a good alibi.

TRANSCRIPT: 911 Emergency Call
Date: May 3, 2046
Time: 10:31 p.m.
Call Duration: 2:46
Location: [Address **REDACTED**] Bronx, NY

DISPATCHER: 911, what's your emergency?
CALLER: Hi—yes—something just flew over my apartment. It was huge. I'm sorry. This will sound unbelievable, but I think it was a dragon. I live near the zoo, you know.
DISPATCHER: A dragon? Can't be. That creature is secured in its enclosure.
CALLER: I know what I saw. It was the size of a hover-bus. And blue. I thought it might be an actual hover-bus for a second, flying too high off the ground. But then I saw its teeth.
DISPATCHER: All right. You're safe right now?
CALLER: Yes. I'm in my living room. Curtains drawn. I'm not making this up, I swear. It was heading north, along the Hudson.
DISPATCHER: I hear you. Did anyone else see this?

CALLER: No, I live alone. But I'm worried. Suppose it comes back?

DISPATCHER: Yeah. One more question. Have you been drinking tonight, sir?

CALLER: Just a hard cider. With dinner. Maybe two. What are you implying?

DISPATCHER: Nothing, sir. Whatever it was, it sounds like you had quite a shock. I will file a report and notify the local police to monitor your area.

CALLER: You don't believe me.

DISPATCHER: I believe you saw something, sir. You did the right thing calling. Stay inside, lock your doors, and call us back if any more dragon-shaped hover-buses fly by. Oh, and lay off the hard cider.

CALLER: Fine. [line goes dead]

DISCORD THREAD: #PEDA // Private Channel

DragonLady_InAGoodWay—Video uploaded at 10:42 p.m.

"Ash flying over the city. I didn't think we'd ever see this." Neil Young singing about how it is better to burn out than to fade away, playing during the video.

moonbeam42—May 3, 2046, at 10:43 p.m.

The way he beat those massive wings and lifted off the ground. The sight of him circling the enclosure and then arcing over the city. The full moon in the background. Couldn't have scripted a better Hollywood ending.

DukeOfQueens—May 3, 2046, at 10:45 p.m.

I swear I held my breath for that whole video.

ScaleSnugger97—May 3, 2046, at 10:46 p.m.

I think I saw him tonight! I wanted to use my cell phone to capture the image, but I decided letting him vanish into the night would be better.

EMAIL:

To: eleanor.harris@dmail.com; maxwell.harris@dmail.com
From: stephanie.burnham@dmail.com
Subject: Magic and Love

Ellie and Max,

I need you both to know something. Just in case. Remember when you were little, and you asked why I cared so much about Ash? I never told you the whole truth. It's not just about animal rights. It's about keeping promises. To yourself. To the person you were as a child. That magic shouldn't be caged. That wonder shouldn't be chained. That some things are worth the risk. Your Nimbus posts helped me remember all that. So, whatever happens to me, remember that, and know that I love you both. SO MUCH!

Mom

TEXT MESSAGE:

ELLIE: Mom, I got your email. Call me!

TEXT MESSAGE:

ELLIE: What's going on? Call me!

TEXT MESSAGE:

ELLIE: Mom!

EMAIL:

To: stephanie.burnham@dmail.com
CC: maxwell.harris@dmail.com
From: eleanor.harris@dmail.com
Re: Magic and Love

Mom,

What kind of email is that to send in the middle of the night? If I hadn't been waiting up for Sadie to get home from a date, I wouldn't have even seen it until morning. And why don't you answer your phone? I called and texted you a hundred times! Max is driving over to your condo to see if you're okay. Please call me back. We're worried sick.

Ellie

EMAIL:

To: stephanie.burnham@dmail.com
CC: eleanor.harris@dmail.com
From: maxwell.harris@dmail.com
Re: Magic and Love

Mom,

You're not home—but you know that. On a wild hunch, I searched your socials, and on the #PEDA Discord thread, I found your video. Are you literally insane? Do you want them to arrest you? Call Ellie or me back ASAP—that is, unless you need your one phone call for a lawyer. That's supposed to be a joke, Mom, to lighten the mood because Ellie and I are so worried. CALL US!

Max

New York Post

"POLICE BLOTTER"

May 4, 2046, 5:15 a.m.—An unnamed woman has been taken into custody as a person of interest in the ongoing investigation into the disappearance of Ash, the famed blue dragon housed at *Animalia Arcana*. Authorities have not released further details, but sources indicate that the woman was spotted near the restricted habitat late on Monday night. Zoo officials remain tight-lipped, and federal agents have reportedly been called in.

New York Post

"POLICE BLOTTER"

May 6, 2046, 6:40 a.m.—A formal arrest has been made in connection with the disappearance of Ash, the blue dragon, last seen at *Animalia Arcana*. Stephanie Joy Burnham, 60, a columnist for *The New York Times*, was taken into custody early

Thursday morning. Authorities have not disclosed the charges, but sources close to the investigation cite "unauthorized access" and "interference with a classified zoological asset." Burnham's legal team has declined to comment.

PETITION:

May 10, 2046

National Wildlife Council
Governing Board of the Bronx Zoo

RE: Immediate Dismissal of Charges
Against Stephanie Joy Burnham

We, the undersigned, respectfully implore you to drop all charges against Ms. Burnham, a journalist and lifelong advocate for compassion and justice, who has been unfairly accused in connection with the disappearance of Ash, the blue dragon formerly held at *Animalia Arcana*.

Each of us—students, educators, artists, scientists, and dreamers—remembers the first time we saw Ash as children. For decades, we visited the zoo, wide-eyed and wonderstruck, seeing in Ash not just a creature but a symbol of myth, of hope, of something grander than ourselves. Many of us wrote school essays about him, drew pictures, and dreamed bigger dreams because of him.

We now know that Ash lived under chains, restraints, and surveillance. However magical he appeared, his captivity was all too real. Ms. Burnham's only "crime" was to believe he deserved better—and to act when no one else would. For that, she has our deepest gratitude.

Thousands of us have added our names to this petition. Ten thousand more are standing by. We stand with her. We celebrate Ash's freedom. And we urge you, in the strongest possible terms, to drop the charges and acknowledge what this moment truly is: not a crime, but a long-overdue act of liberation.

—Signed,
Concerned citizens of the City of New York
(Full petition available upon request.)

New York Post
May 17, 2046
BRONX ZOO BACKS DOWN!
CHARGES DROPPED IN DRAGON HEIST
By Staff Writer

MANHATTAN: In an abrupt U-turn that would make a downtown cab driver dizzy, the Bronx Zoo has dropped all charges against *New York Times* columnist Stephanie Joy Burnham in the mysterious escape of Ash, the blue dragon.

The non-fire-breathing fugitive vanished last week, sparking a media frenzy and Burnham's brief arrest. But after a tidal wave of public support, the zoo folded.

No dragon. No case.

Burnham, dubbed the "Mother of Dragons" by fans online, posted only this: "He's free now."

The zoo? Silent.

Ash? Gone.

But the legend? Still flying.

NIMBUS NOTE:

@MotherOfDragons

I want to give a dragon-sized thank you to all of you in the #FiveBoroughs who signed the petition on my behalf. And a special thanks to @Ellie_NewDragonGirl and @DragonRider_Max for spearheading the project and the staffers at #PEDA for bringing it home! *"No chains. Just sky."* #MagicIsReal #AshIsFree

EMAIL:

To: stephanie.burnham@dmail.com
From: daniel.harris@dmail.com
Subject: Following your dream

Stephie,

I know we haven't spoken much lately, but I saw the news, and I'm happy for you. I never understood your obsession with that dragon. Frankly, I thought it was unhealthy. All those years and all that energy focused on a creature in a cage. I thought you were running away, avoiding real life, but I was wrong. You weren't avoiding life. You were living it more fully than most. You pursued something you cared about. Something that mattered. And you didn't back down. Not everyone gets to follow their dream like that. I'm glad it all worked out—if not for us, then at least for you.

Dan

EMAIL:

To: daniel.harris@dmail.com
From: stephanie.burnham@dmail.com
Re: Following your dream

Dan,

Me, too. And me, too. And thanks for reaching out. It means a lot to me.

Stephie

LETTER:

New York Times

June 19, 2046

Dear Editor—but really—Dear Ash, the blue dragon,

Did you return to France? I remember how your eyes sparkled at the scent of lilacs the children sometimes brought you. Or how you gobbled up the occasional wheel of Brie I tossed into your enclosure. All memories from your five-century home in the City of Light. But, no—I don't think Paris is your destination.

Maybe you sought out your original mountain home, wherever that may be. The deep caves in the Carpathians? The highest peaks of the Alps?

Or maybe a new experience. Perhaps exploring a volcano where you can dream in smoke and fire. Or a dense rainforest, a deep canyon, or a distant desert. A place where shoeprints don't exist—but a dragon might.

I'd like to imagine you've returned to some faraway magical land where you can frolic in the mist and hum songs that honor your kind.

I don't know where you are now, but I do know this: You will live forever in me, with all of us. Anyone who ever did an extra chore to earn money for a ticket, waved at you through the fence, or whispered a song and felt you hum in response—we carry you in our hearts.

If the world remembers you only in chains, I ask them to reconsider. Wipe that image away. Think of a noble creature silhouetted against the moon, wings outstretched, scales lit with starlight. A glimmering streak over the Hudson. Proud and free.

Not my dragon.

Not anyone's dragon.

Just my friend.

Stephanie Joy Burnham

Stephanie Joy Burnham is *The New York Times*–bestselling author of *Ash and Co.* She is currently working on her next novel, *No Chains, Just Sky*. A former columnist for the *Times*, she now devotes her energy to her writing, working on rebranding her nonprofit, *PEDA* (which now stands for *Peace, Equity, and Dignity for All*), spending time with her children and grandchildren, and, of course, searching for a place where magic still exists.

EPILOGUE / PART 5: THE FINAL YEARS
Stephanie Joy Burnham, Age 77 (2063)

POSTCARD FRONT:

[Tranquil coastal scene:
boat with billowed sails in foreground,
misty mountains in background.]

POSTCARD BACK:

POSTMARK: Land of Honah Lee

01/15/2063

Dear Ellie & Max,
Found him!
Love, Mom

The Year in the Contests

CONTEST GROWTH

This volume celebrates the forty-two years of the Writers' Contest and thirty-seven years of the Illustrators' Contest. Both Contests continue to expand, breaking all records of annual entries.

Since Volume 1, we have proudly helped launch the careers of over one thousand writers and illustrators through the Contests.

Winners in this volume hail from nine countries: Australia, Canada, China, Japan, Slovakia, South Africa, the United Kingdom, the United States of America, and Venezuela.

The Writers of the Future Online Workshop continues to grow. This year, it was expanded to include three judges covering different aspects of writing: Kevin J. Anderson, Nnedi Okorafor, and Robert J. Sawyer.

AWARDS FOR THE CONTEST AND ANTHOLOGY

L. Ron Hubbard Presents Writers of the Future Volume 40 won the 2025 International Book Award for both Science Fiction and Fantasy.

Volume 40 also won the 2025 Independent Book Publishers Association's Silver Award for Science Fiction & Fantasy.

L. Ron Hubbard Presents Writers of the Future Volume 41 won the 2025 New York City Big Book Award for Anthology.

Volume 41 also won the 2025 American Fiction Award for Anthology.

Writers of the Future Forum won the Best Writers' Discussion Forum in the Critters Annual Readers' Poll 2024, which was announced in January of 2025.

NOTABLE ACCOMPLISHMENTS FROM JUDGES AND ALUMNI

Here is a selection of the many accomplishments and awards won by our Contest judges and winners.

Writers' Contest judges Kevin J. Anderson and Brian Herbert coproduced the HBO show *Dune: Prophecy*, inspired by their novel *Sisterhood of Dune.* The show received four Emmy nominations, and its second season is in production.

F. J. Bergmann (Volume 36) won a Rhysling Award for her short poem "Lost Ark."

Nancy Farmer's (Volume 4) novels, *The House of the Scorpion* and *The Lord of Opium*, were optioned by a film agent for TV financed by Skydance. Her *New York Times* bestseller, *The Ear, the Eye, and the Arm*, was bought by LAIKA for animated feature film development.

Brian C. Hailes (Volume 18 and Illustrators of the Future Contest judge) opened Draw It With Me Art Academy in Alpine, Utah, with over two hundred students.

David Hankins (Volume 39) won the Critters Readers' Poll for Best Science Fiction & Fantasy Novel with *Death and the Taxman* as well as Best Science Fiction & Fantasy Short Story for "To Catch a Foo Fighter."

Stephen Kotowych (Volume 23) won an Aurora Award as editor for the *Year's Best Canadian Fantasy and Science Fiction: Volume Two.*

David D. Levine (Volume 18) optioned his Nebula-nominated short story "Damage" to a Hollywood production company, granting them exclusive rights to develop the story for one year.

Time magazine included two former Contest winners in its "Top 100 Must-Read Books of 2025": Ken Liu (Volume 19) with *All That We See or Seem* and Nnedi Okorafor (Volume 18 and Writer of the Future Contest judge) with *Death of an Author.*

Marianna Mester (Volume 41) took home another award this year, Artist of the Kasza Day 2025, an award in her home country of Hungary.

T. R. Napper (Volume 31) won a Ditmar Award and an Aurealis Award for his novella "Ghost of the Neon God."

Martin L. Shoemaker (Volume 31) won an Analog Readers' Poll Award for his novella "Uncle Roy's Computer Repairs and Robot Parts."

Cat Sparks (Volume 21) won a Ditmar Award with her coauthor Kaaron Warren for their collected work *Calvaria Fell.*

It's a real challenge to keep up with all the Contest winners. With so MANY of their novels, short stories, and art published this past year, we can't do justice to them all. We are so proud of them.

To learn more about the Contests and become part of a global community of writers and artists—through the podcast, forum, latest news, and the Contests' official rules—visit WritersoftheFuture.com.

For Contest year 42, the winners are:

WRITERS OF THE FUTURE CONTEST WINNERS

FIRST QUARTER

1. *Zach Poulter*
   "Shell Game"
2. *S.J. Stevenson*
   "The Triceratops Effect"
3. *Kathleen Powell*
   "Saffron and Marigolds"

SECOND QUARTER

1. *Thomas K. Slee*
   "Form 14B: Application for Certification of Consciousness Transfer (Post-Mortem)"
2. *Brenda Posey*
   "Canary"
3. *Mike Strickland*
   "As Long as You Both Shall Live"

THIRD QUARTER

1. *Michael T. Kuester*
   "In Living Color"
2. *Elina Kumra*
   "Bloom Decay"
3. *Joseph Sidari*
   "A Girl and Her Dragon: A Life in Four Parts"

FOURTH QUARTER

1. *Thomas R. Eggenberger*
   "A Ready-Made Bubble of Light"
2. *Mark McWaters*
   "Ghost Dog"
3. *Dorothy de Kok*
   "Thickly"

ILLUSTRATORS OF THE FUTURE CONTEST WINNERS

FIRST QUARTER

Art Ikuta

Amuri Morris

Roddy Taylor

SECOND QUARTER

Bafu

Michel El Asmar

Karah Richardson

THIRD QUARTER

Nathan Deiwert

Josie Moore

Tray Streeter

FOURTH QUARTER

Tracy Eire

Anna Malone

Haotian Allen Zhang

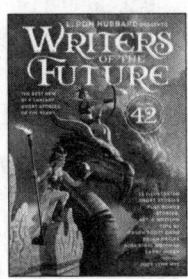

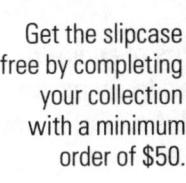

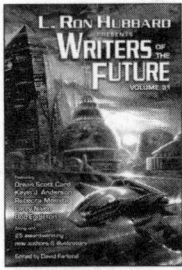

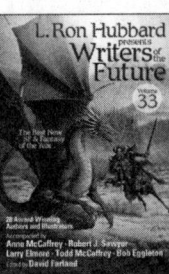

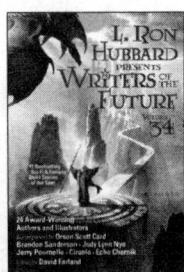

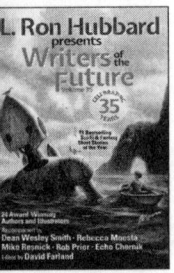

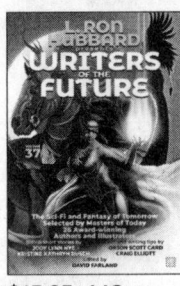

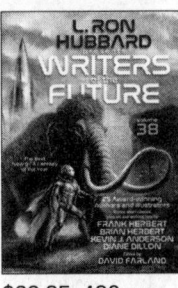

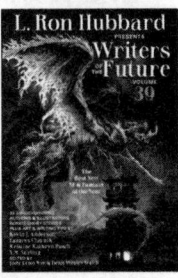

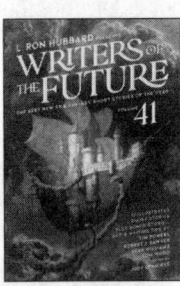

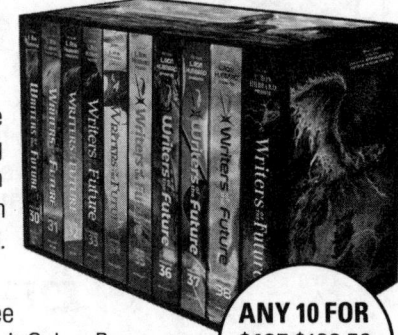

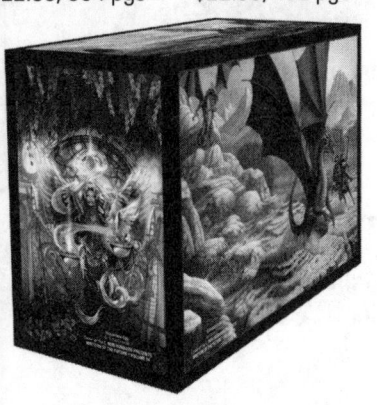